FINITE ELEMENT SOFTWARE FOR PLATES AND SHELLS

FINITE ELEMENT SOFTWARE FOR PLATES AND SHELLS

E. Hinton
D. R. J. Owen
Department of Civil Engineering,
University College, Swansea, U.K.

PINERIDGE PRESS
Swansea, U.K.

First Published, 1984 by
Pineridge Press Limited
91, West Cross Lane, West Cross, Swansea, U.K.

British Library Cataloguing in Publication Data

Finite element software for plates and shells.
1. Plates (Engineering) — Data processing
2. Shells (Engineering) — Data processing
3. Finite element method
I. Hinton, Ernest. II. Owen, D. R. J.
624.1'776'01515353 TA660.P6

CN ISBN 0-906674-24-7

Printed and bound in Great Britain by
Dotesios (Printers) Ltd., Bradford-on-Avon, Wiltshire

PREFACE

The book is concerned with the linear and nonlinear static analysis of plates and shells. Emphasis has been placed on plate and shell solution procedures which allow for transverse shear effects and consequently Mindlin plate and degenerated shell assumptions are used as a basis for the software developed. The first three chapters deal mainly with linear elastic Mindlin plate analysis. In the first chapter, closed form solutions are provided for rectangular, orthotropic simply supported plates on elastic Winkler foundations for various loadings. The displacements and normal rotations are represented by double Fourier series. The solutions obtained from this software may be used to check the finite strip and finite element programs presented later in the book. The second chapter deals with semi-analytical, semi-numerical solution schemes for Mindlin plates, folded plates and axisymmetric shells. A finite strip program for the analysis of Mindlin plates is provided. Chapter 3 is concerned with Mindlin plate bending finite element analysis and a fully-documented program is included for selectively integrated quadratic isoparametric Mindlin plate elements.

The nonlinear finite element analysis of anisotropic plates and shells is described in Chapter 4 and the ultimate load analysis of reinforced concrete plates and shells is covered in Chapter 5. Fully documented programs are presented for each situation.

The juxtaposition of theory, software and applications is deliberate and should allow readers to either implement the programs or alternatively use the programs as a basis for developing software to suit their own local environment and needs. The authors recognise that some readers may not wish to retype the programs and therefore a magnetic tape of the verified programs, together with typical input data, is available from the publishers. The authors believe that the software presented is suitable for commercial as well as research and educational purposes. However, in such an event, the authors cannot be held responsible for any errors resulting from the use of the codes even though every effort has been made to verify them.

The authors are pleased to acknowledge the contributions of E. Onate, H.H. Abdel Rahman, J.A. Figueiras, M. M. Huq, B. Suarez and Y. C. Hoe.

Secretarial and draughting assistance is also duly acknowledged from M. Williams, E. Gower, J. Davies, D. Gabriel and A. McGairl.

E. HINTON

D. R. J. OWEN

JULY 1984

CONTENTS

CHAPTER 1

CLOSED FORM SOLUTIONS FOR SIMPLY SUPPORTED MINDLIN PLATES

E. HINTON

Department of Civil Engineering, University College of Swansea, United Kingdom

1. INTRODUCTION

As Chapters 2 and 3 of this book are concerned with finite strip and finite element analysis of Mindlin plates, it is appropriate to present in this first chapter a closed form solution for simply supported rectangular plates which rest on elastic Winkler foundations. These plates have specially orthotropic form and three different types of loading are considered although others could easily be incorporated. These solutions will be used in Chapters 2 and 3 to demonstrate the accuracy of the finite strip and finite element solutions. In Section 5 of this chapter a program called PLATES is presented for the closed form solution.

To enable a wide variety of plate cross-sections to be considered, Section 4 illustrates how plate rigidities may be evaluated for

(a) homogeneous isotropic plates,

(b) sandwich plates,

(c) symmetrically laminated plates,
and
(d) voided plates.

A small program called RIGID is also provided in Section 5 for the evaluation of the plate rigidities. Some examples illustrating the use of program PLATES are given in Section 6.

2. GOVERNING EQUATIONS

Mindlin plate theory [1] allows for transverse shear deformation effects and thus offers an attractive alternative to classical Kirchhoff thin plate theory [2]. The main assumptions are that:

(a) displacements are small compared with the plate thickness,

(b) the stress normal to midplane of the plates is negligible, and

(c) normals to the midplane before deformation remain straight but not necessarily normal to the midplane after deformation.

From the third assumption, the assumed displacement field can be written as

$$\begin{aligned} U(x,y,z) &= z\theta_x(x,y) \\ V(x,y,z) &= z\theta_y(x,y) \\ W(x,y,z) &= w(x,y) \end{aligned} \qquad (2.1)$$

in which x,y are the rectangular coordinates in the plane of the plate, z is the thickness-direction coordinate measured downwards from the midplane; U,V,W are the displacements in the x,y,z directions respectively, w is the corresponding mid-plane displacement, and θ_x and θ_y are the normal rotations in the xz and yz planes respectively due to bending.

By making use of the assumption that $\sigma_z = 0$, the constitutive equations at a typical point (x,y,z) in a Mindlin plate may be expressed

$$\begin{bmatrix} \sigma_x \\ \sigma_y \\ \tau_{xy} \end{bmatrix} = \begin{bmatrix} Q_{11} & Q_{12} & Q_{16} \\ Q_{12} & Q_{22} & Q_{26} \\ Q_{16} & Q_{26} & Q_{66} \end{bmatrix} \begin{bmatrix} U_{,x} \\ V_{,y} \\ U_{,y} + V_{,x} \end{bmatrix} \qquad (2.2)$$

or $\underline{\sigma} = \underline{Q}\ \underline{\varepsilon}$

where Q_{ij} (i,j = 1,2,6) are the reduced in-plane stiffnesses for plane stress (i.e. assuming that $\sigma_z = 0$) and furthermore

$$\begin{bmatrix} \tau_{yz} \\ \tau_{xz} \end{bmatrix} = \begin{bmatrix} C_{44} & C_{45} \\ C_{45} & C_{55} \end{bmatrix} \begin{bmatrix} V_{,z} + W_{,y} \\ U_{,z} + W_{,x} \end{bmatrix} \qquad (2.3)^*$$

or $\quad \tau = \underline{C}\ \underline{\gamma}$

It should be noted that by using (2.1), the strains on the right hand sides of (2.2) and (2.3) may be written as

$$\begin{bmatrix} U_{,x} \\ V_{,y} \\ U_{,y} + V_{,x} \end{bmatrix} = \begin{bmatrix} z\theta_{x,x} \\ z\theta_{y,y} \\ z(\theta_{x,y} + \theta_{y,x}) \end{bmatrix} \qquad (2.4)$$

or $\quad \underline{\varepsilon} = z\underline{\varepsilon}_f$

and

$$\begin{bmatrix} V_{,z} + W_{,y} \\ U_{,z} + W_{,x} \end{bmatrix} = \begin{bmatrix} \theta_y + w_{,y} \\ \theta_x + w_{,x} \end{bmatrix} \qquad (2.5)$$

or $\quad \underline{\gamma} = \underline{\varepsilon}_s$

The strain energy for a Mindlin plate can be written as

$$\text{S.E.} = \frac{1}{2}\iiint \underline{\sigma}^T\ \underline{\varepsilon}\ dxdydz + \frac{1}{2}\iiint \underline{\tau}^T\ \underline{\gamma}\ dxdydz \qquad (2.6)$$

$$= \frac{1}{2}\iint \underline{\varepsilon}^T\ (z^2\ \underline{Q})\ \underline{\varepsilon}_f\ dxdy + \frac{1}{2}\iint \underline{\varepsilon}_s^T\ \underline{D}_s\ \underline{\varepsilon}_s\ dxdy$$

$$= \frac{1}{2}\iint \underline{\varepsilon}_f^T\ \underline{D}_f\ \underline{\varepsilon}_f\ dxdy + \frac{1}{2}\iint \underline{\varepsilon}_s\ \underline{D}_s\ \underline{\varepsilon}_s\ dxdy$$

and the plate constitutive equations can be written as

$$\begin{bmatrix} M_x \\ M_y \\ M_{xy} \end{bmatrix} = \begin{bmatrix} D_{11} & D_{12} & D_{16} \\ D_{12} & D_{22} & D_{26} \\ D_{16} & D_{26} & D_{66} \end{bmatrix} \begin{bmatrix} \theta_{x,x} \\ \theta_{y,y} \\ \theta_{x,y} + \theta_{y,x} \end{bmatrix} \qquad (2.7)$$

or $\quad \underline{\sigma}_f = \underline{D}_f\ \underline{\varepsilon}_f$

---

*where C_{ij} (i,j=4,5) are appropriate elastic constants

and

$$\begin{bmatrix} Q_y \\ Q_x \end{bmatrix} = \begin{bmatrix} S_{44} & S_{45} \\ S_{45} & S_{55} \end{bmatrix} \begin{bmatrix} \theta_y + w_{,y} \\ \theta_x + w_{,x} \end{bmatrix} \tag{2.8}$$

or $\underline{\sigma}_s = \underline{D}_s \, \underline{\varepsilon}_s$

where

$$D_{ij} = \int_{-t/2}^{t/2} Q_{ij} \, z^2 \, dz \qquad (i,j=1,2,6) \tag{2.9}$$

$$S_{ij} = \int_{-t/2}^{t/2} C_{ij} \, dz \qquad (i,j=4,5) \tag{2.10}$$

Note that for specially orthotropic plates D_{16}, D_{61}, D_{26} and $D_{62} = 0$. For homogeneous plates D_{ij} and S_{ij} may be evaluated from (2.9) and (2.10), whereas for sandwich, laminated and voided plates formulae are given in Section 4. The bending moments in (2.7) may be written as

$$(M_x, M_y, M_{xy}) = \int_{-t/2}^{t/2} z(\sigma_x, \sigma_y, \tau_{xy}) dz \tag{2.11}$$

and the shear forces can be expressed as

$$(Q_x, Q_y) = \int_{-t/2}^{t/2} (\tau_{xz}, \tau_{yz}) dz \tag{2.12}$$

where t is the plate thickness.

Note also that the strain resultants given in (2.7) and (2.8) are the direct curvatures $\theta_{x,x}$ and $\theta_{y,y}$ in the x and y directions respectively, the twisting curvature $(\theta_{x,y} + \theta_{y,x})$, and the shear rotations $(\theta_x + w_{,x})$ and $(\theta_y + w_{,y})$ in the xz and yz planes respectively.

Neglecting body forces, body moments and surface tractions, the equilibrium equations for a Mindlin plate resting on an elastic Winkler foundation of modulus K may be written as

$$Q_{x,x} + Q_{y,y} + Kw = 0$$

$$M_{x,x} + M_{xy,y} - Q_x = 0$$

$$M_{xy,x} + M_{y,y} - Q_y = 0 \qquad (2.13)$$

in which q is the normal pressure, $(.)_{,x}$ denotes $\partial(.)/\ x$.

The closed form solution given later in this chapter is concerned with simply supported rectangular plates of uniform thickness with dimensions a and b for which the boundary conditions are given as

$$w = \theta_{x,x} = 0 \quad \text{at } x = 0,a$$

$$w = \theta_{y,y} = 0 \quad \text{at } y = 0,b \qquad (2.14)$$

By combining (2.7), (2.8) and (2.13) the governing equations may be expressed as

$$S_{55}\theta_{x,x}+S_{55}w_{,xx} + S_{44}\theta_{y,y}+S_{44}w_{,yy} + q + Kw = 0$$

$$D_{11}\theta_{x,xx}+D_{66}\theta_{x,yy} + (D_{12}+D_{66})\theta_{y,xy}-S_{55}\theta_x-S_{55}w_{,x} = 0$$

$$(D_{12}+D_{66})\theta_{x,xy}+D_{66}\theta_{y,xx} + D_{22}\theta_{y,yy}-S_{44}\theta_y-S_{44}w_{,y} = 0 \qquad (2.15)$$

In the closed form solution functions w, θ_x and θ_y are sought which satisfy (2.14) and (2.15). Alternatively, it is possible to find the functions w, θ_x and θ_y which minimise the Total Potential Energy

$$= \frac{1}{2}\int_0^a\int_0^b [D_{11}(\theta_{x,x})^2+D_{66}(\theta_{x,y}+\theta_{y,x})^2+D_{22}(\theta_{y,y})^2$$

$$+ 2D_{12}\theta_{x,x}\theta_{y,y}+S_{55}(\theta_x+w_{,x})^2+S_{44}(\theta_y+w_{,y})^2 +Kw^2-qw]dxdy$$

subject to boundary conditions (2.14).

3. CLOSED FORM SOLUTION

3.1 Evaluation of Fourier Coefficients

Here, the closed form solution given by Dobyns [3] is presented for a simply supported rectangular plate of uniform thickness which has dimensions a and b which rests on an elastic Winkler foundation.

$$\theta_x = A_{mn} \cos(m\pi x/a)\sin(n\pi y/b)$$

$$\theta_y = B_{mn} \sin(m\pi x/a)\cos(n\pi y/b)$$

$$w = C_{mn} \sin(m\pi x/a)\sin(n\pi y/b) \tag{3.1}$$

with the loading function given by the expression

$$q = q_{mn} \sin(m\pi x/a)\sin(n\pi y/b) \tag{3.2}$$

in which q_{mn} are the terms in the Fourier series representation of the applied load. Upon substitution of (3.1) and (3.2) into the equilibrium equations of (2.15) it is possible to obtain the matrix equation

$$\begin{bmatrix} P_{11} & P_{12} & P_{13} \\ P_{12} & P_{22} & P_{23} \\ P_{13} & P_{23} & P_{33} \end{bmatrix} \begin{bmatrix} A_{mn} \\ B_{mn} \\ C_{mn} \end{bmatrix} = \begin{bmatrix} 0 \\ 0 \\ q_{mn} \end{bmatrix} \tag{3.3}$$

where

$$P_{11} = D_{11} (m\pi/a)^2 + D_{66} (n\pi/b)^2 + S_{55}$$

$$P_{12} = (D_{12} + D_{66}) (m\pi/a)^2 (n\pi/b)$$

$$P_{13} = S_{55} (m\pi/a)^2$$

$$P_{22} = D_{66} (m\pi/a)^2 + D_{22} (n\pi/b)^2 + S_{44}$$

$$P_{23} = S_{44} (n\pi/b)^2$$

$$P_{33} = S_{55} (m\pi/a)^2 + S_{44} (n\pi/b)^2 + K$$

or solving for the unknown coefficients A_{mn}, B_{mn} and C_{mn} gives

$$A_{mn} = \frac{(L_{12}L_{23} - L_{22}L_{13})q_{mn}}{det}$$

$$B_{mn} = \frac{(L_{12}L_{13} - L_{11}L_{23})q_{mn}}{det}$$

$$C_{mn} = \frac{(L_{11}L_{22} - L_{12}^{2})q_{mn}}{det} \qquad (3.4)$$

in which det is the determinant of the matrix in (3.3).

It is now possible to evaluate the displacements, curvatures (and hence bending moments from (2.5)), and shear forces at any point in the plate using the expressions

$$w = \sum_m \sum_n A_{mn} \sin(m\pi x/a)\sin(n\pi y/b)$$

$$\theta_{x,x} = -\sum_m \sum_n A_{mn}(m\pi/a)\sin(m\pi x/a)\sin(n\pi y/b)$$

$$\theta_{y,y} = -\sum_m \sum_n B_{mn}(n\pi/b)\sin(m\pi x/a)\sin(n\pi y/b)$$

$$\theta_{x,y}+\theta_{y,x} = -\sum_m \sum_n (A_{mn}(n\pi/b) + B_{mn}(m\pi/a)\cos(m\pi x/a)\cos(n\pi y/b)$$

$$Q_x = \sum_m \sum_n S_{55}(C_{mn}(m\pi/a) + A_{mn})\cos(m\pi x/a)\sin(n\pi y/b)$$

$$Q_y = \sum_m \sum_n S_{44}(C_{mn}(n\pi/b) + B_{mn})\sin(m\pi x/a)\cos(n\pi y/b)$$

(3.5)

where $\sum_m$ implies a summation for m = 1 to ∞.

3.2 Fourier Series Representation of the Loads

The Fourier coefficient for the applied loading q(x,y) is given as

$$q_{mn} = (4/ab)\int_0^a \int_0^b q(x,y)\sin(m\pi x/a)\sin(n\pi y/b)dxdy \qquad (3.6)$$

and thus the following loadings may be considered

(a) uniform load q(x,y) = q for which

$$q_{mn} = (4q/mn\pi^2)\ (1-\cos m\pi)\ (1-\cos n\pi) \qquad (3.7)$$

$= (16q/mn\pi^2)$ for $m,n = 1,3,5,\ldots$

$= 0$ for $m,n = 2,4,6,\ldots$

(b) concentrated load acting at $x = \xi$, $y = \eta$ for which

$$q_{mn} = (4P/ab)\sin(m\pi\xi/a)\sin(n\pi\eta/b) \tag{3.8}$$

(c) a uniform load of total intensity P acting over rectangular area u,v with its centre at $x = \xi$, $y = \eta$ for which

$$q = (16P/\pi^2 mnuv)\sin(m\pi\xi/a)\sin(n\pi\eta/b) \tag{3.9}$$

$$* \sin(m\pi u/2a)\sin(n\pi v/2b)$$

(d) cosine loading of total intensity P acting over a rectangular patch u,v with centre at $x = \xi$, $y = \eta$ for which

$$q = \frac{4P\sin(m\pi\eta/b)\sin(m\pi\xi/a)\cos(n\pi v/2b)\cos(m\pi u/2a)}{abu^2\ v^2\ (n/b-1/v)\ (n/b-1/v)\ (m/a-1/u)\ (m/a-1/u)}$$

$$(\text{for } n/b \neq 1/v,\ m/a \neq 1/u)$$

$$= 0\ (\text{for } n/b = 1/v,\ m/a = 1/u) \tag{3.10}$$

(see Figure 3.1)

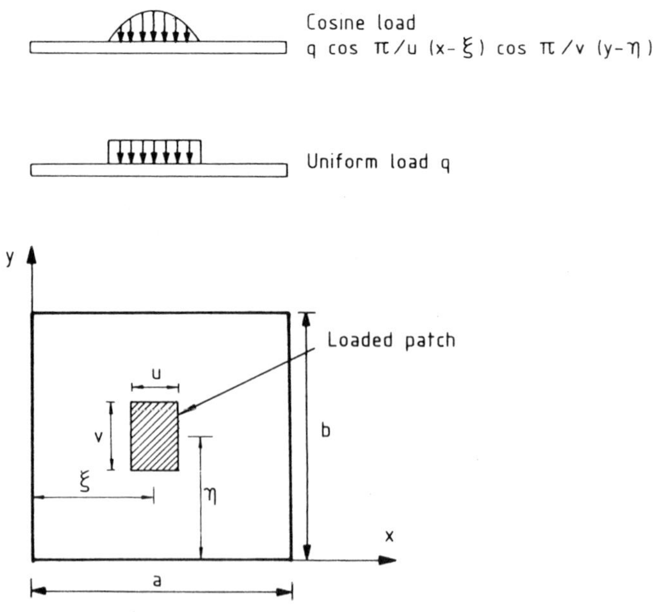

Figure 3.1 Location and shape of loaded area

4. PLATE RIGIDITIES

In this section the evaluation of the plate rigidities is described for homogeneous isotropic, sandwich, laminated and voided plates. In all cases, special orthotropy with respect to the x and y axes is assumed.

4.1 Homogeneous Isotropic Plates

In homogeneous, isotropic plates it is only necessary to specify three constants - Young's modulus E, Poisson's ratio ν and plate thickness t. The rigidities may then be written as

(a) flexural rigidity $D_{11} = D_{22} = Et^3/12(1-\nu^2)$

(b) coupling rigidity $D_{12} = D_{21} = \nu D_{11}$

(c) torsional rigidity $D_{66} = (1-\nu)D_{11}/2$

(d) shear rigidity $S_{44} = S_{55} = \kappa Et/(2+2\nu)$ (4.1)

where κ is the 'shear correction factor' usually taken as 0.83 for homogeneous isotropic plates.

4.2 Sandwich Plates

The equivalent plate rigidities for sandwich plates [4] may be derived from the expressions

$$D_{ij} = \int_{-t/2}^{t/2} Q_{ij}\, z^2\, dz \qquad (i,j = 1,2,6)$$

$$S_{ij} = \int_{-t/2}^{t/2} C_{ij}\, dz \qquad (i,j = 4,5) \qquad (4.2)$$

For sandwich plates with faces of equal thickness and material properties, (see Figure 4.1), there is no flexural-extensional coupling. The elastic rigidities for sandwich plates of isotropic material in both faces and cores may be calculated as follows. Each rigidity has a contribution from the faces of the sandwich plate and also from the core

(a) flexural rigidity $D_{11} = D_{22} = D = [D_{11}]_c + [D_{11}]_f$ (4.3)

(b) coupling rigidity $D_{12} = \nu_c[D_{11}]_c + \nu_f[D_{11}]_f$ (4.4)

(c) torsional rigidity $D_{66} = (1-\nu_c)\ [D_{11}]_c/2 + (1-\nu_f)\ [D_{11}]_f/2$ (4.5)

(d) shear rigidity $S_{44} = S_{55} = S = [S_{44}]_c+[S_{44}]_f$ (4.6)

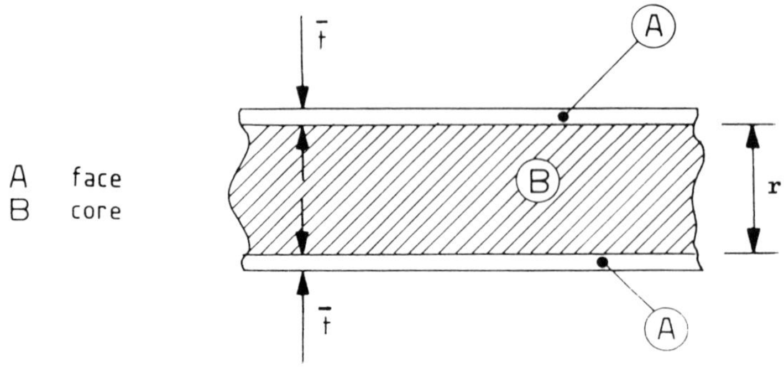

Figure 4.1 Sandwich plate

where r and $\bar{t}$ arc the thicknesses of the core and faces respectively, E_c and E_f are the elastic moduli for the core and faces respectively, ν_c and ν_f are the Poisson's ratios for the core and faces respectively. c and f refer to contributions from the core and faces respectively. Furthermore, the flexural rigidity is written as

$$D = \int_{-r/2}^{r/2} \frac{E_c z^2 dz}{2(1-\nu_c{}^2)} + \int_{-r/2}^{r/2+\bar{t}} \frac{E_f z^2 dz}{2(1-\nu_f{}^2)} + \int_{-r/2-\bar{t}}^{-r/2} \frac{E_f z^2 dz}{2(1-\nu_f{}^2)} \tag{4.7}$$

and the shear rigidity takes the form

$$S = \int_{-r/2}^{r/2} G_c\ dz + \int_{-r/2}^{r/2+\bar{t}} G_f\ dz + \int_{-r/2-\bar{t}}^{-r/2} G_f\ dz \tag{4.8}$$

in which the shear modulus for the core is

$$G_c = E_c/(1+\nu_c) \tag{4.9}$$

and the shear modulus for the faces is

$$G_f = E_f/(1+\nu_f) \tag{4.10}$$

s usual to ignore the contributions of $[D_{11}]_c$ to D_{11} etc. that of $[S_{44}]_f$ to S_{44} etc. Hence

$$D_{11} = D_{22} = \frac{E_f \bar{t}(r^2+2rt+4t^2/3)}{2(1-\nu_f^2)} + \frac{E_f \bar{t}(r+t)^2}{2(1-\nu_f^2)} \qquad (4.11)$$

$$S_{44} = S_{55} = G_c r \qquad (4.12)$$

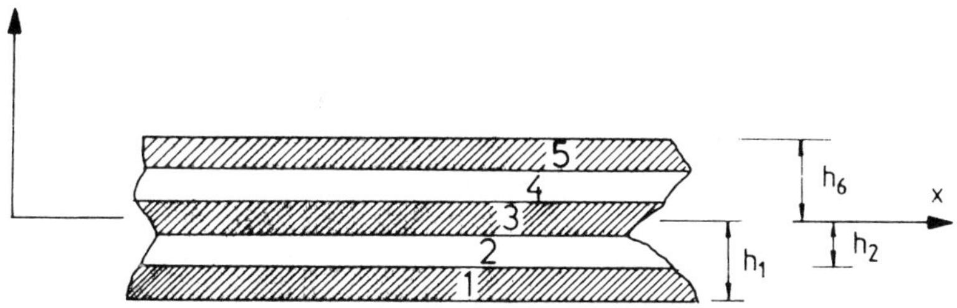

Figure 4.2 Nomenclature for a typical laminate plate

Laminated Plates

Formulae [5] are now presented for the evaluation of dities for laminated plates consisting of m laminates. pical laminated plate is shown in Figure 4.2. The owing information is usually available for each laminate

E_{11}, E_{22} - the elastic moduli in the principal laminate directions 1 and 2 respectively,

ν_{12}, ν_{21} - the Poisson's ratios in the 1-2 plane,

G_{12} - the shear modulus in the 1-2 plane,

G_{31}, G_{23} - the transverse shear moduli in the transverse 3-1 and 2-3 planes respectively, and

(e) $(h_{k+1} - h_k)$ - the thickness of the kth laminate.

The flexural rigidities are then given as

$$D_{ij} = \sum_{k=1}^{m} \int_{h_k}^{h_{k+1}} \bar{Q}^{h}_{ij} \; z^2 \; dz \qquad (i,j = 1,2,6) \qquad (4.13)$$

$$= \sum_{k=1}^{m} \bar{Q}^{k}_{ij} \; (h^3_{k+1} - h^3_{k})/3$$

and the shear rigidities are given as

$$S_{ij} = \sum_{k=1}^{m} \int_{h_k}^{h_{k+1}} \bar{C}_{ij} \; dz \qquad (i,j = 4,5) \qquad (4.14)$$

$$= \sum_{k=1}^{m} \bar{C}_{ij} \; (h_{k+1} - h_k)$$

where $\underline{\bar{Q}}_k$ are obtained by suitably transforming the reduced in-plane stiffnesses for plane stress for each laminate. The principal laminate directions are in the 1-2 directions shown in Figure 4.3. It is therefore necessary to transform from the 1-2 directions to the x-y directions prior to the evaluation of D_{ij}.

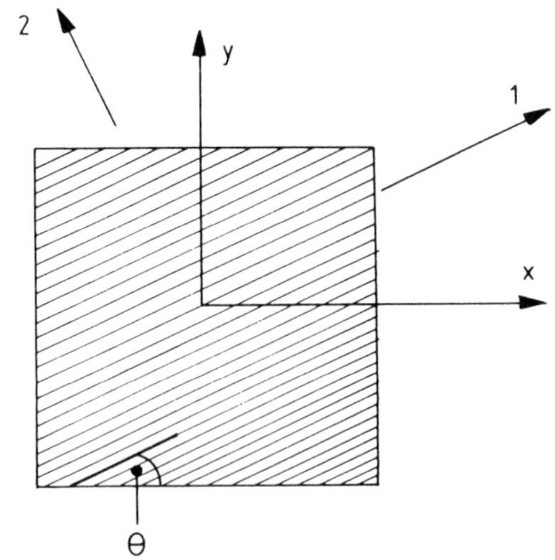

Figure 4.3 Principal laminate directions

The coefficients $\bar{\underline{Q}}_k$ are obtained from the matrix $\underline{Q}_k$ which is given as

$$\bar{\underline{Q}}_k = \underline{T}^{-1}\ \underline{Q}_k\ \underline{T} = \begin{bmatrix} \bar{Q}^k_{11} & \bar{Q}^k_{12} & \bar{Q}^k_{16} \\ \bar{Q}^k_{12} & \bar{Q}^k_{22} & \bar{Q}^k_{26} \\ \bar{Q}^k_{16} & \bar{Q}^k_{26} & \bar{Q}^k_{66} \end{bmatrix} \tag{4.15}$$

where

$$\underline{T} = \begin{bmatrix} c^2 & s^2 & 2sc \\ s^2 & c^2 & -2sc \\ -sc & sc & c^2-s^2 \end{bmatrix}$$

in which $s = \sin\theta$ and $c = \cos\theta$. Further, the reduced in-plane stiffnesses for plane stress may be written as

$$\underline{Q}_k = \begin{bmatrix} Q^k_{11} & Q^k_{12} & 0 \\ Q^k_{12} & Q^k_{22} & 0 \\ 0 & 0 & Q^k_{66} \end{bmatrix}$$

$$Q^k_{11} = E_{11}/(1-\nu_{12}\nu_{21})$$

$$Q^k_{22} = E_{22}/(1-\nu_{12}\nu_{21})$$

$$Q^k_{12} = Q^k_{21} = \nu_{21}E_{11}/(1-\nu_{12}\nu_{21}) = \nu_{12}E_{22}(1-\nu_{12}\nu_{21})$$

$$Q^k_{66} = G_{12}$$

Thus, the components of $\bar{\underline{Q}}_k$ may be written as

$$\bar{Q}^k_{11} = Q^k_{11}\ c^4 + 2(Q^k_{12} + Q^k_{66})c^2s^2 + Q^k_{22}\ s^4$$

$$\bar{Q}^k_{22} = Q^k_{11}\ s^4 + 2(Q^k_{12} + Q^k_{66})c^2s^2 + Q^k_{22}\ c^4$$

$$\bar{Q}_{12} = (Q^k_{11} + Q^k_{22} - 2Q^k_{66})s^2c^2 + Q^k_{12}\ (s^4 + c^4)$$

$$\bar{Q}_{66} = (Q^k_{11} + Q^k_{22} - 2Q^k_{12} - Q^k_{66})s^2c^2 + Q^k_{33}\ (s^4 + c^4)$$

$$\bar{Q}^k_{16} = (Q^k_{11} - Q^k_{12} - Q^k_{66})cs^3 + (Q^k_{12} - Q^k_{12} + Q^k_{66})c^3s$$

$$\bar{Q}^k_{26} = (Q^k_{11} - Q^k_{12} - Q^k_{66})s^3c + (Q^k_{12} - Q^k_{22} + Q^k_{66})sc^3 \qquad (4.16)$$

The matrix of shear stiffnesses $\underline{\bar{C}}_k$ given in (4.14) may be obtained simply by suitably transforming the matrix $\underline{C}_k$ using the expression

$$\underline{\bar{C}}_k = \underline{R}^{-1}\ \underline{C}_k\ \underline{R} = \begin{bmatrix} \bar{C}^k_{44} & \bar{C}^k_{45} \\ \bar{C}^k_{45} & \bar{C}^k_{55} \end{bmatrix} \qquad (4.17)$$

where

$$\underline{R} = \begin{bmatrix} c & -s \\ s & c \end{bmatrix}$$

and

$$\bar{C}^k_{44} = C^k_{44}\ c^2 + C^k_{55}\ s^2$$

$$\bar{C}^k_{55} = C^k_{44}\ s^2 + C^k_{55}\ c^2$$

$$\bar{C}^k_{45} = (C^k_{55} - C^k_{44})sc$$

Note that

$$C^k_{44} = G_{31}$$

and

$$C^k_{55} = G_{23} \qquad (4.18)$$

Note that shear modification factors may also be calculated for laminated plates - see reference [5].

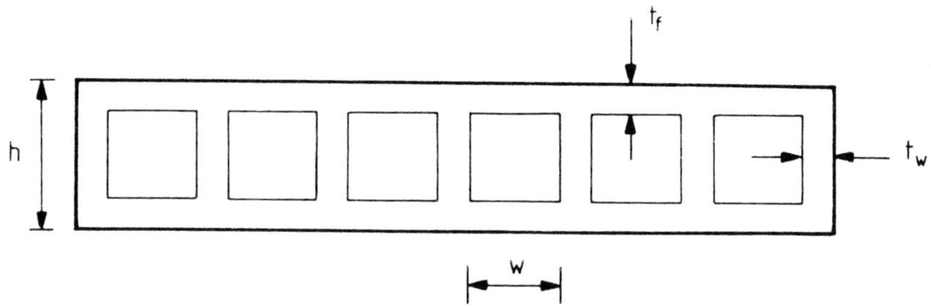

Figure 4.4 Voided plate cross-section

4.4 Voided Plates

The formulate suggested by Basu and Dawson [6] for the evaluation of the rigidities for voided plates of the type shown in Figure 4.4 are now summarised

(a) transverse flexural rigidity

$$D_{11} = \frac{Et_f \ h^2}{2(1-\nu^2)} \tag{4.19}$$

(N.B. the flexural rigidity of the flanges only is considered),

(b) longitudinal flexural rigidity

$$D_{22} = D_{11} \ [1 + (t_w h/6_f w)] \tag{4.20}$$

(N.B. the flexural rigidity of the flanges and the webs are considered.)

(c) coupling rigidity

$$D_{12} = D_{21} = \nu D_{11} \tag{4.21}$$

(d) torsional rigidity

$$D_{66} = Gt_f h^2/2 \tag{4.22}$$

(N.B. If there is a sufficient number of cells (say 4 or more as required for orthotropic plate theory to hold), and the cells are identical or almost identical, then the contribution of the interior webs can be neglected in calculating the torsional rigidity of the multi-cellular plate. The rigidity is thus equal to that of the outside box alone and is given as $Gt_f h^2$).

(e) transverse shear rigidity

$$S_{44} = \frac{2Et_f^3}{w^2[1 + 2(h/w)(t_f/t_w)^3]\ (1-\nu^2)} \tag{4.23}$$

The transverse shearing rigidity S_{44} is provided by the flexure of the flange and the webs on the transverse cross-sections of the plate, as in the case of a Vierendeel girder, and is obtained on the assumption that there are points of contraflexure in the flanges between the vertical webs.

(f) longitudinal shear rigidity

$$S_{55} = Gt_f h(1 + t_f/h)/(t_f w/t_w) \tag{4.24}$$

The longitudinal shear rigidity S_{55} is obtained on the assumption that the vertical shearing force is resisted by the webs alone and that the distribution of shear stress in the webs is uniform.

N.B.
- t_f - flange thickness
- t_w - web thickness
- h - centre to centre distance between flanges
- w - web spacing
- a - width of plate

5. PROGRAMS

In this section two programs are described

(a) PLATES which performs the closed form analysis described in Section 3 for simply supported rectangular Mindlin plates.

(b) RIGID which evaluates the plate rigidities D_{11}, D_{12}, D_{22}, D_{66}, S_{44} and S_{55} for homogeneous isotropic, sandwich, laminated and voided plates.

5.1 Program PLATES

In this section program PLATES which performs the closed form plate analysis is presented. This program consists of a main segment and five subroutines as shown in Figure 5.1. The subroutines are now briefly described:-

MASTER is the main subroutine and it calls all of the other subroutines in order.

DATA reads in all of the input data values.

INIT zeroes all initial values apart from the input values.

COEF calculates the coefficients P_{11}, P_{12}, P_{13}, etc.

CONS calculates the constant values A_{mn}, B_{mn} and C_{mn} for a given loading function.

OUTP prints out the results from PLATES.

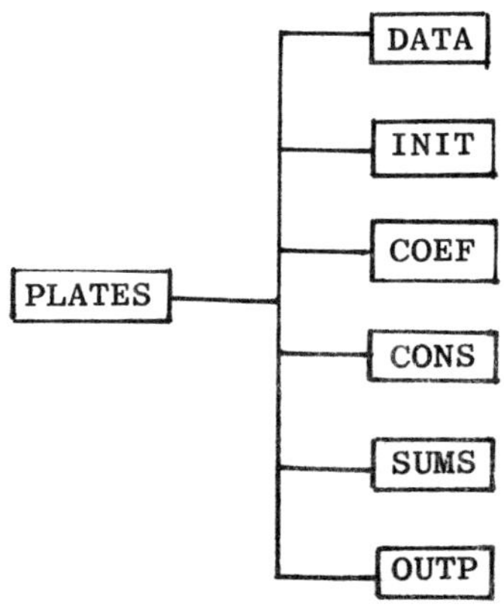

Fig.5.1 Flow diagram for program PLATES

5.1.1 Master Segment

```
C     PROGRAM PLATES
C
C*** CLOSED FORM SOLUTION FOR ORTHOTROPIC MINDLIN PLATES
C
      IMPLICIT REAL*8(A-H,O-Z)
      DIMENSION  BMX(140),BMY(140),BMXY(140),CX(140),CY(140),CXY(140),
     .           QX(140),QY(140),TITLE(20),W(140),WUX(140),WUY(140),
     .           XCORD(140),YCORD(140)
      DATA MPON/140/
C
      CALL DATA (AA,BB,D11,D22,D21,D12,D66,ETA,ETB,G1,
     .           GFS,IQ,IT,MPON,NPON,NUM,PI,PZ,S44,S55,
     .           TITLE,U,V,XCORD,YCORD)
      CALL INIT (BMX,BMY,BMXY,CX,CY,CXY,MPON,NPON,QX,QY,W,WUX,WUY)
      DO 5 M=1,NUM,IT
      DO 5 N=1,NUM,IT
      CALL COEF (AA,BB,D11,D22,D12,D66,DET,G1,GFS,M,N,
     .           P11,P12,P13,P22,P23,PI,S44,S55)
      CALL CONS (AA,AMN,BB,BMN,CMN,DET,ETA,ETB,IQ,M,N,PI,PZ,P11,P12,
     .           P13,P22,P23,U,V)
      CALL SUMS (AA,AMN,BB,BMN,BMX,BMY,BMXY,CMN,CX,CY,CXY,D11,D22,
     .           D12,D66,G1,M,MPON,N,NPON,QX,QY,PI,S44,S55,W,WUX,
     .           WUY,XCORD,YCORD)
    5 CONTINUE
      CALL OUTP (AA,BB,BMX,BMY,BMXY,D11,D22,D21,D12,D66,ETA,ETB,
     .           GFS,G1,IQ,IT,MPON,NPON,NUM,QX,QY,PZ,S44,
     .           S55,TITLE,U,V,W,WUX,WUY,XCORD,YCORD)
      STOP
      END
```

5.1.2 Subroutine DATA

```
      SUBROUTINE DATA (AA,BB,D11,D22,D21,D12,D66,ETA,ETB,
     .                 G1,GFS,IQ,IT,MPON,NPON,NUM,PI,PZ,S44,
     .                 S55,TITLE,U,V,XCORD,YCORD)
C
C***  THIS SUBROUTINE READS IN ALL DATA
C
      IMPLICIT REAL*8(A-H,O-Z)
      DIMENSION        TITLE(20),XCORD(MPON),YCORD(MPON)
C
      READ(5,900)TITLE
  900 FORMAT(20A4)
      READ(5,905)AA,BB
  905 FORMAT(2F10.5)
      READ(5,910)D11,D22,D12,D66
  910 FORMAT(4F10.5)
      D21=D12
      READ(5,915)S44,S55
  915 FORMAT(2F10.5)
      READ(5,920)NUM,NPON
  920 FORMAT(2I5)
      READ(5,925)PZ,GFS,G1
  925 FORMAT(3F10.5)
      DO 5 IPON=1,NPON
      READ(5,930)JPON,XCORD(IPON),YCORD(IPON)
  930 FORMAT(I5,2F10.5)
    5 CONTINUE
      READ(5,935)IQ,IT
```

```
 935  FORMAT(2I5)
      IF(IQ.LT.2) GO TO 10
      READ(5,940)ETA,ETB
 940  FORMAT(2F10.5)
      IF(IQ.LT.3) GO TO 10
      READ(5,945)U,V
 945  FORMAT(2F10.5)
  10  CONTINUE
      PI=3.141592654
      RETURN
      END
```

5.1.3 Subroutine INIT

```
      SUBROUTINE INIT (BMX,BMY,BMXY,CX,CY,CXY,MPON,
     .                 NPON,QX,QY,W,WUX,WUY)
C
C*** THIS SUBROUTINE INITIALISES VARIOUS ARRAYS
C
      IMPLICIT REAL*8(A-H,O-Z)
      DIMENSION        BMX(MPON),BMY(MPON),BMXY(MPON),
     .                 CX(MPON),CY(MPON),CXY(MPON),
     .                 QX(MPON),QY(MPON),WUX(MPON),
     .                 WUY(MPON),W(MPON)
C
      DO 5 IPON=1,NPON
      W(IPON)=0.0
      WUX(IPON)=0.0
      WUY(IPON)=0.0
      CX(IPON)=0.0
      CY(IPON)=0.0
      CXY(IPON)=0.0
      BMX(IPON)=0.0
      BMY(IPON)=0.0
      BMXY(IPON)=0.0
      QX(IPON)=0.0
      QY(IPON)=0.0
   5  CONTINUE
      RETURN
      END
```

5.1.4 Subroutine CONS

```
      SUBROUTINE CONS (AA,AMN,BB,BMN,CMN,DET,ETA,ETB,IQ,M,N,PI,PZ,P11,
     .                 P12,P13,P22,P23,U,V)
      IMPLICIT REAL*8(A-H,O-Z)
C
C*** THIS SUBROUTINE CALCULATES THE VALUES
C*** AMN, BMN AND CMN FOR A GIVEN LOADING FUNCTION
C
      GO TO(5,10,15,20),IQ
   5  QMN=(16.0*PZ)/(PI*PI*FLOAT(M)*FLOAT(N))
      GO TO 25
  10  QMN=((4.0*PZ)/(AA*BB))*SIN((FLOAT(M)*PI*ETA)/AA)*
     .SIN((FLOAT(N)*PI*ETB)/BB)
      GO TO 25
  15  QMN=((16.0*PZ)/(PI*PI*FLOAT(M)*FLOAT(N)*U*V))*
     .SIN((FLOAT(M)*PI*ETA)/AA)*
     .SIN((FLOAT(N)*PI*ETB)/BB)*SIN((FLOAT(M)*PI*U)/(2.0*AA))*
     .SIN((FLOAT(N)*PI*V)/(2.0*BB))
      GO TO 25
  20  QMN=(4.0*PZ*SIN((FLOAT(N)*PI*ETB)/BB)*SIN((FLOAT(M)*PI*ETA)/AA)
     .*COS((FLOAT(N)*PI*V)/(2.0*BB))*COS((FLOAT(M)*PI*U)/(2.0*AA)))/
     .((AA*BB*U*U*V*V)*((FLOAT(N)/BB)-(1.0/V))*
```

```
     .((FLOAT(N)/BB)+(1.0/V))*
     .((FLOAT(M)/AA)-(1.0/U))*((FLOAT(M)/AA)+(1.0/U)))
 25   AMN=(P12*P23-P22*P13)*QMN/DET
      BMN=(P12*P13-P11*P23)*QMN/DET
      CMN=(P11*P22-P12*P12)*QMN/DET
      RETURN
      END
```

5.1.5 Subroutine COEF

```
       SUBROUTINE COEF (AA,BB,D11,D22,D12,D66,DET,G1,GFS,M,
     .                  N,P11,P12,P13,P22,P23,PI,S44,S55)
      IMPLICIT REAL*8(A-H,O-Z)
C
C*** THIS SUBROUTINE CALCULATES THE COEFFICIENTS
C*** (P11,P12,P13,P22,P23,P33) AND THE DETERMINANT
C
      P11=D11*(FLOAT(M)*PI/AA)*(FLOAT(M)*PI/AA)
     .+D66*(FLOAT(N)*PI/BB)*(FLOAT(N)*PI/BB)+G1*S55
      P12=(D12+D66)*(FLOAT(M)*PI/AA)*(FLOAT(N)*PI/BB)
      P13=G1*S55*(FLOAT(M)*PI/AA)
      P22=D66*(FLOAT(M)*PI/AA)*(FLOAT(M)*PI/AA)
     .+D22*(FLOAT(N)*PI/BB)*(FLOAT(N)*PI/BB)+G1*S44
      P23=G1*S44*(FLOAT(N)*PI/BB)
      P33=(G1*S55)*(FLOAT(M)*PI/AA)*(FLOAT(M)*PI/AA)
     .+(G1*S44)*(FLOAT(N)*PI/BB)*(FLOAT(N)*PI/BB)+GFS
      DET=P11*(P22*P33-P23*P23)-P12*(P12*P33-P23*P13)
     .+P13*(P12*P23-P22*P13)
      RETURN
      END
```

5.1.6 Subroutine SUMS

```
      SUBROUTINE SUMS (AA,AMN,BB,BMN,BMX,BMY,BMXY,CMN,CX,CY,CXY,D11,
     .                 D22,D12,D66,G1,M,MPON,N,NPON,QX,QY,PI,S44,S55,W,
     .                 WUX,WUY,XCORD,YCORD)
C
C*** THIS SUBROUTINE SUMS THE VARIOUS FOURIER SERIES
C
      IMPLICIT REAL*8(A-H,O-Z)
      DIMENSION        BMX(MPON),BMY(MPON),BMXY(MPON),CX(MPON),CY(MPON),
     .                 CXY(MPON),QX(MPON),QY(MPON),W(MPON),WUX(MPON),
     .                 WUY(MPON),XCORD(MPON),YCORD(MPON)
C
      DO 5 IPON=1,NPON
      X=XCORD(IPON)
      Y=YCORD(IPON)
C
C*** DEFLECTION W(IPON)
C
      W(IPON)=W(IPON)+CMN*SIN((FLOAT(M)*PI*X)/AA)*
     .SIN((FLOAT(N)*PI*Y)/BB)
C
C*** SLOPE WUX(IPON)
C
      WUX(IPON)=WUX(IPON)+AMN*COS((FLOAT(M)*PI*X)/AA)*
     .SIN((FLOAT(N)*PI*Y)/BB)
C
C*** SLOPE WUY(IPON)
C
      WUY(IPON)=WUY(IPON)+BMN*SIN((FLOAT(M)*PI*X)/AA)*
     .COS((FLOAT(N)*PI*Y)/BB)
C
C*** CURVATURE CX(IPON)
C
```

```
      CX(IPON)=CX(IPON)+AMN*(FLOAT(M)*PI/AA)*SIN((FLOAT(M)*PI*X)/AA)*
     .SIN((FLOAT(N)*PI*Y)/BB)
C
C*** CURVATURE CY(IPON)
C
      CY(IPON)=CY(IPON)+BMN*(FLOAT(N)*PI/BB)*SIN((FLOAT(M)*PI*X)/AA)*
     .SIN((FLOAT(N)*PI*Y)/BB)
C
C*** CURVATURE CXY(IPON)
C
      CXY(IPON)=CXY(IPON)+((AMN*(FLOAT(N)*PI/BB))+
     .(BMN*(FLOAT(M)*PI/AA)))*
     .COS((FLOAT(M)*PI*X)/AA)*COS((FLOAT(N)*PI*Y)/BB)
C
C*** BENDING MOMENT BMX(IPON)
C
      BMX(IPON)=-D11*CX(IPON)-D12*CY(IPON)
C
C*** BENDING MOMENT BMY(IPON)
C
      BMY(IPON)=-D22*CY(IPON)-D12*CX(IPON)
C
C*** TWISTING MOMENT BMXY(IPON)
C
      BMXY(IPON)=D66*CXY(IPON)
C
C*** SHEAR FORCE QX(IPON)
C
      QX(IPON)=QX(IPON)+G1*S55*((CMN*FLOAT(M)*PI/AA)+AMN)*
     .COS((FLOAT(M)*PI*X)/AA)*SIN((FLOAT(N)*PI*Y)/BB)
C
C*** SHEAR FORCE QY(IPON)
C
      QY(IPON)=QY(IPON)+G1*S44*((CMN*FLOAT(N)*PI/BB)+BMN)*
     .SIN((FLOAT(M)*PI*X)/AA)*COS((FLOAT(N)*PI*Y)/BB)
    5 CONTINUE
      RETURN
      END
```

5.1.7 Subroutine OUTP

```
      SUBROUTINE OUTP (AA,BB,BMX,BMY,BMXY,D11,D22,D21,D12,D66,
     .                 ETA,ETB,GFS,G1,IQ,IT,MPON,NPON,NUM,QX,
     .                 QY,PZ,S44,S55,TITLE,U,V,W,WUX,WUY,
     .                 XCORD,YCORD)
C
C*** THIS SUBROUTINE PRINTS OUT THE RESULTS
C
      IMPLICIT REAL*8(A-H,O-Z)
      DIMENSION        BMX(MPON),BMY(MPON),BMXY(MPON),QX(MPON),
     .                 QY(MPON),WUX(MPON),WUY(MPON),W(MPON),
     .                 TITLE(20),XCORD(MPON),YCORD(MPON)
C
      WRITE(6,900)
  900 FORMAT(//,1X,'**** CLOSED FORM SOLUTION ****',/)
      WRITE(6,905)TITLE
  905 FORMAT(/,20A4)
      WRITE(6,910)
  910 FORMAT(1X,50('*'))
      WRITE(6,915)AA,BB
  915 FORMAT(//,1X,'DIMENSION OF PLATE IN X DIRECTION.','(AA)=',E12.5,
     ./,1X,'DIMENSION OF PLATE IN Y DIRECTION.','(BB)=',E12.5,/)
      WRITE(6,920)D11,D22,D21,D66
  920 FORMAT(1X,'FLEXURAL RIGIDITY....','(D11)=',E15.8,/,
     .       1X,'FLEXURAL RIGIDITY....','(D22)=',E15.8,/,
     .       1X,'COUPLING RIGIDITY....','(D12)=',E15.8,/,
     .       1X,'TORSIONAL RIGIDITY...','(D66)=',E15.8,/)
```

```
      WRITE(6,925)S44,S55
  925 FORMAT(1X,'TRANSVERSE SHEAR RIGIDITY.....','(S44)=',E15.8,/,
     .       1X,'TRANSVERSE SHEAR RIGIDITY.....','(S55)=',E15.8,/)
      WRITE(6,930)G1,GFS,PZ
  930 FORMAT(1X,'SHEAR CORRECTION FACTOR.........','(G1)=',E15.8,/,
     .       1X,'MODULUS FOR ELASTIC FOUNDATION.','(GFS)=',E15.8,/,
     .       1X,'LOAD INTENSITY.................','(PZ)=',E15.8,/)
      WRITE(6,935)NUM,NPON
  935 FORMAT(1X,'THE NO. OF TERMS IN FOURIER SERIES',1X,' (NUM)=',I5,
     ./,1X,'THE NO. OF POINTS ON PLATE',8X,' (NPON)=',I5)
      IF (IQ .EQ. 1) GO TO 5
      WRITE(6,940)ETA,ETB
  940 FORMAT(1X,'X-COORD. ',5X,'(ETA)=',E15.8,
     ./,1X,'Y-COORD.',5X,'(ETB)=',E15.8)
      IF (IQ .EQ. 2) GO TO 5
      WRITE(6,945)U,V
  945 FORMAT(1X,'DIMENS. OF LOADED AREA IN X DIRECT.',1X,'(U)=',E15.8,
     ./,1X,'DIMENS. OF LOADED AREA IN Y DIRECT.',2X,'(V)=',E15.8)
    5 WRITE(6,950)
  950 FORMAT(//,50('*'))
      WRITE(6,955)
  955 FORMAT(///,1X,'TABLE SHOWS THE COMPUTED VALUES FOR'/,
     .1X,'SPECIFIED X AND Y COORDINATES.')
      IF (IQ .NE. 1) GO TO 10
      WRITE(6,960)
  960 FORMAT(/,1X,'FOR UNIFORM LOADING.')
   10 IF (IQ .NE. 2) GO TO 15
      WRITE(6,965)
  965 FORMAT(/,1X,'FOR CONCENTRATED LOADING.')
   15 IF (IQ .NE. 3) GO TO 20
      WRITE(6,970)
  970 FORMAT(/,1X,'FOR UNIFORM LOADING OVER RECTANGULAR AREAS (U.V).')
   20 IF (IQ .NE. 4) GO TO 25
      WRITE(6,975)
  975 FORMAT(/,1X,'FOR COSINE LOADING.')
   25 IF(IT.GT.1) GO TO 30
      WRITE(6,980)
  980 FORMAT(/,1X,'NON-SYMMETRIC LOADING.')
   30 IF (IT .NE. 2) GO TO 35
      WRITE(6,985)
  985 FORMAT(/,1X,'SYMMETRIC LOADING.')
   35 WRITE(6,990)
  990 FORMAT(//,9X,'X',9X,'Y',13X,'W',12X,'WUX',11X,'WUY')
      DO 40 IPON=1,NPON
      WRITE(6,995)XCORD(IPON),YCORD(IPON),W(IPON),WUX(IPON),
     .WUY(IPON)
  995 FORMAT(/,1X,2E12.5,3E14.5)
   40 CONTINUE
      WRITE(6,1000)
 1000 FORMAT(//,9X,'X',9X,'Y',13X,'MX',12X,'MY',12X,'MXY')
      DO 45 IPON=1,NPON
      WRITE(6,1005) XCORD(IPON),YCORD(IPON),BMX(IPON),BMY(IPON),
     .BMXY(IPON)
 1005 FORMAT(/,1X,2E12.5,3E14.5)
   45 CONTINUE
      WRITE(6,1010)
 1010 FORMAT(//,9X,'X',9X,'Y',13X,'QX',13X,'QY')
      DO 50 IPON=1,NPON
      WRITE(6,1015) XCORD(IPON),YCORD(IPON),QX(IPON),QY(IPON)
 1015 FORMAT(/,1X,2E12.5,2E14.5)
   50 CONTINUE
      RETURN
      END
```

5.2 Input Instructions for PLATES

Input instructions for program PLATES are now provided

Data set 1	PROBLEM TITLE	(20A4)	one record
Cols. 1 - 80	TITLE	title of problem	
Data set 2	PLATE DIMENSIONS	(2F10.5)	one record
Cols. 1 - 10	AA	side length parallel to x axis	
11 - 20	BB	side length parallel to y axis	
Data set 3	FLEXURAL PROPERTIES	(4F10.5)	one record
Cols. 1 - 10	D11	flexural rigidity D_{11}	
11 - 20	D22	flexural rigidity D_{22}	
21 - 30	D21	coupling rigidity D_{21}	
31 - 40	D66	torsional rigidity D_{66}	
Data set 4	SHEAR PROPERTIES	(2F10.5)	one record
Cols. 1 - 10	S44	shear rigidity S_{44}	
11 - 20	S55	shear rigidity S_{55}	
Data set 5	CONTROL DATA	(2I5)	one record
Cols. 1 - 5	NUM	number of Fourier terms	
6 - 10	NPON	number of output points	
Data set 6	FURTHER PLATE PARAMETERS	(3F10.5)	one record
Cols. 1 - 10	PZ	Load intensity	
11 - 20	GFS	foundation modulus	
21 - 30	G1	shear modification factor	
Data set 7	SAMPLING POINTS	(I5,2F10.5)	NPON records
Cols. 1 - 5	JPON	point number	
6 - 15	XCORD(.)	x-coordinate	
21 - 30	YCORD(.)	y-coordinate	
Data set 8	LOADING TYPE	(2I5)	one record
Cols. 1 - 5	IQ	=1 uniform loading 2 concentrated loading 3 uniform loading over rectangular patch uv 4 cosine loading over rectangular patch uv	

6 - 10	IT	=1 non-symmetric loads 2 symmetric loads

Data set 9 LOADING POSITION (2F10.5) one record

Cols. 1 - 10	ETA	x-coordinate at centre of rectangular patch
11 - 20	ETB	y-coordinate at centre of rectangular patch

[N.B. Omit this data set if IQ < 2.]

Data set 10 LOAD PATCH SIZE (2F10.5) one record

Cols. 1 - 10	U	side length of patch parallel to x axis
11 - 20	V	side length of patch parallel to y axis

[N.B. Omit this data set if IQ < 2.]

5.3 Glossary of variable names used in PLATES

AA,BB length of plates in the x,y directions resp.

AMN, BMN, CMN unknown Fourier coefficients A_{mn}, B_{mn}, C_{mn}

BMX, BMY, BMXY bending moments M_x, M_y, M_{xy}

CX, CY, CXY curvatures $\theta_{x,x}$, $\theta_{y,y}$, $(\theta_{x,y} + \theta_{y,x})$

D11, D22, D12 D21, D66 flexural rigidities D_{11}, D_{22}, D_{12}, D_{21}, D_{66}

DET determinant of matrix $\underline{P}$

ETA, ETB x,y coordinates of centroid of load

G1 shear modification factor

GFS elastic foundation modulus K

IT =1 for non-symmetric loads
2 for symmetric loads

IQ =1 for uniform loads
2 for concentrated loads
3 for uniform patch loads
4 for cosine patch loads

MPON maximum number of output points (140)

NPON number of output points

NUM number of Fourier terms

P11, P22, etc. coefficients used to evaluate A_{mn}, B_{mn}, C_{mn}

PI π

PZ load intensity

QX, QY shear forces

S44, S55 shear rigidities S_{44}, S_{55}

TITLE problem title

U, V dimensions of patch load

W, WUX, WUY w, $w_{,x}$, $w_{,y}$

XCORD,YCORD x,y coordinates of output points

5.4 Program RIGID

The program RIGID is now described. RIGID evaluates the plate rigidities for isotropic, sandwich, laminated and voided plates using the formulae presented in Section 4. RIGID consists of a main segment and six other subroutines as shown in Figure 5.2.

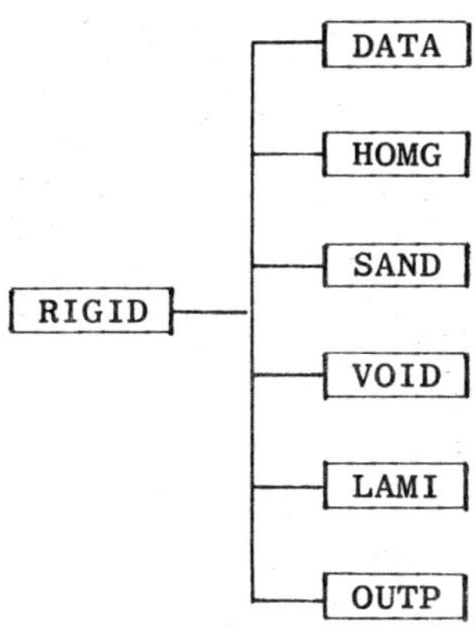

Figure 5.2 Flow diagram for program RIGID

These routines are now briefly described and then listed.

RIGID is the main routine and it calls all of the other subroutines in order.

DATA reads in all of the main input values.

HOMG calculates the rigidities for isotropic homogeneous plates E, ν and t.

SAND calculates the rigidities for a sandwich plate given E_f, G_c, $\bar{t}$, r and ν_f.

VOID calculates the rigidities of a voided plate given E, ν, t_f, h, w and t_w.

LAMI calculates the rigidities of a laminated plate given the number of layers and properties for each typical layer k - E^k_{11}, E^k_{22}, ν^k_{12}, ν^k_{21}, G^k_{12}, G^k_{23}, G^k_{31}, θ^k and $(h_{k+1} - h_k)$.

OUTP outputs the results D_{11}, D_{22}, D_{12}, D_{66}, S_{44}, S_{55}.

5.4.1 Main segment

```
C      PROGRAM RIGID
C
C*** CALCULATES RIGIDITIES FOR
C*** HOMOGENEOUS,SANDWICH,VOIDED AND LAMINATED
C*** PLATES
C
      IMPLICIT REAL*8(A-H,O-Z)
      DIMENSION A(10),D(6,6),H1(11),TITLE(20)
C
      CALL DATA (A,C,E,E11,E22,EF,GC,G12,G23,H,H1,IQ,
     .           M,T,TF,TITLE,TW,V,V12,VF,W)
      GO TO (5,10,15,20),IQ
    5 CALL HOMG (D11,D22,D12,D21,D66,E,S44,S55,T,V)
      GO TO 25
   10 CALL SAND (C,D11,D22,D12,D21,D66,EF,GC,S44,S55,T,VF)
      GO TO 25
   15 CALL VOID (D11,D22,D12,D21,D66,E,H,S44,S55,TF,TW,V,W)
      GO TO 25
   20 CALL LAMI (A,D,E11,E22,G12,G23,H1,M,V12)
   25 CALL OUTP (C,D,D11,D22,D12,D21,D66,E,E11,E22,EF,
     .           GC,G12,G23,H,IQ,S44,S55,T,TF,TITLE,
     .           TW,V,V12,VF,W)
      STOP
      END
```

5.4.2 Subroutine DATA

```
      SUBROUTINE DATA (A,C,E,E11,E22,EF,GC,G12,G23,H,H1,
     .                 IQ,M,T,TF,TITLE,TW,V,V1,VF,W)
C
C*** READS ALL DATA
C
      IMPLICIT REAL*8(A-H,O-Z)
      DIMENSION        A(10),H1(11),TITLE(20)
C
      READ(5,900)TITLE
  900 FORMAT(20A4)
      READ(5,905)IQ
  905 FORMAT(I5)
      GO TO (5,10,15,20),IQ
C
    5 READ(5,910)V,T,E
  910 FORMAT(3F10.5)
      RETURN
C
   10 READ(5,915)VF,GC,EF
  915 FORMAT(3F10.5)
      READ(5,920)C,T
  920 FORMAT(2F10.5)
      RETURN
C
   15 READ(5,925)V,E
  925 FORMAT(2F10.5)
      READ(5,930)TF,TW,H,W
  930 FORMAT(4F10.5)
      RETURN
C
   20 READ(5,935)E11,E22,V12
  935 FORMAT(3F10.5)
      READ(5,940)G12,G23
  940 FORMAT(2F10.5)
      READ(5,945)M
  945 FORMAT(I5)
      J=M+1
      DO 25 I=1,J
      READ(5,950)H1(I)
  950 FORMAT(F10.5)
   25 CONTINUE
      DO 30 K=1,M
      READ(5,955)A(K)
  955 FORMAT(F10.5)
   30 CONTINUE
      RETURN
      END
```

5.4.3 Subroutine HOMG

```
      SUBROUTINE HOMG (D11,D22,D12,D21,D66,E,S44,S55,T,V)
C
C*** CALCULATES RIGIDITIES FOR HOMOGENEOUS PLATES
C
      IMPLICIT REAL*8(A-H,O-Z)
      D11=(E*(T**3))/(12.0*(1.0-V*V))
      D22=D11
      D12=V*D11
      D21=D12
      D66=((1.0-V)*D11)/2.0
      S44=(E*T)/(2.0*(1.0+V)*(1.2))
      S55=S44
      RETURN
      END
```

5.4.4 Subroutine SAND

```
      SUBROUTINE SAND (C,D11,D22,D12,D21,D66,EF,GC,S44,
     .                  S55,T,VF)
C
C*** CALCULATES RIGIDITIES FOR SANDWICH PLATES
C
      IMPLICIT REAL*8(A-H,O-Z)
      D11=(EF*T*((C+T)**2.0))/(2.0*(1.0-VF*VF))
      D22=D11
      D12=VF*D11
      D21=D12
      D66=((1.0-VF)*D11)/2.0
      S44=C*GC/1.2
      S55=S44
      RETURN
      END
```

5.4.5 Subroutine VOID

```
      SUBROUTINE VOID (D11,D22,D12,D21,D66,E,H,S44,S55,
     .                  TF,TW,V,W)
C
C*** CALCULATES RIGIDITIES FOR VOIDED PLATES
C
      IMPLICIT REAL*8(A-H,O-Z)
      G=E/(2.0*(1.0+V))
      D11=(E*TF*H*H)/(2.0*(1.0-V*V))
      D22=D11*(1.0+(TW*H)/(6.0*TF*W))
      D12=V*D11
      D21=D12
      D66=(G*TF*H*H)/2.0
      S44=(2.0*E*(TF**3))/(W*W*(1.0+(2.0*(H/W))*((TF/TW)
     .**3))*(1.0-V*V))
      S55=(G*TF*H*(1.0+(TF/H)))/((TF/TW)*W)
      RETURN
      END
```

5.4.6 Subroutine LAMI

```
      SUBROUTINE LAMI (A,D,E11,E22,G12,G23,H1,M,V12)
C
C*** CALCULATES RIGIDITIES FOR LAMINATED PLATES
C
      IMPLICIT REAL*8(A-H,O-Z)
      DIMENSION       A(10),AA(6,6),A2(6,6),C1(6,6),
     .                D(6,6),H1(11),Q(6,6)
C
      V21=V12*E22/E11
      G31=G23
C
      Q(1,1)=E11/(1.0-V12*V21)
      Q(1,2)=E11*V12/(1.0-V12*V21)
      Q(1,6)=0.0
      Q(2,1)=Q(1,2)
      Q(2,2)=E22/(1.0-V12*V21)
      Q(2,6)=0.0
      Q(4,4)=G23
      Q(5,5)=G31
      Q(6,1)=0.0
      Q(6,2)=0.0
      Q(6,6)=G12
```

```
C
      DO 15 I=1,6
      DO 10 J=1,6
      D(I,J)=0.0
      DO 5 K=1,M
      ANGLE=3.1415927*A(K)/180.0
      CC=COS(ANGLE)
      SS=SIN(ANGLE)
      C1(1,1)=0.0
      C1(1,2)=0.0
      C1(1,6)=0.0
      C1(2,2)=0.0
      C1(2,6)=0.0
      C1(2,1)=0.0
      C1(6,6)=0.0
      C1(6,1)=0.0
      C1(6,2)=0.0
C
      C1(1,1)=(Q(1,1)*(CC**4))+(2.0*(Q(1,2)+Q(6,6))*CC*CC
     .*SS*SS)+(Q(2,2)*(SS**4))
      C1(1,2)=((Q(1,1)+Q(2,2)-2.0*Q(6,6))*SS*SS*CC*CC)
     .+(Q(1,2)*(CC**4+SS**4))
      C1(1,6)=((Q(1,1)-Q(1,2)-Q(6,6))*SS*CC**3)+
     .((Q(1,2)-Q(2,2)+Q(6,6))*(SS**3)*CC)
      C1(2,2)=(Q(1,1)*SS**4)+(2.0*(Q(1,2)+Q(6,6))*CC*CC*SS*SS)
     .+(Q(2,2)*CC**4)
      C1(2,6)=((Q(1,1)-Q(1,2)-Q(6,6))*(SS**3)*CC)+((Q(1,2)-Q(2,2)
     .+Q(6,6))*SS*(CC**3))
      C1(6,6)=((Q(1,1)+Q(2,2)-2.0*Q(1,2)-Q(6,6))*CC*CC*SS*SS)
     .+(Q(6,6)*(CC**4+SS**4))
      C1(2,1)=C1(1,2)
      C1(6,1)=C1(1,6)
      C1(6,2)=C1(2,6)
      D(I,J)=D(I,J)+(C1(I,J)*(((H1(K+1))**3)-((H1(K))**3)))/3.0
    5 CONTINUE
   10 CONTINUE
   15 CONTINUE
C
      DO 30 I=4,5
      DO 25 J=4,5
      AA(I,J)=0.0
      DO 20 K=1,M
      ANGLE=3.1415927*A(K)/180.0
      CC=COS(ANGLE)
      SS=SIN(ANGLE)
      A2(4,4)=0.0
      A2(5,5)=0.0
      A2(4,5)=0.0
      A2(5,4)=0.0
      A2(4,4)=(Q(4,4)*CC*CC)+(Q(5,5)*SS*SS)
      A2(5,5)=(Q(4,4)*SS*SS)+(Q(5,5)*CC*CC)
      A2(4,5)=((Q(5,5)-Q(4,4))*CC*SS)
      A2(5,4)=A2(4,5)
      AA(I,J)=AA(I,J)+(A2(I,J)*((H1(K+1))-(H1(K))))
   20 CONTINUE
   25 CONTINUE
   30 CONTINUE
      D(4,4)=AA(4,4)
      D(5,5)=AA(5,5)
      D(4,5)=AA(4,5)
      D(5,4)=AA(5,4)
      RETURN
      END
```

5.4.7 Subroutine OUTP

```
      SUBROUTINE OUTP (C,D,D11,D22,D12,D21,D66,E,E11,E22,
     .                 EF,GC,G12,G23,H,IQ,S44,S55,T,TF,
     .                 TITLE,TW,V,V12,VF,W)
C
C*** OUTPUTS THE RESULTS
C
      IMPLICIT REAL*8(A-H,O-Z)
      DIMENSION        D(6,6),TITLE(20)
C
      WRITE(6,900)
  900 FORMAT(//,1X,'... FLEXURAL AND SHEAR RIGIDITIES...')
      WRITE(6,905)TITLE
  905 FORMAT(20A4)
      WRITE(6,910)
  910 FORMAT(40('*'))
      GO TO (5,10,15,20),IQ
C
    5 WRITE(6,915)V,T,E
  915 FORMAT(/,5X,'V=',E10.5,/,5X,'T=',E10.5,/,5X,'E=',E14.7)
      GO TO 35
C
   10 WRITE(6,920)VF,EF,GC
  920 FORMAT(/,5X,'VF=',E10.5,/,5X,'EF=',E10.5,/,5X,'GC=',E10.5)
      WRITE(6,925)C,T
  925 FORMAT(5X,'C =',E10.5,/,5X,'T =',E10.5)
      GO TO 35
C
   15 WRITE(6,930)V,E,TF
  930 FORMAT(/,5X,'V =',E10.5,/,5X,'E =',E10.5,/,5X,'TF=',E10.5)
      WRITE(6,935)TW,H,W
  935 FORMAT(5X,'TW=',E10.5,/,5X,'H =',E10.5,/,5X,'W =',E10.5)
      GO TO 35
C
   20 WRITE(6,940)E11,E22
  940 FORMAT(/,5X,'E11=',E12.5,/,5X,'E22=',E12.5)
      WRITE(6,945)G12,G23,V12
  945 FORMAT(5X,'G12=',E12.5,/,5X,'G23=',E12.5,/,5X,'V12=',E10.5)
      DO 30 I=1,6
      DO 25 J=1,6
      WRITE(6,950)I,J,D(I,J)
  950 FORMAT(/,1X,'RIGIDITY D(',I2,',',I2,')=',E15.8)
   25 CONTINUE
   30 CONTINUE
      GO TO 40
C
   35 WRITE(6,955)D11
  955 FORMAT(/,1X,'FLEXURAL RIGIDITY [D11]=',E15.5)
      WRITE(6,960)D22
  960 FORMAT(/,1X,'FLEXURAL RIGIDITY [D22]=',E15.5)
      WRITE(6,965)D12,D21
  965 FORMAT(/,1X,'COUPLING RIGIDITY [D12]=',E15.5,
     .//,1X,'COUPLING RIGIDITY [D21]=',E15.5)
      WRITE(6,970)D66
  970 FORMAT(/,1X,'TORSIONAL RIGIDITY[D66]=',E15.5)
      WRITE(6,975)S44,S55
  975 FORMAT(/,1X,'THE SHEAR RIGIDITY    [S44]=',E15.5,
     .//,1X,'SHEAR RIGIDITY    [S55]=',E15.5)
   40 CONTINUE
      RETURN
      END
```

5.5 Input Instructions for program RIGID

The input instructions for program RIGID are now presented.

Data set 1 PROBLEM TITLE (20A4) one record

Cols. 1 - 80	TITLE	title of problem

Data set 2 PLATE INDICATOR (I5) one record

Cols. 1 - 5	IQ	=1 for homogeneous isotropic plates =2 for sandwich plates =3 for voided plates =4 for laminated plates

Data set 3 HOMOGENEOUS PLATE (3F10.5) one record

Cols. 1 - 10	V	Poisson's ratio ν
11 - 20	T	plate thickness, t
21 - 30	E	elastic modulus, E

[N.B. only include if IQ = 1]

Data set 4 SANDWICH PLATE-I (2F10.5) one record

Cols. 1 - 10	VF	Poisson's ratio for flange ν_f
11 - 20	GC	core for shear modulus G_c
21 - 30	EF	elastic modulus for flange E_f

[N.B. only include if IQ = 2]

Data set 5 SANDWICH PLATE-II (2F10.5) one record

Cols. 1 - 10	C	core thickness, r
11 - 20	T	flange thickness, t

[N.B. only include if IQ = 2]

Data set 6 VOIDED PLATES (4F10.5) one record

Cols. 1 - 10	TF	flange thickness, t_f
11 - 20	TW	web thickness, t_w
21 - 30	H	centre-to-centre distance between flanges
31 - 40	W	centre-to-centre distance between webs

[N.B. only include if IQ = 3]

Data set 7 LAMINATED PLATES-I (3F10.5) one record

Cols. 1 - 10	E11	elastic modulus E_{11}
11 - 20	E22	elastic modulus E_{22}
21 - 30	V12	Poisson's ratio ν_{12}

[N.B. only include if IQ = 4]

Data set 8 LAMINATED PLATES-II (2F10.5) one record

Cols.	1 - 10	G12	in-plane shear modulus G_{12}
	11 - 20	G23	transverse shear modulus G_{23}

[N.B. only include if IQ = 4]

Data set 9 LAMINATED PLATES-III (2I5) one record

Cols.	1 - 5	M	number of layers m

[N.B. only include if IQ = 4]

Data set 10 LAMINATE PROPERTY-I (F10.5)

One record for each laminate boundary - m+1 records

Cols.	1 - 10	H1(.)	z-coordinate of boundary of laminate

Data set 11 LAMINATE PROPERTY-II (F10.5)

One record for each laminate - m records

Cols.	1 - 10	A(.)	angle of principal laminate direction

5.6 Glossary of Variables in RIGID

Name	Description
A(10)	angle between principal laminate direction 1 and x axis - one for each laminate
AA(5,5)	matrix of shear rigidities (4.14) S_{ij} [N.B. only i,j = 4,5 is used]
A2(5,5)	matrix of transformed shear moduli (4.17) $\bar{C}^k_{ij}$ for kth laminate [N.B. only i,j=4,5 is used]
C	core thickness r in sandwich plate
C1(6,6)	matrix of transformed moduli (4.15) $\bar{Q}^k_{ij}$ for kth laminate [N.B. only i,j=1,2,6 is used]
E	elastic modulus
E11,E22	elastic moduli in principal laminate directions 1 and 2
EF	elastic modulus for face in sandwich plate

GC	shear modulus for core in sandwich plate
G12,G13,G23	in-plane and transverse shear moduli respectively in laminated plate
H	centre-to-centre distance h between flanges in voided plate - see Figure 4.4
H1(11)	z-coordinate of boundary of laminate
IQ	plate type indicator =1 for homogeneous, isotropic plates =2 for sandwich plates =3 for voided plates =4 for laminated plates
M	number of laminates in laminated plate

6. SAMPLE SOLUTIONS

6.1 Introduction

Some test examples are now presented using program PLATES. Three groups of examples are considered,

(a) homogeneous isotropic rectangular thin plates

(b) rectangular plates on elastic foundations

(c) shear-weak plates

Results obtained using program PLATES are compared with those given by classical Kirchhoff thin plate theory where appropriate. However, for shear-weak plates comparisons with other works are also provided.

6.2 Homogeneous isotropic rectangular thin plates

Results from the analysis of a series of thin (t/a=0.01), simply supported, homogeneous isotropic, rectangular plates are now presented for various loading cases. Comparisons with results obtained from classical Kirchhoff thin plate theory [2] are also presented (in brackets).

(a) uniformly distributed loading q

Table 6.1 summarises the results for a set of thin, uniformly loaded plates. Typical input and output data for a square plate of t/a = 0.1 are given in Figures 6.1 and 6.2 respectively.

(b) central point load of intensity P

In Table 6.2, results for various thin centrally point-loaded plates are presented.

(c) uniformly distributed patch loading of intensity P

Table 6.3 provides the results for a set of plates subjected to a uniformly distributed patch loading of intensity P.

The only differences between the Kirchhoff and Mindlin results are attributable to a different number of Fourier terms being used.

```
INPUT DATA FOR SQUARE UNIFORMLY LOADED PLATE (T/L=0.01)
      1.0         1.0
      1.0         1.0         0.3         0.35
  35000.0     35000.0
 51    3
      1.0         0.0         1.0
  1        0.0         0.0
  2        0.5         0.5
  3        0.5         0.0
  1    2
```

Figure 6.1 Typical input data for square uniformly loaded thin plate

INPUT DATA FOR SQUARE UNIFORMLY LOADED PLATE (T/L=0.01)
****

DIMENSION OF PLATE IN X DIRECTION.(AA)= 0.10000E+01
DIMENSION OF PLATE IN Y DIRECTION.(BB)= 0.10000E+01

FLEXURAL RIGIDITY....(D11)= 0.10000000E+01
FLEXURAL RIGIDITY....(D22)= 0.10000000E+01
COUPLING RIGIDITY....(D12)= 0.30000000E+00
TORSIONAL RIGIDITY...(D66)= 0.35000000E+00

TRANSVERSE SHEAR RIGIDITY.....(S44)= 0.35000000E+05
TRANSVERSE SHEAR RIGIDITY.....(S55)= 0.35000000E+05

SHEAR CORRECTION FACTOR.........(G1)= 0.10000000E+01
MODULUS FOR ELASTIC FOUNDATION.(GFS)= 0.00000000E+00
LOAD INTENSITY..................(PZ)= 0.10000000E+01

THE NO. OF TERMS IN FOURIER SERIES (NUM)= 51
THE NO. OF POINTS ON PLATE (NPON)= 3

****

TABLE SHOWS THE COMPUTED VALUES FOR
SPECIFIED X AND Y COORDINATES.

FOR UNIFORM LOADING.

SYMMETRIC LOADING.

X	Y	W	WUX	WUY
0.00000E+00	0.00000E+00	0.00000E+00	0.00000E+00	0.00000E+00
0.50000E+00	0.50000E+00	0.40645E-02	0.24048E-11	0.24048E-11
0.50000E+00	0.00000E+00	0.00000E+00	0.00000E+00	-0.13482E-01

X	Y	MX	MY	MXY
0.00000E+00	0.00000E+00	0.00000E+00	0.00000E+00	-0.32476E-01
0.50000E+00	0.50000E+00	0.47886E-01	0.47886E-01	-0.10393E-20
0.50000E+00	0.00000E+00	0.00000E+00	0.00000E+00	0.54052E-11

X	Y	QX	QY
0.00000E+00	0.00000E+00	0.00000E+00	0.00000E+00
0.50000E+00	0.50000E+00	-0.31849E-10	-0.31849E-10
0.50000E+00	0.00000E+00	0.00000E+00	0.33372E+00

Figure 6.2 Typical output for square uniformly loaded thin plate

TABLE 6.1 Results for uniformly loaded, simply-supported, rectangular thin plates (Kirchhoff thin plate solution given in brackets)

b/a	$\bar{w}_{max}$	$(\bar{M}_x)_{max}$	$(\bar{M}_y)_{max}$	$(\bar{Q}_x)_{max}$	$(\bar{Q}_y)_{max}$
1.0	0.00406 (0.00406)	0.0479 (0.0479)	0.0479 (0.0479)	0.331 (.0338)	0.331 (.0338)
1.5	0.00772 (0.00772)	0.0812 (0.0812)	0.0498 (0.0498)	0.422 (0.424)	0.361 (0.363)
2.0	0.01013 (0.01013)	0.1017 (0.1017)	0.0464 (0.0464)	0.458 (0.458)	0.357 (0.379)
3.0	0.01223 (0.01223)	0.1189 (0.1189)	0.0406 (0.0406)	0.491 (0.493)	0.373 (0.372)

TABLE 6.2 Central displacements for simply supported rectangular thin plates subjected to a central load (Kirchhoff thin plate solution given in brackets)

b/a	1.0	1.4	1.6	2.0	3.0
	0.01162	0.01488	0.01572	0.01653	0.01694
	(0.01160)	(0.01484)	(0.01570)	(0.01651)	(0.01690)

TABLE 6.3 Maximum bending moment for simply supported rectangular thin plates subjected to a uniformly loaded patch load. (Kirchhoff thin plate solution given in brackets)

v/a \ u/a	0.1	0.5	0.8	0.9
0.1	0.284 (0.284)	0.150 (0.150)	0.108 (0.108)	0.098 (0.098)
0.3	0.224 (0.225)	0.134 (0.134)	0.098 (0.098)	0.081 (0.080)
0.5	0.185 (0.185)	0.118 (0.116)	0.087 (0.087)	0.079 (0.079)

6.3 Plates on elastic foundation

Consideration is now given to the analysis of plates on an elastic Winkler foundation. A useful parameter in the study of plates on elastic foundations is the non-dimensional quantity

$$\beta = \frac{a}{2}\,(k/4D)^{1/4}$$

In the examples which follow, values of β ranging from 0 (corresponding to no elastic foundation) to 3 (corresponding to a very stiff elastic foundation) will be used.

Table 6.4 compares the central deflection predicted by a Mindlin finite strip analysis with those obtained by the closed form solution for square plates for $t/a = 0.1$ with values of β ranging from 0 to 3 and Poisson's ratio $\nu = 0.0$.

Table 6.5 compares the central deflections obtained using both types of analysis for plates with various aspect ratios in which $\beta = 1.0$, $h/a = 0.1$ and $\nu = 0.0$. In all cases there is excellent agreement between the two methods.

Typical input and output for PLATES are shown in Figures 6.3 and 6.4 respectively. In this case $\beta = 1$, $D = 1$, $t/a = 0.01$ and $\nu = 0.3$.

```
SQUARE PLATE ON ELASTIC FOUNDATION K = 64
      1.0         1.0
      1.0         1.0         0.3        0.35
 35000.0     35000.0
 51     3
      1.0        64.0         1.0
  1        0.0         0.0
  2        0.5         0.5
  3        0.5         0.0
  1     2
```

Figure 6.3 Typical input data for plate on elastic foundation

```
SQUARE PLATE ON ELASTIC FOUNDATION K = 64
**************************************************

DIMENSION OF PLATE IN X DIRECTION.(AA)= 0.10000E+01
DIMENSION OF PLATE IN Y DIRECTION.(BB)= 0.10000E+01

FLEXURAL RIGIDITY....(D11)= 0.10000000E+01
FLEXURAL RIGIDITY....(D22)= 0.10000000E+01
COUPLING RIGIDITY....(D12)= 0.30000000E+00
TORSIONAL RIGIDITY...(D66)= 0.35000000E+00

TRANSVERSE SHEAR RIGIDITY.....(S44)= 0.35000000E+05
TRANSVERSE SHEAR RIGIDITY.....(S55)= 0.35000000E+05

SHEAR CORRECTION FACTOR.........(G1)= 0.10000000E+01
MODULUS FOR ELASTIC FOUNDATION.(GFS)= 0.64000000E+02
LOAD INTENSITY..................(PZ)= 0.10000000E+01

THE NO. OF TERMS IN FOURIER SERIES  (NUM)=   51
THE NO. OF POINTS ON PLATE          (NPON)=   3

**************************************************

TABLE SHOWS THE COMPUTED VALUES FOR
SPECIFIED X AND Y COORDINATES.

FOR UNIFORM LOADING.

SYMMETRIC LOADING.
```

X	Y	W	WUX	WUY
0.00000E+00	0.00000E+00	0.00000E+00	0.00000E+00	0.00000E+00
0.50000E+00	0.50000E+00	0.34776E-02	0.20286E-11	0.20286E-11
0.50000E+00	0.00000E+00	0.00000E+00	0.00000E+00	-0.11635E-01

X	Y	MX	MY	MXY
0.00000E+00	0.00000E+00	0.00000E+00	0.00000E+00	-0.28402E-01
0.50000E+00	0.50000E+00	0.40395E-01	0.40395E-01	-0.87025E-21
0.50000E+00	0.00000E+00	0.00000E+00	0.00000E+00	0.45763E-11

X	Y	QX	QY
0.00000E+00	0.00000E+00	0.00000E+00	0.00000E+00
0.50000E+00	0.50000E+00	-0.24584E-10	-0.24584E-10
0.50000E+00	0.00000E+00	0.00000E+00	0.29707E+00

Figure 6.4 Typical output for plate on elastic foundation

TABLE 6.4 Central deflection of uniformly-loaded, simply supported square plates on elastic foundations (t/a = 0.01, Poisson's ratio ν = 0.0)

Foundation parameter	Central deflection w qa^4/D closed form solution	finite strip solution
0	.00406	.00406
0.25	.00406	.00406
0.50	.00402	.00402
0.75	.00386	.00386
1.00	.00348	.00348
1.25	.00287	.00287
1.50	.00218	.00218
1.75	.00155	.00155
2.00	.00106	.00106
2.25	.00072	.00072
2.50	.00048	.00048
2.75	.00033	.00033
3.00	.00023	.00023

TABLE 6.5 Central deflection of uniformly loaded rectangular plates on elastic foundations (t/a=0.1, Poisson's ratio ν = 0.0)

Foundation parameter	Central deflection w qa^4/D closed form solution	finite strip solution
1.0	.00358	.00358
2.0	.00710	.00710
3.0	.00783	.00783
4.0	.00793	.00792
5.0	.00794	.00793

6.4 Shear weak plates

6.4.1 Modcrately thick plate

Results for a uniformly-loaded, moderately thick ($t/a = 0.1$) simply supported, square plate are compared in Table 6.6 with results based on Kirchhoff plate theory. Typical input and output for PLATES are shown in Figures 6.5 and 6.6 respectively. A useful parameter for shear weak plates is $\alpha = Sa^2/D$. For the present plate $\alpha = 350.0$ whereas for the thin shear stiff plate ($t/a = 0.01$) considered in Section 6.2, $\alpha = 35000.0$.

```
INPUT DATA FOR SQUARE UNIFORMLY LOADED PLATE (T/L=0.1)
        1.0          1.0
        1.0          1.0          0.3        0.35
     350.00       350.00
    51    3
        1.0          0.0          1.0
     1         0.0          0.0
     2         0.5          0.5
     3         0.5          0.0
     1    2
```

Figure 6.5 Typical input data for square uniformly loaded moderately thick plates

```
INPUT DATA FOR SQUARE UNIFORMLY LOADED PLATE (T/L=0.1)
****************************************************


DIMENSION OF PLATE IN X DIRECTION.(AA)= 0.10000E+01
DIMENSION OF PLATE IN Y DIRECTION.(BB)= 0.10000E+01

FLEXURAL RIGIDITY....(D11)= 0.10000000E+01
FLEXURAL RIGIDITY....(D22)= 0.10000000E+01
COUPLING RIGIDITY....(D12)= 0.30000000E+00
TORSIONAL RIGIDITY...(D66)= 0.35000000E+00

TRANSVERSE SHEAR RIGIDITY.....(S44)= 0.35000000E+03
TRANSVERSE SHEAR RIGIDITY.....(S55)= 0.35000000E+03

SHEAR CORRECTION FACTOR.........(G1)= 0.10000000E+01
MODULUS FOR ELASTIC FOUNDATION.(GFS)= 0.00000000E+00
LOAD INTENSITY..................(PZ)= 0.10000000E+01

THE NO. OF TERMS IN FOURIER SERIES  (NUM)=   51
THE NO. OF POINTS ON PLATE         (NPON)=    3


****************************************************


TABLE SHOWS THE COMPUTED VALUES FOR
SPECIFIED X AND Y COORDINATES.

FOR UNIFORM LOADING.

SYMMETRIC LOADING.

       X           Y            W            WUX          WUY

0.00000E+00 0.00000E+00  0.00000E+00  0.00000E+00  0.00000E+00

0.50000E+00 0.50000E+00  0.42728E-02  0.24048E-11  0.24048E-11

0.50000E+00 0.00000E+00  0.00000E+00  0.00000E+00 -0.13482E-01


       X           Y            MX           MY           MXY

0.00000E+00 0.00000E+00  0.00000E+00  0.00000E+00 -0.32476E-01

0.50000E+00 0.50000E+00  0.47886E-01  0.47886E-01 -0.10393E-20

0.50000E+00 0.00000E+00  0.00000E+00  0.00000E+00  0.54052E-11


       X           Y            QX           QY

0.00000E+00 0.00000E+00  0.00000E+00  0.00000E+00

0.50000E+00 0.50000E+00 -0.31849E-10 -0.31849E-10

0.50000E+00 0.00000E+00  0.00000E+00  0.33372E+00
```

Figure 6.6 Typical output for square uniformly loaded moderately thick plate

TABLE 6.6 Results for uniformly loaded, moderately-thick ($t/a = 0.1$), simply supported, square plate

	w_{max} qa^4/D	$(M_x)_{max}$ qa^2	$(M_{xy})_{max}$ qa^2	$(Q_x)_{max}$ qa
Closed form solution	0.00427	0.0479	-0.0325	0.334 *
Kirchhoff solution	0.00406	0.0479	-0.0325	0.338

*This value converges to 0.338 qa when more harmonics are considered.

6.4.2 Sandwich plates

A series of uniformly loaded, rectangular, simply supported sandwich plates are analysed using PLATES. Figure 6.7 shows the central deflections for plates with various values of the parameter α^{-1}. The agreement with the results of Plantema [4] is excellent. Note that for the square plate the central displacement may be approximated by the expression

$$w_{max} = (0.00406 + 0.07344/\alpha)\ qa^4/D$$

6.4.3 Fibre reinforced plates

In this example, consideration is given to the flexure of uniformly loaded, square boron/epoxy and glass/epoxy composite simply supported plates. Figure 6.8 gives some general details of these plates and Table 6.7 provides the necessary elastic rigidities when the plates are treated as homogeneous orthotropic plates. Results obtained by Hussainy and Srinivas [9] are compared with those obtained using PLATES in Table 6.8.

TABLE 6.7 Equivalent elastic moduli of boron/epoxy and glass/epoxy composite plates

Property	Boron/epoxy			Glass/epoxy	
	30°	45°	60°	30°	45°
	30°	45°	60°	30°	45°
Q_{11}	2.487037×10^7	1.238941×10^7	4.832576×10^6	5.269582×10^6	3.499728×10^6
Q_{22}/Q_{11}	0.1943106	1.0	5.1463985	0.4524497	1.0
Q_{12}/Q_{11}	0.3180775	0.8372123	1.6369534	0.2788481	0.5133482
Q_{66}/Q_{11}	0.3375236	0.8762567	1.7370310	0.3478621	0.6172642
C_{44}/Q_{11}	0.02799501	0.06459594	0.1871394	0.1148516	0.1963975
C_{55}/Q_{11}	0.03636318	0.06459594	0.1440735	0.1460184	0.1963975

N.B. The above moduli are derived by assuming the Young's modulus and Poisson's ratio of epoxy, boron fibre and glass fibre respectively as 0.5×10^6, 0.36; 60×10^6, 0.2; 10.6×10^6, 0.22.

TABLE 6.8 Results for uniformly loaded, square, boron-epoxy and glass-epoxy composite, simply supported plates

Orientation	Results at centre of plate					
	$E_f\ w/tq$		M_x/t^2q		$M_y/t^2 2$	
	Hussainy and Srinivas	Present analysis	Hussainy and Srinivas	Present analysis	Hussainy and Srinivas	Present analysis
			Boron/epoxy			
30°	2124.0	2123.8	5.821	5.818	2.228	2.227
45°	1960.5	1960.5	3.533	3.533	3.533	3.533
60°	2124.0	2123.8	2.228	2.227	5.821	5.818
			Glass/epoxy			
30°	1283.7	1283.7	5.501	5.499	3.072	3.073
45°	1205.0	1205.0	3.953	3.953	3.953	3.953

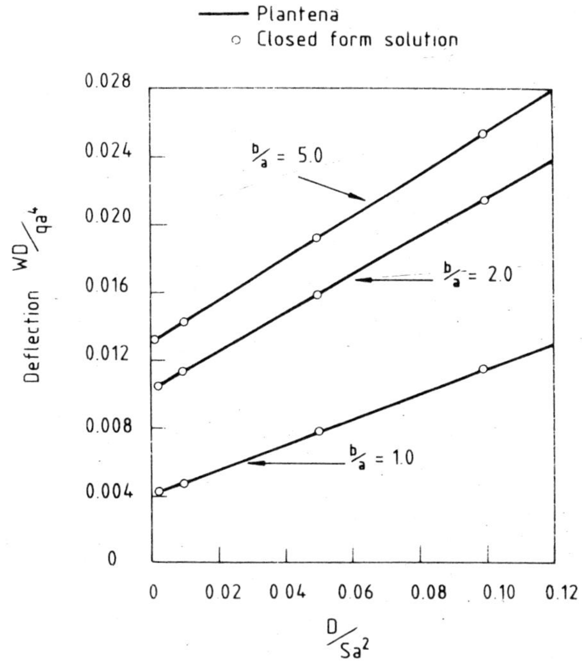

Figure 6.7 Central deflection of uniformly loaded simply supported sandwich plates

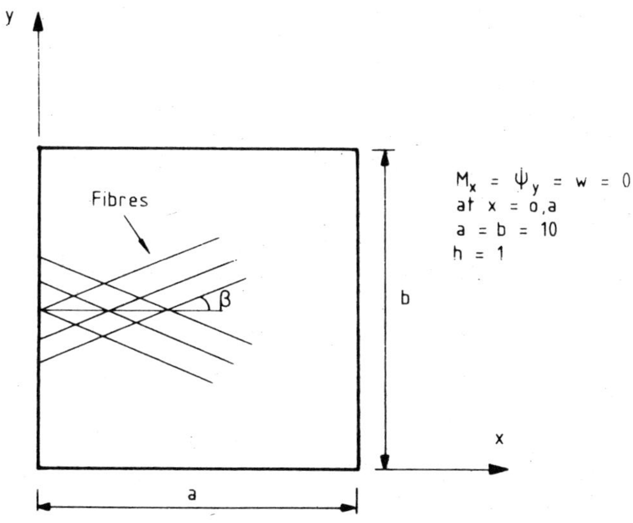

Figure 6.8 Boron-epoxy composite plate

6.4.4 Three ply laminates

Values for the central deflection of uniformly loaded, square, simply supported, three ply laminated plates (shown in Figure 6.9) are presented in Table 6.9. Results are compared with those given by Srinivas and Rao [10] which are based on 'exact' analysis and thin plate theory. Excellent agreement is obtained with the results based on the 'exact' analysis. If the transverse shear rigidities are multiplied by 10^4 then the thin plate theory results are obtained.

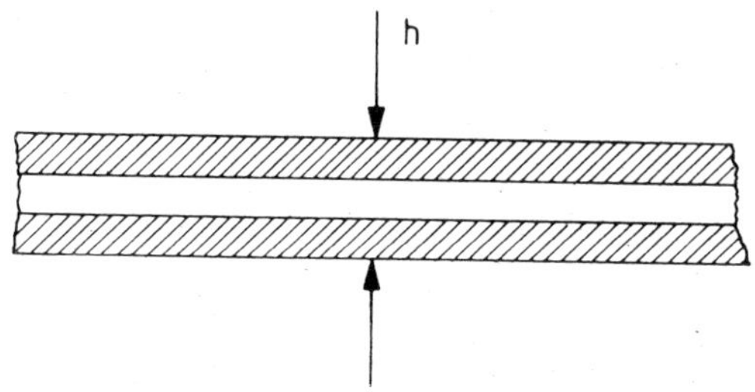

Figure 6.9 Three-ply laminate

TABLE 6.9 Central deflection for three-ply square laminated plates under uniform load q

C	Analytical method	$w\ G_{12}(2)/tq$
1	Exact theory	181.05
	Present analysis ($k_1 = k_2 = 0.83$)	181.35
	Thin plate theory	168.38
	Present analysis ($G_{13}, G_{23} * 10^4$)	168.49
10	Exact theory	41.906
	Present analysis ($k_1 = k_2 = 0.3521$)	42.004
	Thin plate theory	31.241
	Present analysis ($G_{13}, G_{23} * 10^4$)	31.255
50	Exact theory	16.753
	Present analysis ($k_1 = k_2 = 0.0938$)	16.838
	Thin plate theory	6.762
	Present analysis ($G_{13}, G_{23} * 10^4$)	6.764

$(Q_{11} : Q_{12} : Q_{22} : Q_{66} : C_{44} : C_{55})$

$= (3.802 : 0.879 : 1.996 : 1 : 0.608 : 1.015)$

$$a/b = 1, \quad t/a = 0.1; \quad C = \frac{Q_{66}(1)}{Q_{66}(2)} = \frac{Q_{66}(3)}{Q_{66}(2)}$$

Note: k_1 and k_2 are the shear modification factors

REFERENCES

1. MINDLIN, R.D.
"Influence of rotatory inertia and shear or flexural motions of isotropic elastic plates", J. Appl. Mech., 37, 1031-1036, 1951.

2. TIMOSHENKO, S.P. and WOINOWSKY-KRIEGER, S.
"Theory of plates and shells", 2nd ed., McGraw-Hill, New York, 1961.

3. DOBYNS, A.L.
"Analysis of simply-supported orthotropic plates subject to static and dynamic loads", J.AIAA, 19, 642-650, 1982.

4. PLANTEMA, F.J.
"Sandwich Construction", John Wiley, New York, 1966.

5. VINSON, J. R. and CHOU, T. W.
"Composite materials and their use in structures", Applied Science Publishers, London, 1975.

6. BASU, A. K. and DAWSON, J. M.
"Orthotropic sandwich plates", Proc. Instn. Civ. Engrs., Supplement, 87-115, 1970.

7. FREDERICK, D.
"Thick rectangular plates on an elastic foundation", Trans. A.S.C.E., 122, 1069-1085, 1957
[See also discussion by Lorsch, H.G., Trans. A.S.C.E., 122, 1086-1087, 1957].

8. ARIMAN, T.
"On orthotropic thick elastic plates on an elastic foundation", Die Bautechnik, 7, 230-234, 1968.

9. HUSSAINY, S.A. and SRINIVAS, S.
"Flexure of rectangular composite plates", Fibre Science and Technology, 8, 59-76, 1975.

10. SRINIVAS, S. and RAO, A. K.
"Bending, vibration and buckling of simply supported thick orthotropic rectangular plates and laminates", Int. J. Solids Structures, 6, 1463-1481, 1970.

CHAPTER 2

MINDLIN FINITE STRIP AND AXISYMMETRIC FINITE ELEMENT SHELL ANALYSIS

E. Oñate*, B. Suarez* and E. Hinton+

* Universidad Politecnica de Cataluña, Spain
\+ University College of Swansea, U.K.

1. INTRODUCTION

Many engineering structures have constant geometrical properties along a particular direction. Such prismatic structures are very common in plate and folded plate problems, where the transverse cross-section of the structure often remains constant in the longitudinal direction. In axisymmetric shell problems, the cross-section of the shell does not change in the circumferential direction (see Figure 1). If the material properties of the structures are also constant in the same direction, the analysis can be simplified by the combined use of finite elements and Fourier expansions to model the transverse and longitudinal behaviour.

The combination of finite elements and Fourier series is not new and it has for some years been used in the study of un-symmetrically loaded axisymmetric shells and solids by Grafton and Strone[1], Ahmad et al[2], and Wilson[3].

The extension to plate and shell analysis was first developed by Cheung[4] and termed the finite strip method. Since then, further refinements in the method have been made and some of the main contributions are listed in references [5] - [18].

In this chapter the basis of the Mindlin finite strip formulation will be presented for a wide range of prismatic

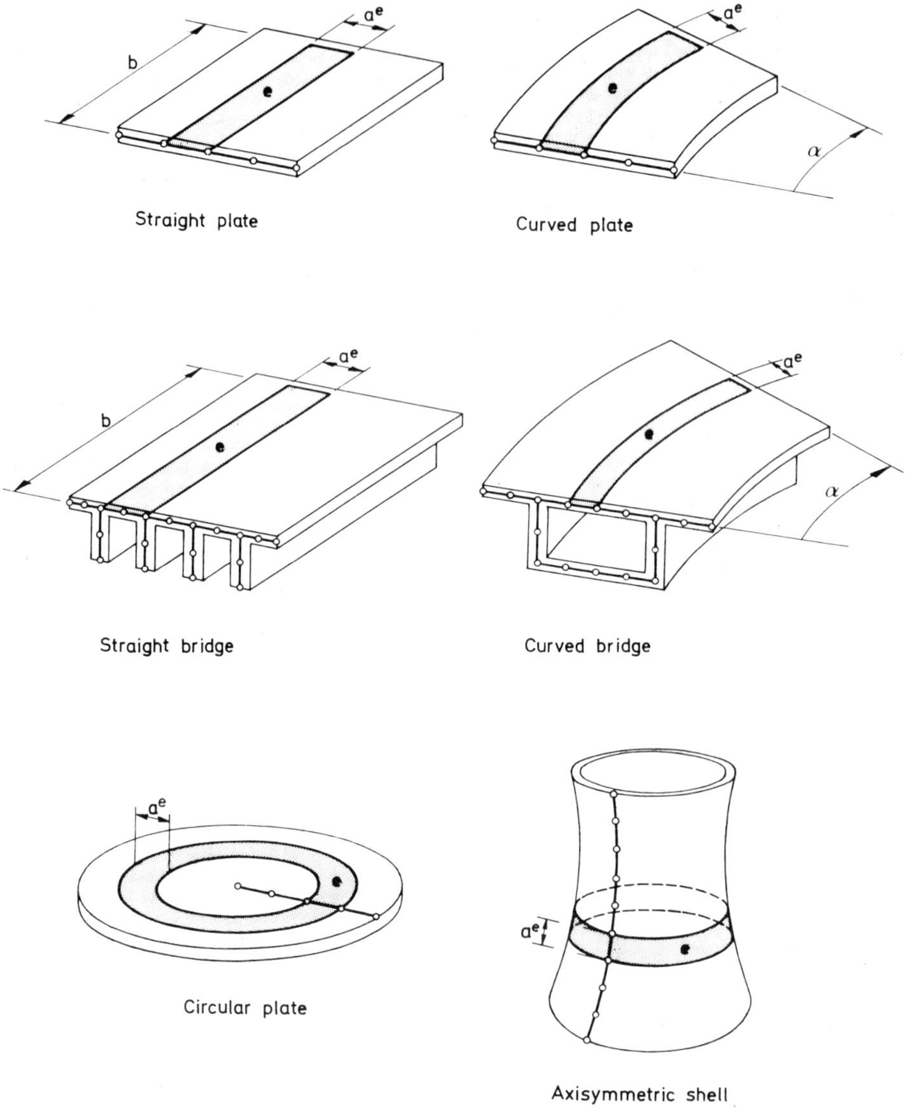

Fig. 1 Different structures which can be analyzed with the finite strip method.

structures. Rectangular plates will be considered first as an introduction to more complex problems and details of the general finite strip formulation will be given. The second part of the chapter deals with folded plate shell type structures. It will be shown that the finite strip formulation for right folded plates and axisymmetric shells can be simply derived as a special case of the more general formulation for folded plates with curved planforms which will be presented first. The last part of the chapter deals with the computer implementation of the finite strip method and full details of a finite strip computer program for the analysis of rectangular and curved plates will be given.

Before discussing the basis of the finite strip method, it is interesting to introduce the reader to the basic concepts of the use of Fourier series for structural analysis. This is done in the next section for the simple case of a beam.

2. ANALYSIS OF A SIMPLY-SUPPORTED BEAM BY FOURIER SERIES

Consider the beam shown in Figure 2 under an arbitrary loading q(y). The Total Potential Energy for a beam in bending can be written as

$$\pi(w) = \frac{EI}{2}\int_o^b \left(\frac{d^2w}{dy^2}\right)^2 dy - \int_o^b qwdy \tag{1}$$

where E and I are the Young's modulus and the modulus of inertia of the beam transverse cross-section, resp., and w is the lateral beam deflection at each point, which must satisfy the following boundary conditions

$$w = \frac{d^2w}{dy^2} = 0 \text{ at } y = 0 \text{ and } y = b \tag{2}$$

The above conditions are satisfied by the following Fourier series

$$w = \sum_{\ell=1}^{\infty} w^{\ell} \sin \ell \frac{\pi y}{b} \tag{3}$$

where b is the beam length and ℓ refers to a particular term, i.e. $\ell = 1,2,3$, etc. and w^{ℓ} is the undetermined deflection amplitude for the ℓth harmonic.

The loading q(y) is defined by Fourier series as

$$q(y) = \sum_{\ell=1}^{\infty} q^{\ell} \sin \ell \frac{\pi y}{b} \tag{4}$$

where q^{ℓ} is the loading amplitude for the ℓth harmonic, which can be obtained using Euler's formula for Fourier series, i.e.

$$q^{\ell} = \frac{\int_{b_0}^{b_1} q(y) \sin \ell \frac{\pi y}{b} dy}{\int_{0}^{b} \sin^2 \ell \frac{\pi y}{b} dy} = \frac{2}{b} \int_{b_0}^{b_1} q(y) \sin \ell \frac{\pi y}{b} dy \tag{5}$$

where the load is applied in the zone from $y = b_0$ to $y = b_1$, as shown in Figure 2.

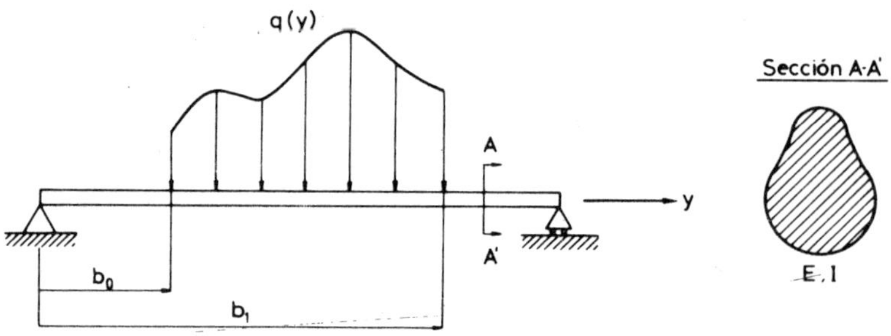

Figure 2 Simple supported beam of constant cross-section under arbitrary distributed loading q(y)

The value of q^{ℓ} is easily obtained provided that the product $q(y) \sin \frac{\ell \pi y}{b}$ is integrable. Therefore, for a given load harmonic the problem is one of finding the unknown amplitude w^{ℓ} which uniquely describes the deflected beam profile for that harmonic. Substituting (3) and (4) into (1) it is possible to write

$$\pi(w) = \sum_{\ell=1}^{\infty} \left(\frac{EI}{4} (w^{\ell})^2 \frac{\ell^4 \pi^4}{b^3} - \frac{b}{2} q^{\ell} w^{\ell}\right) \tag{6}$$

The value of w^{ℓ} is obtained by minimising the Total Potential Energy with respect to w^{ℓ}, i.e.

$$\frac{\partial \pi(w)}{\partial w^{\ell}} = 0 \quad \text{which leads to} \quad w^{\ell} = \frac{q^{\ell} b^4}{EI\ell^4 \pi^4} \tag{7}$$

The first deflected profile is obtained if the summation in (3) is performed and from the deformed shape of the beam the curvatures and hence, the bending moments may be calculated.

As an example of application the beam shown in Figure 3 is subjected to two loading cases: a uniformly distributed loading of intensity q acting along the whole beam length; and a vertical point load, P, acting at midspan. The Fourier coefficient, q^{ℓ}, for each loading can be obtained using (5) as

$$q^{\ell} = \frac{2q}{\ell\pi}(1 - \cos \ell\pi) \quad \text{for the uniform load}$$

$$q^{\ell} = \frac{2P}{b} \sin \frac{\ell\pi}{2} \quad \text{for the point load} \tag{8}$$

Thus the vertical deflection for each load case is obtained from (7) and (3) as

$$w = \frac{2qb^4}{EI\pi^5} \sum_{\ell=1}^{\infty} \frac{1 - \cos \ell\pi}{\ell^5} \sin \frac{\ell\pi y}{b} \quad \text{for the uniform load}$$

$$w = \frac{2Pb^3}{EI\pi^4} \sum_{\ell=1}^{\infty} \frac{1}{\ell^4} \sin \frac{\ell\pi}{2} \sin \frac{\ell\pi y}{b} \quad \text{for the point load} \tag{9}$$

The bending moments are obtained by the expression

$$M = -EI \frac{d^2 w}{dy^2} \tag{10}$$

Thus for each loading case the bending moment can be written as

$$M = \frac{2qb^2}{\pi^3} \sum_{\ell=1}^{\infty} \frac{1}{\ell^3}(1 - \cos \ell\pi) \sin \frac{\ell\pi y}{b} \quad \text{for the uniform load}$$

$$M = \frac{2Pb}{\pi^2} \sum_{\ell=1}^{\infty} \frac{1}{\ell^2} \sin \frac{\ell\pi}{2} \sin \frac{\ell\pi y}{b} \quad \text{for the point load} \tag{11}$$

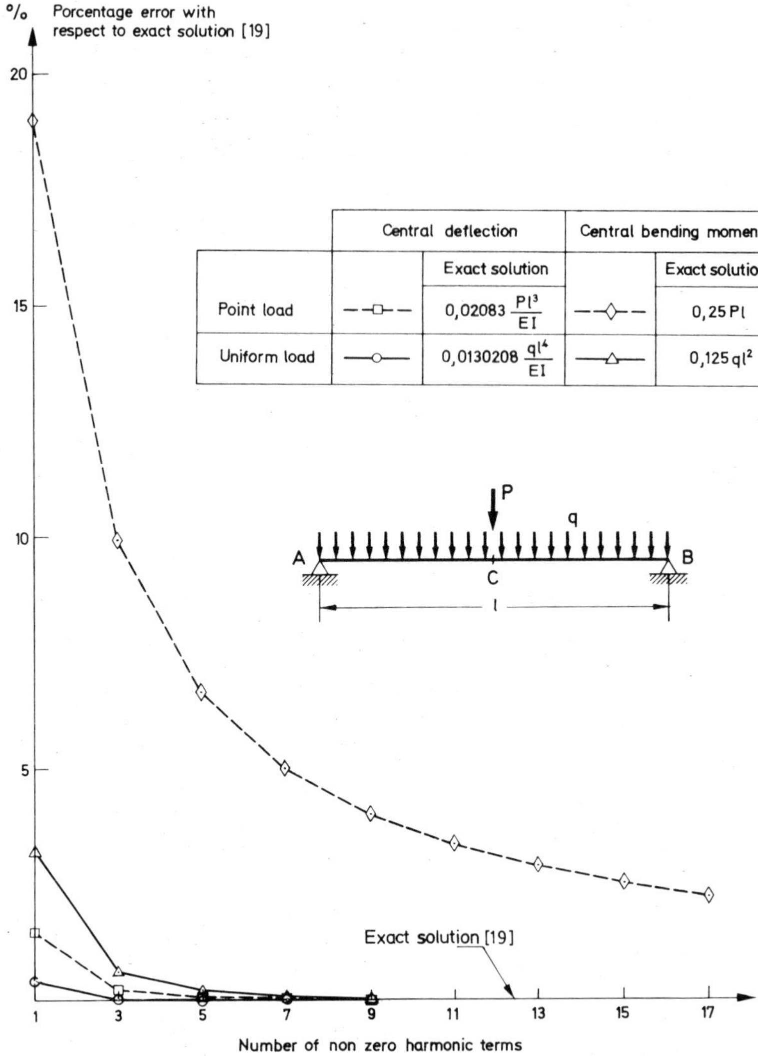

Fig. 3 *Simple supported beam analyzed by Fourier series. Convergence study of central deflection and central bending moment for a point load and an uniform load.*

Note that in (9) and (11) the even harmonic terms are zero. This is due to the symmetry of the loading about the centre of the beam.

Numerical results for the vertical deflection and bending moment at midspan, taking nine non-zero harmonics for the two load cases, are shown in Figure 3, from which the following conclusions can be drawn:

a) Convergence to the theoretical value [19] for the deflection and bending moment is much faster for the uniform loading than for the point load case.

b) In both load cases the convergence of the bending moment is slower than that of the vertical deflection.

The convergence of the solution is therefore loading dependent. As a rule, solutions for uniform loads converge much more rapidly than those for point loads. Also, the number of harmonic terms needed to achieve a given degree of accuracy for the solution is greater for the bending moments than for the displacements.

These practical rules, deduced from a simple case, apply for the general finite strip formulation.

3. FINITE STRIP FORMULATION FOR THE ANALYSIS OF RECTANGULAR MINDLIN PLATES

In this section the Mindlin finite strip formulation for rectangular plate bending analysis will be derived in detail as an introduction to more complex structural problems, e.g. folded plates, shells, which can also be treated with the strip formulation in a similar manner.

3.1 Basic theory for Mindlin plates

Mindlin's assumptions for plate bending can be stated simply as follows:

1) The lateral deflections of the plate, w, are small.
2) Normals to be mid-plane of the plate before deformation

remain straight but no necessarily normal to the mid-plane after deformation.

3) The stress normal to the mid-plane, σ_z, is negligible.

From assumptions 1 and 2 the displacement field at any point of the plate can be expressed as

$$\begin{aligned} u(x,y,z) &= z\ \theta_x(x,y) \\ v(x,y,z) &= z\ \theta_y(x,y) \\ w(x,y,z) &= w(x,y) \end{aligned} \tag{12}$$

where u, v and w are the displacements in the x, y and z directions, respectively, and θ_x and θ_y are the rotations contained in the zx and yz plane, resp. (see Figure 4).

The Total Potential Energy of the plate can be written as [21]

$$\pi = \tfrac{1}{2} \iint \underset{\sim}{\varepsilon}^T \underset{\sim}{\sigma}\ dxdy - \iint q\ w\ dxdy \tag{13}$$

where, for simplicity only, the contribution of a distributed load of intensity q(x y) has been considered. In (13) $\underset{\sim}{\varepsilon}$ and $\underset{\sim}{\sigma}$ are the generalised strain and stress vectors, resp., which can be expressed as

$$\underset{\sim}{\varepsilon} = [\underset{\sim}{\varepsilon}_b^T, \underset{\sim}{\varepsilon}_s^T]^T \ ; \qquad \underset{\sim}{\sigma} = [\underset{\sim}{\sigma}_b^T, \underset{\sim}{\sigma}_s^T]^T \tag{14}$$

with

$$\underset{\sim}{\varepsilon}_b = [\frac{\partial\theta_x}{\partial x}, \frac{\partial\theta_y}{\partial y}, \frac{\partial\theta_x}{\partial y} + \frac{\partial\theta_y}{\partial x}]^T \tag{15.a}$$

$$\underset{\sim}{\varepsilon}_s = [\theta_x + \frac{\partial w}{\partial x}, \theta_y + \frac{\partial w}{\partial y}]^T \tag{15.b}$$

$$\underset{\sim}{\sigma}_b = [M_x, M_y, M_{xy}]^T \tag{16.a}$$

and $$\underset{\sim}{\sigma}_s = [Q_x, Q_y]^T \tag{16.b}$$

where $\underset{\sim}{\varepsilon}_b$, $\underset{\sim}{\sigma}_b$ and $\underset{\sim}{\varepsilon}_s$, $\underset{\sim}{\sigma}_s$ are resp. the generalised strain and stress resultant vectors due to bending and shear effects. In (16), the sign convention for bending moments and shear forces Q_x and Q_y are shown in Figure 5. The relationship between the generalised stresses and strains can be written as

$$\underset{\sim}{\sigma} = \underset{\sim}{D}\ \underset{\sim}{\varepsilon} \tag{17}$$

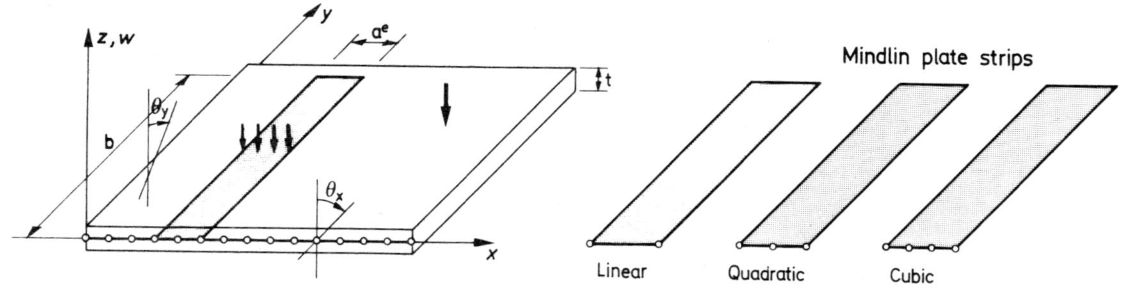

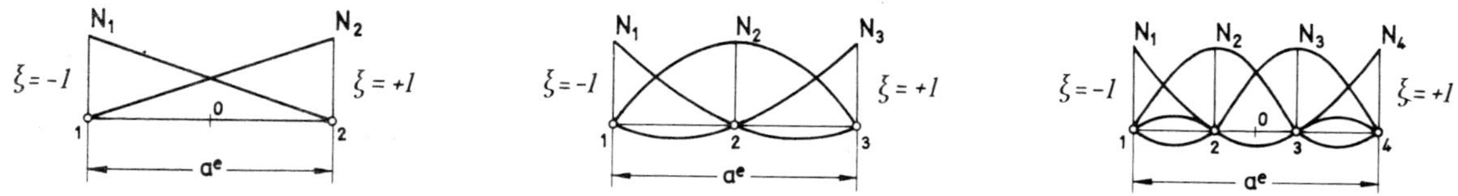

Linear:

$$N_1 = \frac{1-\xi}{2}$$
$$N_2 = \frac{1+\xi}{2}$$
$$\xi = \frac{2(x-x_0)}{a^e}$$

Quadratic:

$$N_1 = \frac{1}{2}\xi(\xi-1)$$
$$N_2 = (1-\xi)(1+\xi)$$
$$N_3 = \frac{1}{2}\xi(\xi+1)$$
$$\xi = \frac{2(x-x_2)}{a^e}$$

Cubic:

$$N_1 = -\frac{9}{16}(\xi^3-\xi^2-\frac{1}{9}\xi+\frac{1}{9})$$
$$N_2 = \frac{27}{16}(\xi^3-\frac{1}{3}\xi^2-\xi+\frac{1}{3})$$
$$N_3 = -\frac{27}{16}(\xi^3+\frac{1}{3}\xi^2-\xi-\frac{1}{3})$$
$$N_4 = \frac{9}{16}(\xi^3+\xi^2-\frac{1}{9}\xi-\frac{1}{9})$$
$$\xi = \frac{2(x-x_0)}{a^e}$$

Fig. 4 Linear, Quadratic and Cubic Mindlin plate strips and their shape functions.

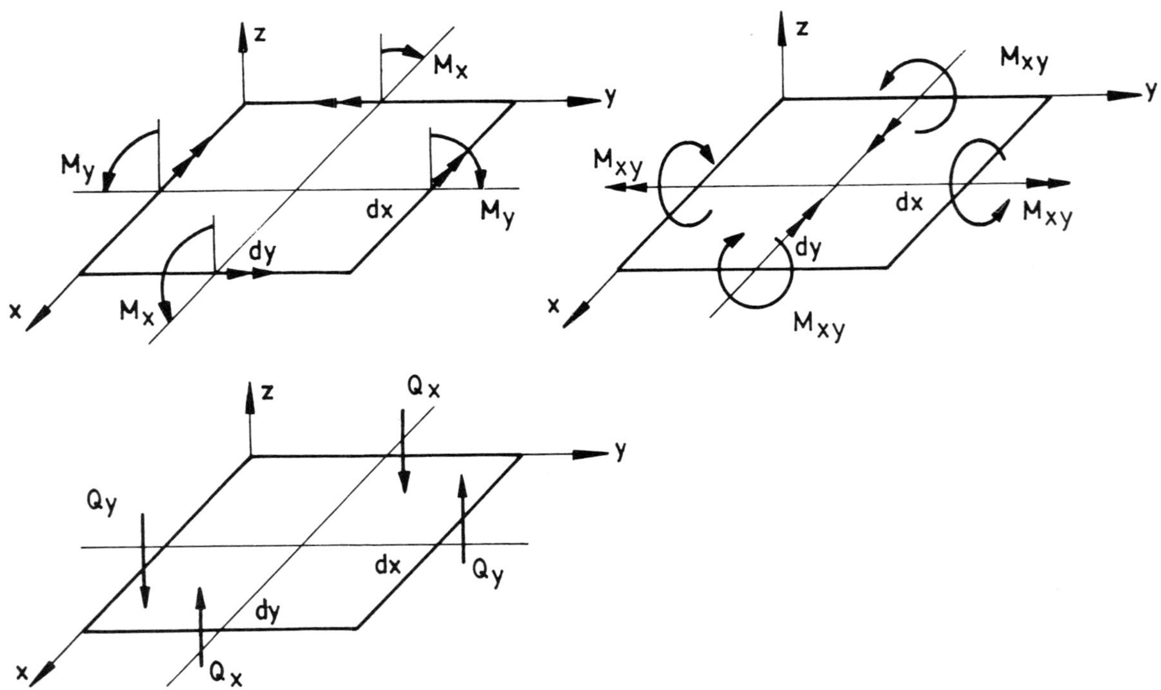

Fig. 5 Sign convention for moments and shear forces.

where for an isotropic plate

$$\underset{\sim}{D} = \begin{bmatrix} \underset{\sim}{D}_b & 0 \\ 0 & \underset{\sim}{D}_s \end{bmatrix} \tag{18}$$

in which

$$\underset{\sim}{D}_b = \frac{Et^3}{12(1-\nu^2)} \begin{bmatrix} 1 & \nu & 0 \\ \nu & 1 & 0 \\ 0 & 0 & \frac{(1-\nu)}{2} \end{bmatrix} ; \quad \underset{\sim}{D}_s = \frac{Et}{2(1+\nu)} \begin{bmatrix} \gamma & 0 \\ 0 & \gamma \end{bmatrix} \tag{19}$$

are resp. the bending and shear contributions to the elasticity matrix $\underset{\sim}{D}$; E and ν are the Young's modulus and Poisson's ratio resp., t is the plate thickness, and γ is a shear modification coefficient to take into account the warping of the section [19]. (Note that $\gamma = \frac{5}{6}$ for rectangular sections.)

3.2 Finite strip formulation for Mindlin plates

The finite strip formulation for the analysis of Mindlin plates follows similar steps to those involved in the simple beam problem of Section 2. Thus, the displacements are now expanded in terms of truncated Fourier series along direction, y, in which both the material and geometrical properties of the plate are taken to be constant, i.e.

$$\begin{aligned} w(x,y) &= \sum_{\ell=1}^{\infty} w^{\ell}(x) \sin \frac{\ell\pi}{b} y \\ \theta_x(x,y) &= \sum_{\ell=1}^{n} \theta_x^{\ell}(x) \sin \frac{\ell\pi}{b} y \\ \theta_y(x,y) &= \sum_{\ell=1}^{n} \theta_y^{\ell}(x) \cos \frac{\ell\pi}{b} y \end{aligned} \tag{20}$$

where b is the plate length, w^{ℓ}, θ_x^{ℓ} and θ_y^{ℓ} are the displacement amplitudes for the ℓth harmonic term and n is the number of harmonic terms used in the analysis.

The next step is to discretise the displacement amplitudes (which are a function of the x coordinate only) using a standard finite element representation [20] along the transverse direction of the plate. Thus, within an element e the displacement amplitudes can be written as

$$w^{\ell}(x) = \sum_{i=1}^{n_e} N_i(x)\, w_i^{\ell}$$

$$\theta_x^{\ell}(x) = \sum_{i=1}^{n_e} N_i(x)\, \theta_{xi}^{\ell} \qquad (21)$$

$$\theta_y^{\ell}(x) = \sum_{i=1}^{n_e} N_i(x)\, \theta_{yi}^{\ell}$$

where w_i^{ℓ}, θ_{xi}^{ℓ} and θ_{yi}^{ℓ} are the nodal amplitudes of the ith node of element e, $N_i(x)$ is the (one-dimensional) shape function of node i of element e, and n_e is the number of nodes in element e.

Thus, the process is equivalent to dividing the plate into longitudinal elements (or strips) so that each strip has a certain number of nodes (or more accurately, nodal lines) associated with its transverse direction. The displacement field is defined longitudinally by the Fourier expansion of (21) and transversely by the finite element discretisation of (22). (See Figure 4.)

Substituting (21) into (20) it is possible to write

$$\underset{\sim}{u} = \sum_{\ell=1}^{n} \sum_{i=1}^{n_e} \underset{\sim}{N}_i^{\ell}\, \underset{\sim}{a}_i^{\ell} \qquad (22)$$

where $\underset{\sim}{u} = [w, \theta_x, \theta_y]^T$

$$\underset{\sim}{N}_i^{\ell} = \begin{bmatrix} N_i S_{\ell} & 0 & 0 \\ 0 & N_i S_{\ell} & 0 \\ 0 & 0 & N_i C_{\ell} \end{bmatrix}$$

and $\underset{\sim}{a}_i^{\ell} = [w_i^{\ell}, \theta_{xi}^{\ell}, \theta_{yi}^{\ell}]^T$ (23)

with $S_{\ell} = \sin \frac{\ell\pi}{b} y$ and $C_{\ell} = \cos \frac{\ell\pi}{b} y$

The harmonic expansions chosen satisfy the following boundary conditions

$$\left.\begin{array}{l} w = \theta_x = 0 \\ \dfrac{\partial w}{\partial x} = \dfrac{\partial \theta_x}{\partial x} = \dfrac{\partial \theta_y}{\partial y} = 0 \end{array}\right\} \text{ for } y = 0 \text{ and } y = b$$

which imply that the plate is simply supported at the two opposite ends. Other boundary conditions can be reproduced by an appropriate selection of the Fourier-like expansions of (20). However, the expansions chosen here are the most simple and usual in practice for economy reasons as will be explained later.

The discretised expressions for the strains and stresses within a strip can be simply obtained by substituting (22) into (14) and (18) resp. Thus, after substitution, the following expressions are obtained

$$\underset{\sim}{\varepsilon} = \sum_{\ell=1}^{n} \sum_{i=1}^{n_e} \underset{\sim}{B}{}_i^{\ell} \, \underset{\sim}{a}{}_i^{\ell} \tag{24}$$

and

$$\underset{\sim}{\sigma} = \sum_{\ell=1}^{n} \sum_{i=1}^{n_e} \underset{\sim}{D} \, \underset{\sim}{B}{}_i^{\ell} \, \underset{\sim}{a}{}_i^{\ell} \tag{25}$$

where $\underset{\sim}{B}{}_i^{\ell}$ is the strain-displacement matrix of the ith node of strip e for the ℓth harmonic term, which is written as

$$\underset{\sim}{B}{}_i^{\ell} = \begin{Bmatrix} \underset{\sim}{B}{}_{bi}^{\ell} \\ \underset{\sim}{B}{}_{si}^{\ell} \end{Bmatrix} \tag{26}$$

where

$$\underset{\sim}{B}{}_{bi}^{\ell} = \begin{bmatrix} 0 & \dfrac{\partial N_i}{\partial x} S_\ell & 0 \\ 0 & 0 & -N_i \dfrac{\ell\pi}{b} S_\ell \\ 0 & N_i \dfrac{\ell\pi}{b} C_\ell & \dfrac{\partial N_i}{\partial x} C_\ell \end{bmatrix} \tag{27}$$

and

$$\underset{\sim}{B}_{si}^{\ell} = \begin{bmatrix} \frac{\partial N_i}{\partial x} S_\ell & N_i S_\ell & 0 \\ N_i \frac{\ell\pi}{b} C_\ell & 0 & N_i C_\ell \end{bmatrix}$$

in which $\underset{\sim}{B}_{bi}^{\ell}$ and $\underset{\sim}{B}_{si}^{\ell}$ are the contributions to the strain matrix of node i due to bending and shear, resp. for the ℓth harmonic term.

Expanding the forces in the same way as the displacements, the distributed vertical load can be represented by the expression

$$q = \sum_{\ell=1}^{n} q^{\ell} \sin \frac{\ell\pi}{b} y \tag{28}$$

Thus, substituting (25) and (26) and using (21) and (28) the expression for the Total Potential Energy of the plate can be expressed as the sum of the contributions π^e from each strip

$$\pi = \sum_e \pi^e = \sum_e \left\{ \frac{1}{2} \iint_{A_e} \left(\sum_{\ell=1}^{n} \sum_{i=1}^{n_e} \underset{\sim}{B}_i^{\ell} \underset{\sim}{a}_i^{\ell} \right)^T \underset{\sim}{D} \left(\sum_{m=1}^{m} \sum_{j=1}^{n_e} \underset{\sim}{B}_j^{m} \underset{\sim}{a}_j^{m} \right) dA \right.$$
$$\left. - \iint_{A^e} \left(\sum_{\ell=1}^{n} \sum_{i=1}^{n_e} N_i \, S_\ell \, w^{\ell} \right) \left(\sum_{m=1}^{n} q^m \, S_m \right) dA \right\} \tag{29}$$

where A^e is the area of the transverse cross-section of strip e. Taking into account the orthogonal properties of the functions S_ℓ and C_ℓ, i.e.

$$\left. \begin{aligned} \int_0^b S_\ell \, S_m \, dy \\ \int_0^b C_\ell \, C_m \, dy \end{aligned} \right\} \begin{aligned} &= \frac{b}{2} \quad \text{if} \quad \ell = m \\ &= 0 \quad \text{if} \quad \ell \neq m \end{aligned} \tag{30}$$

The expression for the Total Potential Energy of the strip, π^e, of (29) can be written as

$$\pi^e = \frac{1}{2} \sum_{\ell=1}^{n} \sum_{m=1}^{n} \sum_{i=1}^{n_e} \sum_{j=1}^{n_e} (\underset{\sim}{a}_i^{\ell})^T [\underset{\sim}{K}_{ij}^{\ell m}]^e \underset{\sim}{a}_j^m - \sum_{\ell=1}^{n} \sum_{i=1}^{n_e} (\underset{\sim}{a}_i^{\ell})^T [\underset{\sim}{f}_i^{\ell}]^e \tag{31}$$

in which

$$[\underset{\sim}{K}_{ij}^{\ell m}]^e \begin{cases} = \frac{b}{2} \int_0^{a^e} (\bar{\underset{\sim}{B}}_i)^T \underset{\sim}{D} \bar{\underset{\sim}{B}}_j \, dx \text{ for } \ell = m \\ = 0 \text{ for } \ell \neq m \end{cases} \tag{32}$$

is the stiffness matrix of the strip e of width a^e connecting nodes i and j for the ℓth harmonic term, and

$$[\underset{\sim}{f}_i^{\ell}]^e = \frac{b}{2} \int_0^{a^e} [0, N_i q^{\ell}, 0]^T \, dx \tag{33}$$

is the vector of forces of node i of element e for the ℓth harmonic term.

Matrix $\bar{\underset{\sim}{B}}_i^{\ell}$ of (32) is obtained from (27) by simply making $S_\ell = C_\ell = 1$.

From (32) and (33) it can be seen that there is no coupling between the different harmonic terms and, therefore, the stiffness matrix and load vectors of the strip can be computed separately for each harmonic.

The discretised equilibrium equations can be easily obtained by minimising the Total Potential Energy of the plate, π, with respect to all the nodal amplitudes, i.e.

$$\frac{\partial \pi}{\partial \underset{\sim}{a}_i^{\ell}} = 0 \tag{34}$$

Equation (34) leads to a system of linear equations of the form

$$\begin{bmatrix} \underset{\sim}{K}^{11} & & & \underset{\sim}{0} \\ & \underset{\sim}{K}^{22} & & \\ & & \ddots & \\ \underset{\sim}{0} & & & \underset{\sim}{K}^{nn} \end{bmatrix} \begin{Bmatrix} \underset{\sim}{a}^{1} \\ \underset{\sim}{a}^{2} \\ \vdots \\ \underset{\sim}{a}^{n} \end{Bmatrix} = \begin{Bmatrix} \underset{\sim}{f}^{1} \\ \underset{\sim}{f}^{2} \\ \vdots \\ \underset{\sim}{f}^{n} \end{Bmatrix} \qquad (35)$$

Therefore the global stiffness $\underset{\sim}{K}^{\ell\ell}$ and the load vector $\underset{\sim}{f}^{\ell}$ can be computed separately for each harmonic by assembling the contributions from the different strip matrices in the standard manner [20]. The system of equations for the ℓth harmonic term can then be solved independently for the nodal amplitudes $\underset{\sim}{a}^{\ell}$. By repeating this process for all the harmonics, the different nodal amplitude parameters can be obtained. Subsequently, the displacements at each point in the structure can be computed by (22), whereas the strains and stresses can be evaluated using (24) and (25) resp.

It is worth pointing out here that the decoupling of the stiffness and load matrices for each harmonic term is a direct consequence of the Fourier expansions chosen for the displacement field in (20). If any other expansion is used to reproduce a different type of boundary condition at the plate ends, then decoupling does not occur due to the appearance of products $S_{\ell}C_{m}$ which do not satisfy the orthogonality condition [12]. In such cases the stiffness matrix is a full matrix and special iterative techniques have to be used in order to make the finite strip method competitive by comparison with the more general finite element procedures [12].

3.3 Numerical evaluation of the integrals

In the present implementation all integrals in the transverse direction, such as those of (33) and (34), are evaluated numerically using one-dimensional Gauss-Legendre quadrature [20]. Thus, each integral is evaluated as follows

$$\int_0^{a^e} f(x)\, dx = \int_{-1}^{+1} g(\xi)\, d\xi = \sum_{i=1}^{n_g} g(\xi_i)\, W_i \qquad (36)$$

where ξ_i and W_i are the coordinate and weight factor of the ith Gaussian point, resp. and n_g is the number of integration points used. A table with the values of the Gaussian coordinates and weights can be found in reference [20].

3.4 The reduced integration family of Mindlin finite strips

Mindlin finite strips must be used with caution when dealing with very thin plates. The problem is similar to that experienced with Mindlin plate finite elements. As the plate thickness becomes small, the influence of the shear terms tends to dominate the numerical solution and unrealistic overstiff results (locking) can be obtained unless some precautions are taken.

The stiffness matrix associated with the ℓth harmonic can be written in terms of bending and shear contribution $\tilde{K}_b^{\ell\ell}$ and $\tilde{K}_s^{\ell\ell}$ resp. as

$$\tilde{K}^{\ell\ell} = \tilde{K}_b^{\ell\ell} + \tilde{K}_s^{\ell\ell} \qquad (37)$$

One of the simplest ways to ensure the singularity of matrix $\tilde{K}_s^{\ell\ell}$ and hence avoid locking is to relax the constraint imposed by the shear terms by under-integrating the coefficients of $\tilde{K}_s^{\ell\ell}$ in the numerical integration of the integrals which appear in the stiffness matrix $\tilde{K}^{\ell\ell}$. The rest of the stiffness matrix can be exactly integrated and thus the process is called "selective integration", or else it can also be under-integrated which is usually termed "reduced integration".

The number of integrating points to exactly integrate the strip matrices, obviously depends on the degree of the shape function polynomials of each particular strip. Figure 4 shows the shape functions for the linear, quadratic and cubic Mindlin strip elements. Table 1 gives the number of Gaussian

TABLE 1: Gaussian integration rules for matrices $\underset{\sim}{K}_b^{\ell\ell}$ and $\underset{\sim}{K}_s^{\ell\ell}$ for various Mindlin strip elements.

GAUSSIAN INTEGRATION RULES						
STRIP ELEMENT	FULL (F)		SELECTIVE (S)		REDUCED (R)	
	$\underset{\sim}{K}_b^{\ell\ell}$	$\underset{\sim}{K}_s^{\ell\ell}$	$\underset{\sim}{K}_b^{\ell\ell}$	$\underset{\sim}{K}_s^{\ell\ell}$	$\underset{\sim}{K}_b^{\ell\ell}$	$\underset{\sim}{K}_s^{\ell\ell}$
o—o	2	2	2	1	1	1
o—o—o	3	3	3	2	2	2
o—o—o—o	4	4	3	3	3	3
	NUMBER OF INTEGRATION POINTS					

integrating points [20] needed for the full, reduced and selective integration of $\underset{\sim}{K}_b^{\ell\ell}$ and $\underset{\sim}{K}_s^{\ell\ell}$.

It has been shown by Oñate and Suarez [18] that singularity of matrix $\underset{\sim}{K}_s^{\ell\ell}$ for the linear, quadratic and cubic strips is ensured for most practical cases if reduced <u>or</u> selective integration is used. The linear strip with full integration behaves badly and it gives overstiff, unrealistic results even for the case of moderately thick plates. The quadratic strip with full integration is somewhat unreliable. Although deflections and bending moments agree in most cases with theoretical values, the shear forces oscillate and smoothing is recommended. The cubic strip, however, performs well with full integration. Nevertheless, reduced integration can be recommended for economy reasons.

Some of the properties mentioned above will be shown in the examples which are now presented.

3.5 Numerical examples

3.5.1 Example 1: Convergence with number of harmonics

Figure 6 shows the convergence of the central deflection and central bending moment, M_x, for a simply supported square thick plate under uniform loading. Numerical results have been obtained for the meshes of linear, quadratic and cubic elements shown in the same figure. The convergence rate for reduced, selective or full integration is the same for all strip elements. Five non-zero harmonics (the even harmonic terms are zero due to the symmetry of loading) are required to obtain an error of less than 0.3% in both deflection and bending moment compared with the converged solution.

3.5.2 Example 2: Convergence with number of strips

Figure 7 shows the convergence of the central deflection and central bending moment, M_x, with the number of strips for the plate considered in Section 3.5.1 for two different thickness/span ratios of $t/L = 0.1$ and 0.01 resp. The percentage error with respect to the theoretical "exact" solution [21] is shown for the linear, quadratic and cubic strip elements with full, selective and reduced integration.

It can be seen that results for the linear element with selective and reduced integration are extremely good in both cases when compared with the quadratic or cubic strips.

3.5.3 Example 3: Thin plate behaviour study

Figure 8 shows the value of the central deflection of the square plate of Figure 6 for a wide range of thickness/span ratios. It can be seen that all elements behave well with reduced or selective integration and give the correct solution for thick, thin and very thin plates.

It is interesting to look at the distribution of shear forces along the centre line of the plate for different thicknesses. The shear forces have been plotted in Figure 9 for

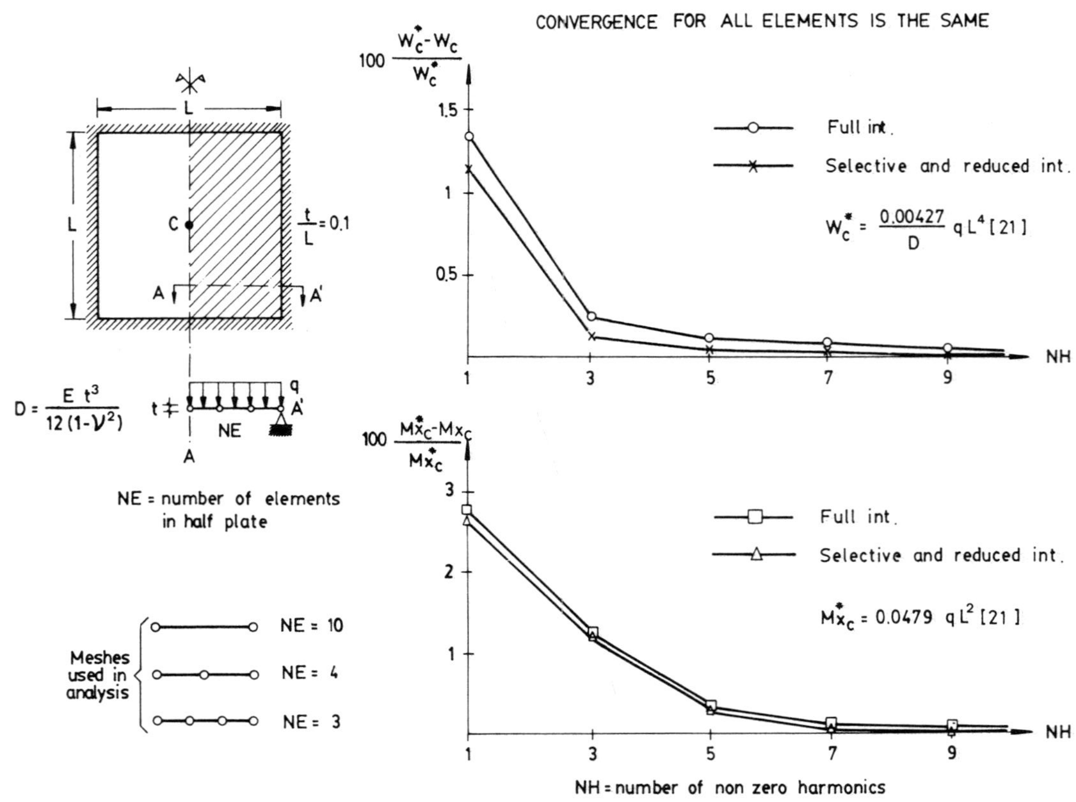

Fig 6 Simply supported square under uniform loading. Convergence of central deflection, W_c, and central bending moment, M_c, with number of harmonics for the linear quadratic and cubic strip elements.

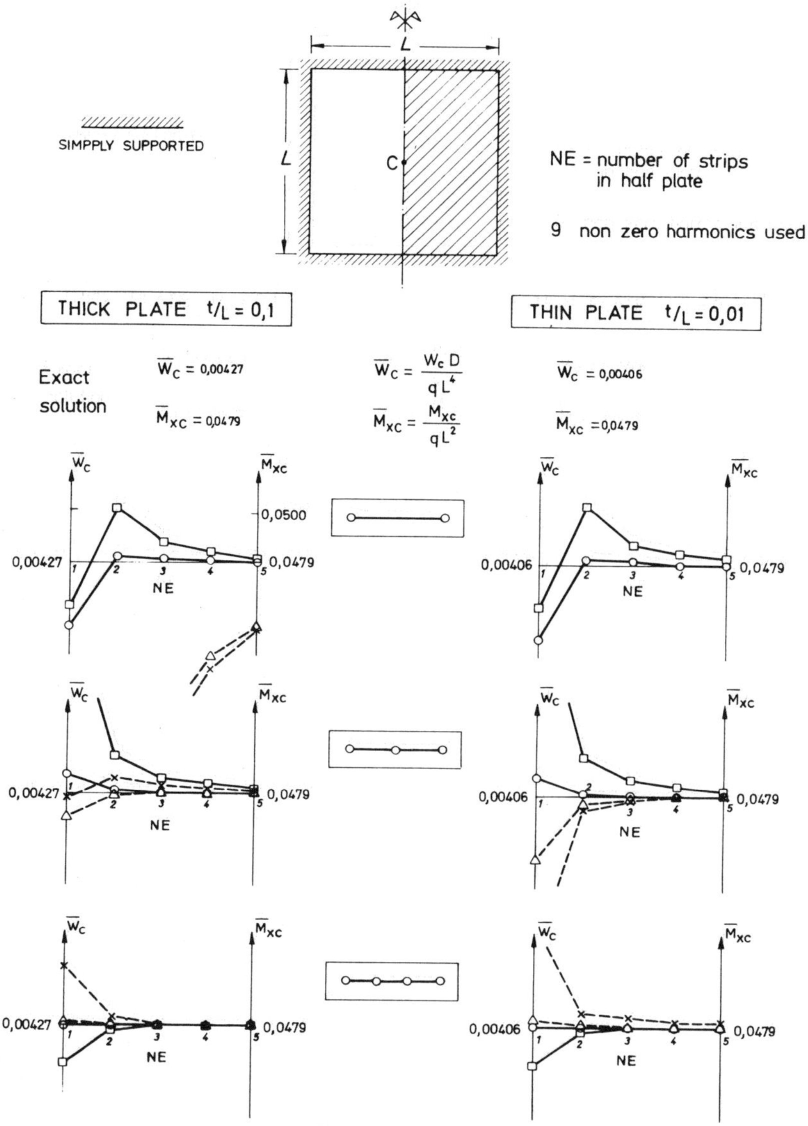

	Full integration	Select. and Red. integration
$\overline{W}_C$	——△——	——○——
$\overline{M}_{XC}$	——×——	——□——

Fig. 7 Square plate under uniform loading. Convergence study of the deflection and bending moment, M_x, at the plate center for the linear, quadratic and cubic strip elements with full, reduced and selective integration.

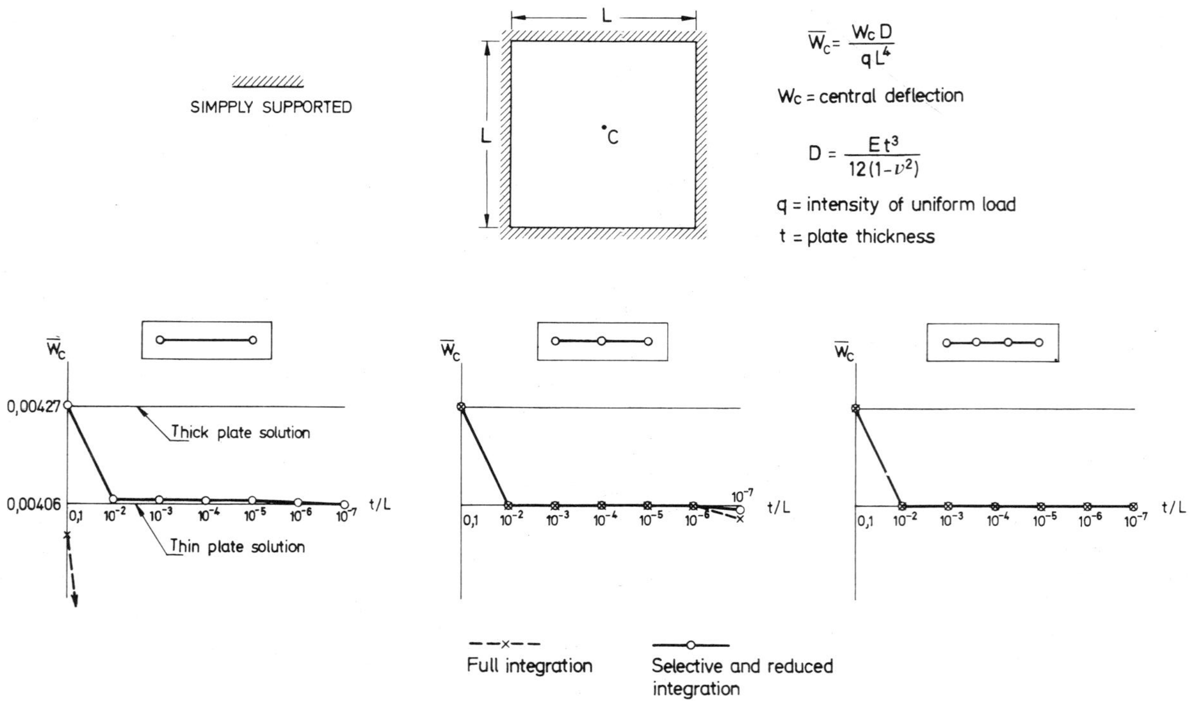

Fig. 8 Square plate under uniformly distributed load. Central deflection versus t/L for the linear, quadratic and cubic strip elements with full, reduced and selective integration.

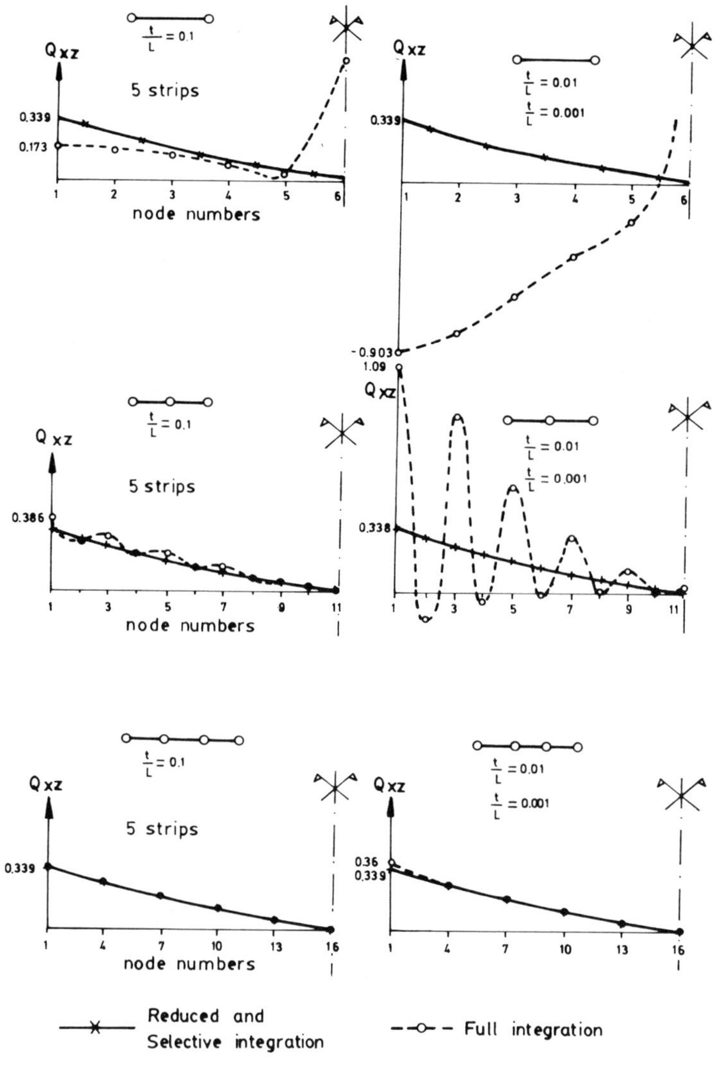

Fig. 9 Distribution of shear forces along the center of the plate for different strip elements and integration rules.

two thickness/span ratios of t/L = 0.1 and 0.01. We can see that wrong results are obtained with the linear element and full integration for the two cases, whereas selective and reduced integration give the correct results. Results for the quadratic element with full integration oscillate, as already mentioned. These oscillations are eliminated if selective or reduced integration is used. More information about this example can be found in reference [18].

3.5.4 Example 4: Square plate subjected to localised edge load

This square plate is simply supported on two opposite edges, clamped on one edge and free on the remaining edge. The centre of the free edge is subjected to a line load of varying intensity, as shown in Figure 10. In the vicinity of the load, the transverse stresses are large compared to the in-plane stresses with the result that shear deformation, in these regions, contributes significantly to the total deformation. The results obtained with the linear, quadratic and cubic Mindlin strips with reduced integration are compared with solutions based on classical plate theory [21] and with results of the work of Alwar and Ramachandran in which Reissner's theory is compared with experimental results [22]. Figure 10 clearly shows the importance of allowing for shear deformation. As the intensity of the line load increases, the numerical results obtained with classical plate theory, seriously overestimate the maximum tensile stress at the centre of the free edge. This situation worsens for thicker plates. The numerical results obtained for all three Mindlin strip elements are in good agreement with the experimental results. Figure 11 shows the localised nature of this overestimate and how thin plate theory is satisfactory in regions not too distant from the load.

3.5.5 Conclusions

For economical solutions which do not exhibit locking

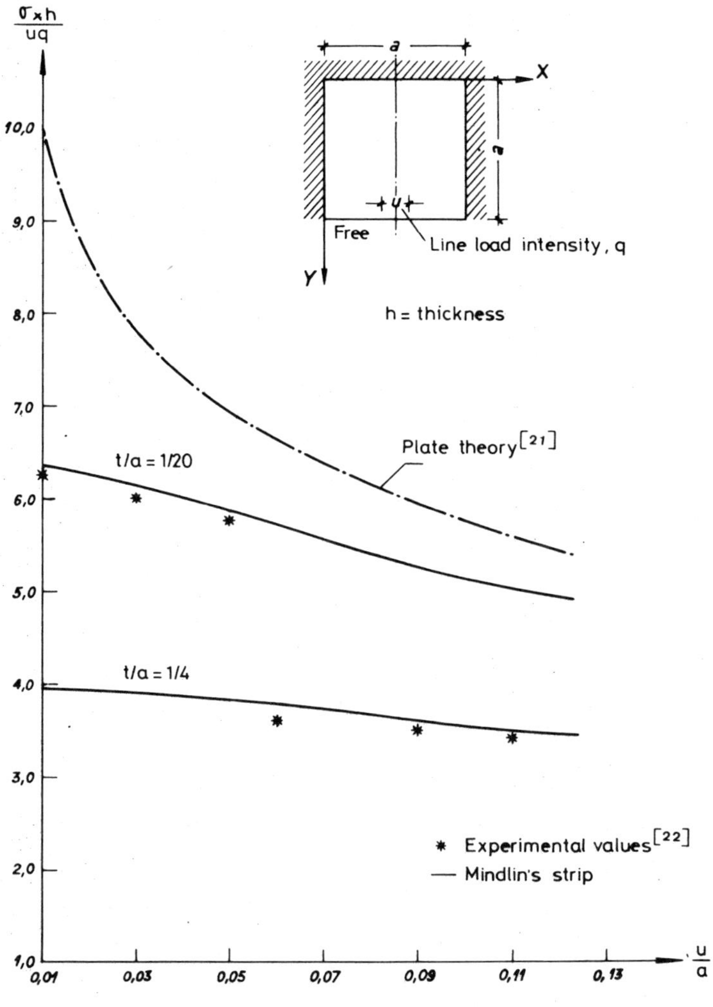

Fig. 10 Square plate simply supported at three edges under localised line load acting at free edge. Maximum tensile stress.

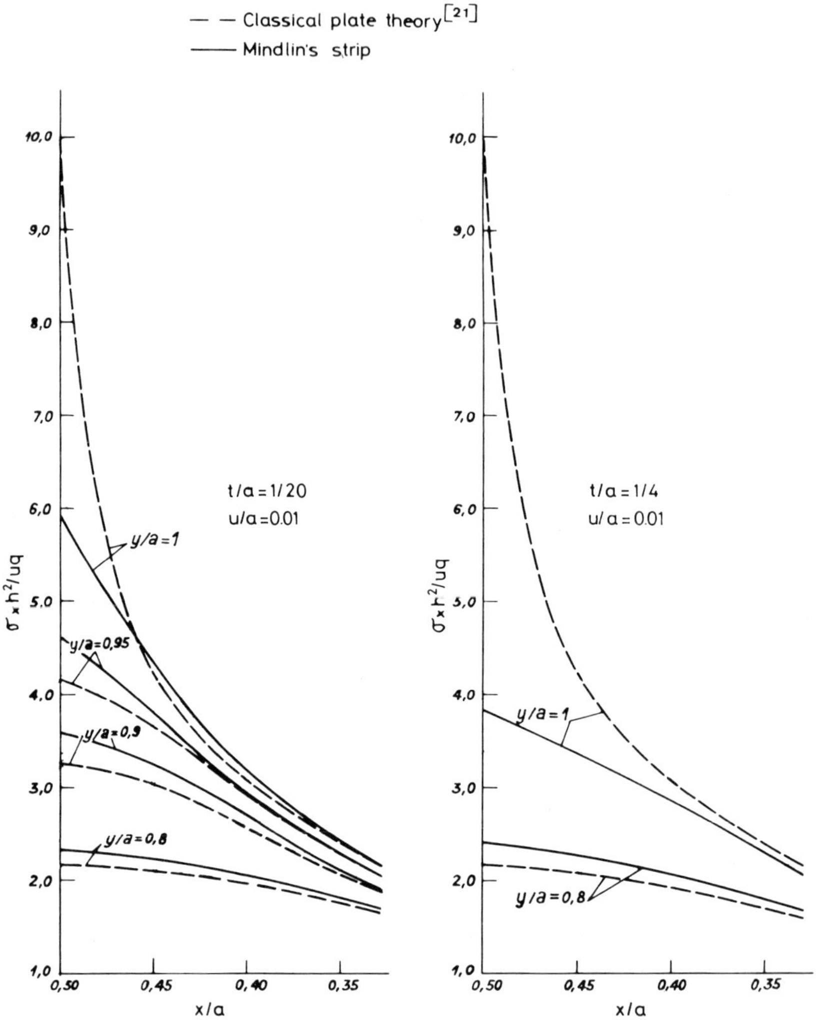

Fig. 11 Square plate simply supported at three edges under localised line load acting at free edge. Maximum tensile stress at various points.

reduced integration should be used in all strip elements for practical plate problems. In addition, the linear strip with reduced integration seems to be the best value element due to its excellent behaviour for thick and thin plate analysis and its simplicity. Moreover, an explicit form of the element matrices is simply obtained by evaluating the integrals at the strip mid-point [18].

4. ANALYSIS OF FOLDED PLATES WITH CURVED PLANFORMS

The Mindlin finite strip formulation for folded plate analysis closely follows the pattern presented for the analysis of plates in the previous sections.

The most general case of curved folded plates with circular plan shape is considered. It will also be shown later that the formulations for folded plates with rectangular planforms and axisymmetric shells can be obtained as particular cases of the formulation for curved folded plates presented in the next sections.

4.1 Basic theory

4.1.1 Displacement field

In the shell element shown in Figure 12, the three displacements, u, v, w, of a typical point can be expressed in terms of the three mid-plane displacements u_o, v_o and w_o and the two normal rotations θ_s and θ_θ as

$$\begin{aligned} u(s,\theta,n) &= u_o(s,\theta) + n\,\theta_s(s,\theta) \\ v(s,\theta,n) &= v_o(s,\theta) + n\,\theta_t(s,\theta) \qquad (38) \\ w(s,\theta,n) &= w_o(s,\theta) \end{aligned}$$

In (38) θ_s and θ_t are the normal rotations contained in planes sn and tn resp. These rotations can be expressed as the sum of the change in slope of the middle surface and an extra average rotation due to shear, so that

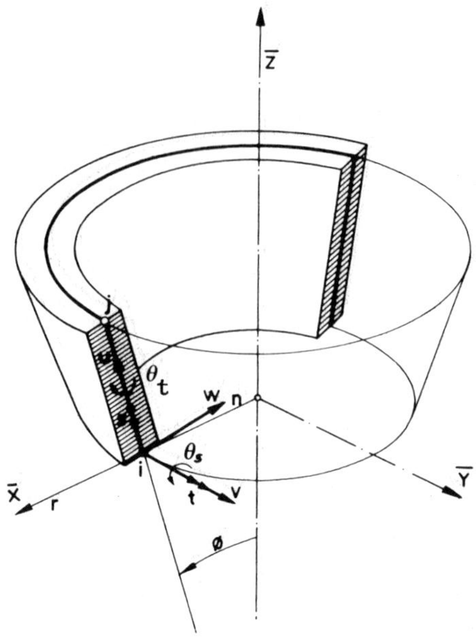

Fig. 12 Sign convention for displacements in a troncoconical shell.

$$\theta_s = -\frac{\partial w_o}{\partial s} + \phi_s$$

$$\theta_t = -\frac{1}{r}\frac{\partial w_o}{\partial \theta} + \phi_t \tag{39}$$

The displacement vector at a typical point may now be defined as

$$\underset{\sim}{u} = [u_o, v_o, w_o, \theta_s, \theta_t]^T \tag{40}$$

4.1.2 Strains

The relevant strains in the local system (s, t, n) which is illustrated in Figure 12, are defined as

$$\underset{\sim}{\varepsilon} = \begin{Bmatrix} \varepsilon_s \\ \varepsilon_t \\ \gamma_{st} \\ \gamma_{sn} \\ \gamma_{tn} \end{Bmatrix} = \begin{bmatrix} \frac{\partial u}{\partial s} \\ \frac{1}{r}\frac{\partial v}{\partial \theta} + \frac{u}{r}\sin\phi - \frac{w}{R_t} \\ \frac{1}{r}\frac{\partial u}{\partial \theta} + \frac{\partial v}{\partial s} - \frac{v}{r}\sin\phi - \frac{n}{R_t}\frac{\partial v}{\partial s} \\ \theta_s + \frac{\partial w}{\partial s} \\ \theta_t + \frac{1}{r}\frac{\partial w}{\partial \theta} + \frac{v}{R_t} \end{bmatrix} \tag{41}$$

Upon substitution of (38) into (41) the strain vector can be written as

$$\underset{\sim}{\varepsilon} = \underset{\sim}{\varepsilon}_m + \begin{Bmatrix} n\underset{\sim}{\varepsilon}_b \\ \underset{\sim}{\varepsilon}_s \end{Bmatrix} \tag{42}$$

where

$$\underset{\sim}{\varepsilon}_m = \begin{bmatrix} \frac{\partial u_o}{\partial s} \\ \frac{1}{r}\frac{\partial v_o}{\partial \theta} + \frac{u_o}{r}\sin\phi - \frac{w_o}{r}\cos\phi \\ \frac{\partial v_o}{\partial s} + \frac{1}{r}\frac{\partial u_o}{\partial \theta} - \frac{v_o}{r}\sin\phi \\ 0 \\ 0 \end{bmatrix} ;$$

$$\underset{\sim}{\varepsilon}_b = \begin{bmatrix} \frac{\partial \theta_s}{\theta s} \\ \frac{1}{r}\frac{\partial \theta_t}{\partial \theta} + \frac{\theta_s}{r}\sin\phi \\ \frac{\partial \theta_t}{\partial s} + \frac{1}{r}\frac{\partial \theta_s}{\partial \theta} - \frac{\theta_t}{r}\sin\phi - \frac{\cos\phi}{r}\frac{\partial v_o}{\partial s} \end{bmatrix}$$

and

$$\underset{\sim}{\varepsilon}_s = \begin{bmatrix} \theta_s + \dfrac{\partial w_o}{\partial s} \\ \dfrac{1}{r}\dfrac{\partial w_o}{\partial \theta} + \dfrac{v_o}{r}\cos\phi \end{bmatrix} \tag{43}$$

are the generalised strain vectors due to membrane, bending and shear effects, resp. In obtaining (43) the following assumptions have been made

$$(1 + \frac{n}{R_t}) = 1, \qquad \frac{n^2}{R_t}\frac{\partial \theta_t}{\partial s} = 0 \tag{44}$$

and $r = R_t \cos\phi$.

The generalised strain vector is now defined as

$$\underset{\sim}{\varepsilon} = [\underset{\sim}{\varepsilon}_m^T, \underset{\sim}{\varepsilon}_b^T, \underset{\sim}{\varepsilon}_s^T]^T \tag{45}$$

4.1.3 Stresses

The vector of stress resultants corresponding to the generalised strain vector of (45) can be written as

$$\underset{\sim}{\sigma} = [\underset{\sim}{\sigma}_m^T, \underset{\sim}{\sigma}_b^T, \underset{\sim}{\sigma}_s^T]^T \tag{46}$$

where

$$\begin{aligned} \underset{\sim}{\sigma}_m &= [N_s, N_t, N_{st}]^T \\ \underset{\sim}{\sigma}_b &= [M_s, M_t, M_{st}]^T \\ \underset{\sim}{\sigma}_s &= [Q_s, Q_t]^T \end{aligned} \tag{47}$$

are the stress resultant vectors due to membrane, bending and shear effects, resp. For the sign convention see Figure 13.

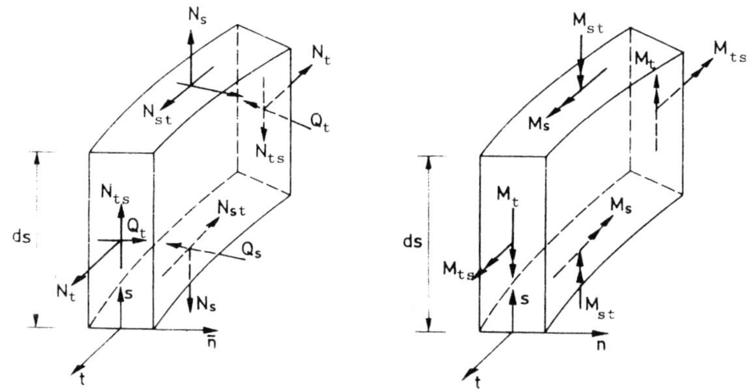

Fig. 13 Sign convention for displacements and resultant stresses in a troncoconical shell.

4.1.4 Stress-strain relationship

For an elastic material the relationship between generalised strains and stress resultants can be written as

$$\underset{\sim}{\sigma} = \underset{\sim}{D}\,\underset{\sim}{\varepsilon} \tag{48}$$

with

$$\underset{\sim}{D} = \begin{bmatrix} \underset{\sim}{D}_m & \underset{\sim}{0} & \underset{\sim}{0} \\ \underset{\sim}{0} & \underset{\sim}{D}_b & \underset{\sim}{0} \\ \underset{\sim}{0} & \underset{\sim}{0} & \underset{\sim}{D}_s \end{bmatrix} \tag{49}$$

where for an isotropic material the membrane elasticity matrix may be written as

$$\underset{\sim}{D}_m = \frac{Et}{1-\nu^2}\begin{bmatrix} 1 & \nu & 0 \\ 0 & 1 & 0 \\ 0 & 0 & \frac{1-\nu}{2} \end{bmatrix} \tag{50}$$

and $\underset{\sim}{D}_b$ and $\underset{\sim}{D}_s$ have the same meaning as those in (19).

4.1.5 Total Potential Energy of the shell

It can be shown [25] that the Total Potential Energy of the shell can be written in a form equivalent to that for a plate in (13) as

$$\pi = \tfrac{1}{2} \iint_A \underset{\sim}{\varepsilon}^T \underset{\sim}{\sigma} \, dA - \iint_A \underset{\sim}{u}^T \underset{\sim}{b} \, dA - \iint_A \underset{\sim}{u}^T \underset{\sim}{t} \, dA - \int_\Gamma \underset{\sim}{u}^T \underset{\sim}{P} \, d\Gamma \qquad (51)$$

where $\underset{\sim}{u}$ is the displacement vector, $\underset{\sim}{b}$ and $\underset{\sim}{t}$ are the body force and distributed loading vectors acting per unit area, $\underset{\sim}{P}$ is the vector of point loads acting along a line Γ, and A the area of the shell mid-surface.

4.2 Finite strip formulation for curved folded plates

The folded plate is divided into longitudinal strips as shown in Figure 14.

If k is the number of nodes within a particular strip e, the displacement field within the strip is expressed as

$$\underset{\sim}{u} = \sum_{\ell=1}^{n} \sum_{i=1}^{k} \underset{\sim}{N}_i^{\ell} \underset{\sim}{a}_i^{\ell} \qquad (52)$$

where $\underset{\sim}{N}_i^{\ell}$ and $\underset{\sim}{a}_i^{\ell}$ are resp. the generalised shape function matrix and the vector of nodal displacement amplitudes associated with node i for the ℓth harmonic term. These matrices have the following form

$$\underset{\sim}{N}_i^{\ell} = \begin{bmatrix} N_i S_\ell & 0 & 0 & 0 & 0 \\ 0 & N_i C_\ell & 0 & 0 & 0 \\ 0 & 0 & N_i S_\ell & 0 & 0 \\ 0 & 0 & 0 & N_i S_\ell & 0 \\ 0 & 0 & 0 & 0 & N_i C_\ell \end{bmatrix} \qquad (53)$$

$$\underset{\sim}{a}_i^{\ell} = [u_{oi}^{\ell}, \ v_{oi}^{\ell}, \ w_{oi}^{\ell}, \ \theta_{si}^{\ell}, \ \theta_{ti}^{\ell}]^T \qquad (54)$$

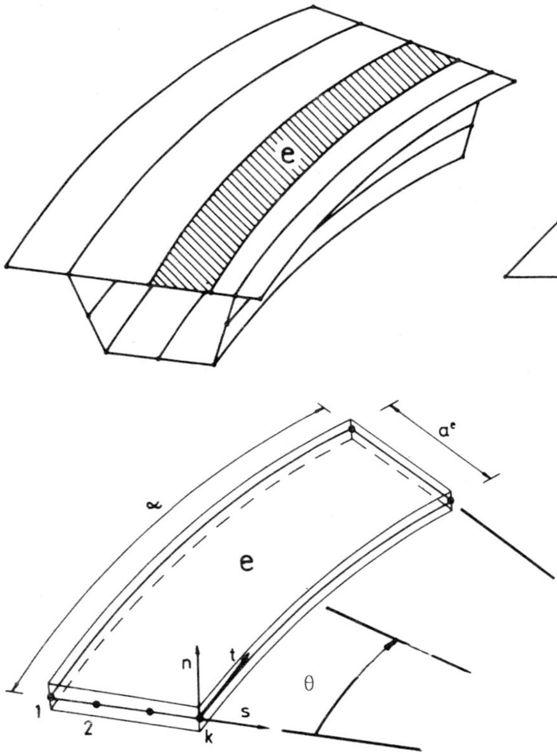

14a: Curved bridge strip

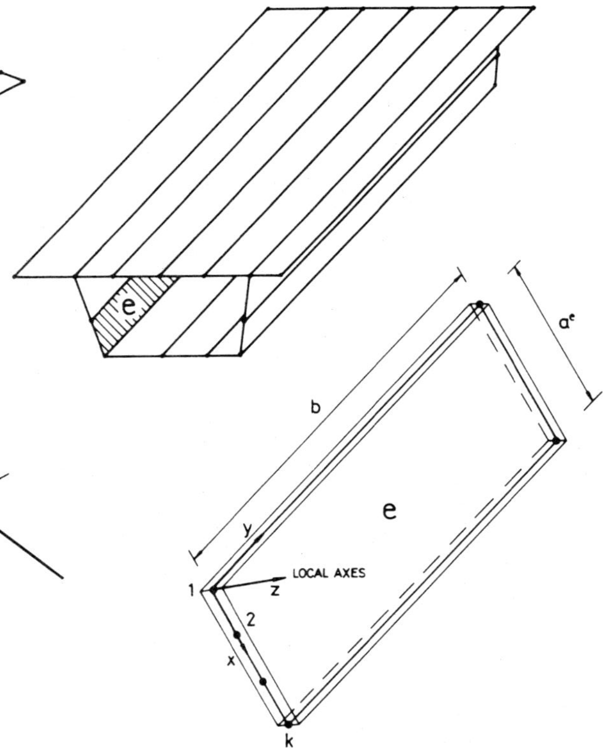

14b: Straight bridge strip

FIG. 14: Curved and straight bridges. Finite strip discretisations and local element coordinate axes.

where $S_\ell = \sin \frac{\ell\pi}{\alpha}\theta$, $C_\ell = \cos \frac{\ell\pi}{\alpha}\theta$ and the angle α is defined in Figure 14.

It is easy to check that the chosen harmonic expansions satisfy the simple support conditions at $\theta = 0$ and $\theta = \alpha$. This formulation is thus valid for simply supported folded plates with rigid diaphragms at the two ends.

The generalised strain vector $\underset{\sim}{\varepsilon}$ of (41) can be simply obtained at any point within the strip in terms of the nodal displacement amplitues by substituting (52) into (41) to give

$$\underset{\sim}{\varepsilon} = \sum_{\ell=1}^{n} \sum_{i=1}^{k} \underset{\sim}{B}_i^\ell \, \underset{\sim}{a}_i^\ell \tag{55}$$

where $\underset{\sim}{B}_i^\ell$ can now be written as

$$\underset{\sim}{B}_i^\ell = [\underset{\sim}{B}_{mi}^\ell, \underset{\sim}{B}_{bi}^\ell, \underset{\sim}{B}_{si}^\ell]^T \tag{56}$$

with

$$\underset{\sim}{B}_{mi}^\ell = \begin{bmatrix} \frac{\partial N_i}{\partial s} S_\ell & 0 & 0 & 0 & 0 \\ \frac{N_i}{r} \sin\phi \, S_\ell & -\frac{N_i \ell\pi}{b'} S_\ell & -\frac{N_i}{r} \cos\phi \, S_\ell & 0 & 0 \\ \frac{N_i \ell\pi}{b'} C_\ell & (\frac{\partial N_i}{\partial s} - \frac{N_i}{r} \sin\phi) C_\ell & 0 & 0 & 0 \end{bmatrix}$$

$$\underset{\sim}{B}_{bi}^\ell = \begin{bmatrix} 0 & 0 & 0 & \frac{\partial N_i}{\partial s} S_\ell & 0 \\ 0 & 0 & 0 & \frac{N_i}{r} \sin\phi \, S_\ell & -\frac{N_i \ell\pi}{b'} S_\ell \\ 0 & -\frac{\partial N_i}{\partial s} \frac{\cos\phi}{r} C_\ell & 0 & \frac{N_i \ell\pi}{b'} C_\ell & (\frac{\partial N_i}{\partial s} - \frac{N_i}{r} \sin\phi) C_\ell \end{bmatrix}$$

$$\underset{\sim}{B}_{si}^\ell = \begin{bmatrix} 0 & 0 & \frac{\partial N_i}{\partial s} S_\ell & N_i S_\ell & 0 \\ 0 & \frac{N_i}{r} \cos\phi \, C_\ell & \frac{N_i \ell\pi}{b'} C_\ell & 0 & N_i C_\ell \end{bmatrix} \tag{57}$$

where $b' = r\alpha$ and $\underset{\sim}{B}_{mi}^{\ell}$, $\underset{\sim}{B}_{bi}^{\ell}$ and $\underset{\sim}{B}_{si}^{\ell}$ are resp. the generalised strain matrices due to membrane, bending and shear effects for node i and the ℓth harmonic term.

If the force vectors are represented by the same kind of expansion as the one used for the displacement field, then it is possible to write

$$[\underset{\sim}{b},\ \underset{\sim}{t},\ \underset{\sim}{p}] = \sum_{\ell=1}^{n} [\underset{\sim}{S}_{\ell}\ \underset{\sim}{b}^{\ell},\ \underset{\sim}{S}_{\ell}\ \underset{\sim}{t}^{\ell},\ \underset{\sim}{S}_{\ell}\ \underset{\sim}{p}^{\ell}] \tag{58}$$

where

$$\underset{\sim}{S}_{\ell} = \begin{bmatrix} S_{\ell} & 0 & 0 & 0 & 0 \\ 0 & C_{\ell} & 0 & 0 & 0 \\ 0 & 0 & S_{\ell} & 0 & 0 \\ 0 & 0 & 0 & S_{\ell} & 0 \\ 0 & 0 & 0 & 0 & C_{\ell} \end{bmatrix} \tag{59}$$

and $\underset{\sim}{b}^{\ell}$, $\underset{\sim}{t}^{\ell}$ and $\underset{\sim}{p}^{\ell}$ are the force amplitude vectors for the ℓth term of the series.

Upon substitution of (52), (56) and (48) into the Total Potential Energy expression of (51) and by taking into account the orthogonal properties of functions S_{ℓ} and C_m it is possible to obtain, after an identical process to that followed in (30) - (32) for the plate bending case, the expression for the strip stiffness matrix and load vectors which have the following form

$$[K_{ij}^{\ell m}] = \begin{cases} \dfrac{\alpha}{2} \displaystyle\int_0^{a_e} [\bar{\underset{\sim}{B}}_i^{\ell}]^T \underset{\sim}{D}\ \bar{\underset{\sim}{B}}_j^{\ell}\ r ds & \text{for } \ell = m \\ 0 & \text{for } \ell \neq m \end{cases} \tag{60}$$

and

$$\underset{\sim}{f}_i^{\ell} = \iint_A (\underset{\sim}{N}_i^{\ell})^T \underset{\sim}{b}\ dA + \iint_A (\underset{\sim}{N}_i^{\ell})^T \underset{\sim}{t}\ dA + \int_{\Gamma} (\underset{\sim}{N}_i^{\ell})^T \underset{\sim}{p}\ d\Gamma \tag{61}$$

Further details on the load vector for different loading cases are given in Section 8. Matrix $\bar{\underset{\sim}{B}}_i^{\ell}$ of (60) can be obtained from (57) by simply making $S_{\ell} = C_{\ell} = 1$.

The discretised equilibrium equations for the whole structure are obtained by minimising the Total Potential Energy. This process leads to a set of uncoupled systems of equations which are similar to those of (3) and which can be solved separately for each harmonic term.

4.3 Assembly of the stiffness matrices and coordinate transformation

One of the main differences between the plate and folded plate strip formulations is that in the plate bending case the strips are all in the same plane, which coincides with the plate middle surface, whereas in the folded plate case the strips usually meet at different angles. Thus, to assemble the complete stiffness matrix of the structure from the individual strip stiffness matrices all nodal forces and displacements must be expressed in a common, and uniquely defined, coordinate system. The nodal displacements defined in the local strip coordinate system, shown in Figure 12, contain only two rotation components θ_s and θ_t. The third rotation, θ_n, about the n axis does not appear in the strain definition and is therefore not required to model the structural behaviour of each individual strip. However, when several strips meeting at a common node lie in different planes it is necessary to include $\theta_{\bar{z}}$, the rotation about the global $\bar{Z}$ axis, for a consistent transformation of displacements and forces from the local to the global coordinate system.

Thus, it is possible to write that

$$\bar{\underset{\sim}{a}}_i^{\ell} = \underset{\sim}{T}^{(e)} \underset{\sim}{a}_i^{\ell} \tag{62}$$

and
$$\bar{\underset{\sim}{f}}_i^{\ell} = \underset{\sim}{T}^{(e)} \underset{\sim}{f}_i^{\ell} \tag{63}$$

where

$$\begin{aligned} \bar{\underset{\sim}{a}}_i^{\ell} &= [\bar{u}_i^{\ell}, \bar{v}_i^{\ell}, \bar{w}_i^{\ell}, \theta_{\bar{x}i}^{\ell}, \theta_{\bar{y}i}^{\ell}, \theta_{\bar{z}i}^{\ell}]^T \\ \bar{\underset{\sim}{f}}_i^{\ell} &= [F_{\bar{x}i}^{\ell}, F_{\bar{y}i}^{\ell}, F_{\bar{z}i}^{\ell}, M_{\theta_{\bar{x}i}}^{\ell}, M_{\theta_{\bar{y}i}}^{\ell}, M_{\theta_{\bar{z}i}}^{\ell}]^T \end{aligned} \tag{64}$$

are the displacement and force vectors at node i of element e in the global coordinate system $\bar{x}$, $\bar{y}$, $\bar{z}$ where $\bar{y}$ is parallel to t and $\bar{z}$ is the vertical axis as shown in Figure 12. Note also that

$$\underset{\sim}{a}_i^{\ell} = [u_i^{\ell}, v_i^{\ell}, w_i^{\ell}, \theta_{si}^{\ell}, \theta_{ti}^{\ell}, 0]^T \tag{65}$$

$$\underset{\sim}{f}_i^{\ell} = [F_{si}^{\ell}, F_{ti}^{\ell}, F_{ni}^{\ell}, M_{\theta_{si}}^{\ell}, M_{\theta_{ti}}^{\ell}, 0]^T \tag{66}$$

are the same vectors in the local strip coordinate system. The matrix

$$\underset{\sim}{T}^{(e)} = \begin{bmatrix} \sin\phi & 0 & -\cos\phi & 0 & 0 & 0 \\ 0 & 1 & 0 & 0 & 0 & 0 \\ \cos\phi & 0 & \sin\phi & 0 & 0 & 0 \\ 1 & 0 & 0 & 1 & 0 & 0 \\ 0 & 0 & 0 & 0 & \sin\phi & \cos\phi \\ 0 & 0 & 0 & 0 & -\cos\phi & \sin\phi \end{bmatrix} \tag{67}$$

is the coordinate transformation matrix of element e, and ϕ is the angle between axes s and $\bar{z}$, as shown in Figure 12.

The strip stiffness matrix in the global system can be written in the standard form as

$$[\bar{\underset{\sim}{K}}_{ij}^{\ell\ell}]^e = \underset{\sim}{T}^{(e)} \underset{\sim}{K}_{ij}^{\prime\ell\ell} [\underset{\sim}{T}^e]^T \tag{68}$$

with

$$\underset{(6\times6)}{\underset{\sim}{K}_{ij}^{\prime\ell\ell}} = \begin{bmatrix} \underset{(5\times5)}{\underset{\sim}{K}_{ij}^{\ell\ell}} & 0 \\ 0 & 0 \end{bmatrix} \tag{69}$$

The sixth row and column of $\underset{\sim}{K}_{ij}^{\prime\ell\ell}$ contains only zeros. This is done in (69) to facilitate the transformation from the local to the global system. Equation (69) can be written in a more practical form as

$$[\bar{\underset{\sim}{K}}_{ij}^{\ell\ell}]^e = \frac{\alpha}{2} \int [\underset{\sim}{B}_i^*]^T \underset{\sim}{D} \underset{\sim}{B}_j^* \, r \, ds \tag{70}$$

with

$$\underset{\sim}{B}_i^* = \bar{\underset{\sim}{B}}_i \, \underset{\sim}{T}^T \tag{71}$$

Matrix $\underset{\sim}{B}_i^*$ allows the direct evaluation of the local stress resultants from the global displacements using (55) and (49). Thus, the local stress resultants can be written as

$$\underset{\sim}{\sigma} = \sum_{\ell=1}^{n} \sum_{i=1}^{K} \underset{\sim}{D} \, \underset{\sim}{B}_i^{*\ell} \, \underset{\sim}{a}_i^{\ell} \tag{72}$$

In general, matrix $\bar{\underset{\sim}{K}}_{ij}^{\ell\ell}$ will be fully populated and thus θ_z will be an independent degree of freedom. A problem, however, arises if all strips associated with a particular node happen to lie in the same plane because the resultant diagonal stiffness coefficient corresponding to θ_z, after assembly, will be zero. Such a node is termed a coplanar node and examples can be seen in Figure 14. This singularity of the global stiffness matrix has been avoided, in practice, by assembling the equations corresponding to the three rotations at a coplanar node, in the local system s,t,n in which s,n lie in the same common plane containing all the adjacent strips. Any arbitrary number is then put in the sixth leading diagonal location of matrix $\bar{\underset{\sim}{K}}_{ij}^{\ell\ell}$. This implies that the sixth equation is a pseudo-equation. However, this does not affect the solution process since such an equation is uncoupled from the rest of the stiffness equations. This artifice, first suggested by Clough and Wilson[23] implies that all coplanar and non-coplanar nodes have six degrees of freedom. This is very convenient if the equation solution system does not allow for varying numbers of degrees of freedom at different nodes.

5. CURVED PLATES

The Mindlin strip formulation for the analysis of curved plates can be derived directly from the formulation given for curved folded plates presented in the previous section by simply neglecting the membrane behaviour of the structure in all equations. Details of the formulation follow precisely the same steps as those explained previously and they will not be repeated here. The stiffness and load vectors for each harmonic term are obtained from (60) and (61) resp. However, $\tilde{B}_i^{\ell}$ and $\tilde{D}$ are now written as

$$\tilde{B}_i^{\ell} = \begin{bmatrix} 0 & \frac{\partial N_i}{\partial s} S_\ell & 0 \\ 0 & \frac{N_i}{r}\sin\phi\, S_\ell & -\frac{N_i \ell\pi}{b'} S_\ell \\ 0 & \frac{N_i \ell\pi}{b'} C_\ell & (\frac{\partial N_i}{\partial s} - \frac{N_i}{r}\sin\phi) C_\ell \\ \frac{\partial N_i}{\partial s} S_\ell & N_i S_\ell & 0 \\ \frac{N_i \ell\pi}{b'} C_\ell & 0 & N_i C_\ell \end{bmatrix} \quad (73)$$

and

$$\tilde{D} = \begin{bmatrix} \tilde{D}_b & \tilde{0} \\ 0 & \tilde{D}_s \end{bmatrix} \quad (74)$$

in which $b' = r\alpha$ and $\tilde{a}_i^{\ell}$ of (54) is now expressed as

$$a_i^{\ell} = [w_i^{\ell} , \theta_{si}^{\ell} , \theta_{ti}^{\ell}]^T$$

By making r very large so that $\frac{1}{r} \simeq 0$ and by letting $b' = b$ it is possible to obtain the strain-displacement matrix for a straight (or right) plate given in (27).

6. FOLDED PLATES WITH RECTANGULAR PLANFORM

The formulation for folded plates with rectangular planform can be easily derived from the formulation for their curved counterparts presented in Section 4. Thus, only details of the main differences will be given here.

6.1 Displacement field

For a plane "shell" element the three displacements of a point across the thickness can be expressed in terms of the displacements of the middle surface as

$$
\begin{aligned}
u(x,y,z) &= u_o(x,y) + z\,\theta_x(x,y) \\
v(x,y,z) &= v_o(x,y) + z\,\theta_y(x,y) \\
w(x,y,z) &= w_o(x,y)
\end{aligned}
\tag{75}
$$

where all terms have the same meaning of those of (38) except that the fixed local axes x,y,z replace the curvilinear ones s,t,n (see Figure 14(b)).

6.2 Strain field

From standard elasticity theory the strain vector can be written as

$$
\underset{\sim}{\varepsilon} = \begin{Bmatrix} \underset{\sim}{\varepsilon}_m \\ \underset{\sim}{0} \end{Bmatrix} + \begin{Bmatrix} z\underset{\sim}{\varepsilon}_b \\ \underset{\sim}{\varepsilon}_s \end{Bmatrix}
\tag{76}
$$

where

$$
\underset{\sim}{\varepsilon}_m = \begin{Bmatrix} \dfrac{\partial u_o}{\partial x} \\[2mm] \dfrac{\partial v_o}{\partial y} \\[2mm] \dfrac{\partial u_o}{\partial y} + \dfrac{\partial v_o}{\partial x} \end{Bmatrix} \quad
\underset{\sim}{\varepsilon}_b = \begin{Bmatrix} \dfrac{\partial \theta_x}{\partial x} \\[2mm] \dfrac{\partial \theta_y}{\partial y} \\[2mm] \dfrac{\partial \theta_x}{\partial y} + \dfrac{\partial \theta_y}{\partial x} \end{Bmatrix} \quad
\underset{\sim}{\varepsilon}_s = \begin{Bmatrix} \theta_x + \dfrac{\partial w_o}{\partial x} \\[2mm] \theta_y + \dfrac{\partial w_o}{\partial y} \end{Bmatrix}
\tag{77}
$$

are the corresponding generalised strain vectors for membrane, bending and shear. It is worth noting the decoupling between membrane and flexural effects at element level, which did not occur in the curved bridge formulation (see (43)).

6.3 Stresses

The expressions for the stress resultant vectors are identical to those for curved bridges (see (47)) with indices x, y and s replacing s, θ and n resp. The sign convention given in Figure 13 is also followed. The stress/strain relationship is identical to the one given in (48).

6.4 Finite strip formulation for folded plates with rectangular planform

The displacement and strain fields are expressed in exactly the same way as for curved bridges (see (52) and (56)) with the bridge length b and the coordinate y replacing the angles α and ε resp. (see Figure 14). The strain matrix $\underset{\sim}{B}_i^{\ell}$ is obtained from (77) and (56), and it now takes the form

$$\underset{\sim}{B}_i^{\ell} = [\underset{\sim}{B}_{mi}^{\ell}\ ,\ \underset{\sim}{B}_{bi}^{\ell}\ ,\ \underset{\sim}{B}_{si}^{\ell}]^T \tag{78}$$

where

$$\underset{\sim}{B}_{mi}^{\ell} = \begin{bmatrix} \frac{\partial N_i}{\partial x} S_\ell & 0 & 0 & 0 & 0 \\ 0 & -N_i \frac{\ell\pi}{b} S_\ell & 0 & 0 & 0 \\ N_i \frac{\ell\pi}{b} C_\ell & \frac{\partial N_i}{\partial x} C_\ell & 0 & 0 & 0 \end{bmatrix}$$

$$\underset{\sim}{B}_{bi}^{\ell} = \begin{bmatrix} 0 & 0 & 0 & \frac{\partial N_i}{\partial x} S_\ell & 0 \\ 0 & 0 & 0 & 0 & -N_i \frac{\ell\pi}{b} S_\ell \\ 0 & 0 & 0 & N_i \frac{\ell\pi}{b} C_\ell & \frac{\partial N_i}{\partial x} C_\ell \end{bmatrix} \tag{79}$$

$$\underset{\sim}{B}_{si}^{\ell} = \begin{bmatrix} 0 & 0 & \frac{\partial N_i}{\partial x} S_\ell & N_i S_\ell & 0 \\ 0 & 0 & N_i \frac{\ell\Pi}{b} C_\ell & 0 & N_i C_\ell \end{bmatrix}$$

are resp. the membrane, bending and shear strain matrix for node i and the ℓth harmonic term.

It is worth noting that matrix $\underset{\sim}{B}_i^{\ell}$ for rectangular folded plates can be directly obtained from the expression for curved folded plates by simply making r very large so that $1/r \to 0$ and by setting $b' = b$ in (57).

These useful analogies allow all of the relevant matrices for rectangular folded plates to be derived from the corresponding expressions for their curved counterparts[17].

7. AXISYMMETRIC SHELLS

The Mindlin formulation for axisymmetric shells closely follows the steps of the formulation for folded plates with curved planforms. In fact, for axisymmetric shells under arbitrary loading the expressions for the displacement, strain and stress fields are identical to those of (38) - (50). Moreover, the Total Potential Energy of the shell can be written in a form which is almost identical to that of (51) - the only difference being that all integrals are taken now over a whole circumference.

Thus, the Mindlin finite strip formulation for axisymmetric shell problems can be considered as another particular case of the curved folded plate formulation (as was the case for rectangular and folded plates) and the basic steps in both formulations are essentially the same.

7.1 Mindlin finite strip formulation for axisymmetric shells under arbitrary loading

The shell is divided into circular strips as shown in Figure 1. For axisymmetric shells under arbitrary loading the displacement vector can be expressed within each strip in terms of the symmetrical and non-symmetrical contribution

$$\underset{\sim}{u} = \sum_{\ell=0}^{n} \sum_{i=1}^{k} (\underset{\sim}{N}_i^{\ell} \underset{\sim}{a}_i^{\ell} + \bar{\underset{\sim}{N}}_i^{\ell} \bar{\underset{\sim}{a}}_i^{\ell}) \qquad (80)$$

where the displacement vector, $\underset{\sim}{u}$, and nodal parameter vectors $\underset{\sim}{a}_i^{\ell}$ and $\bar{\underset{\sim}{a}}_i^{\ell}$ are defined by (40) and (54) resp., and

$$\underset{\sim}{N}_i^{\ell} = \begin{bmatrix} N_iC_\ell & 0 & 0 & 0 & 0 \\ 0 & N_iS_\ell & 0 & 0 & 0 \\ 0 & 0 & N_iC_\ell & 0 & 0 \\ 0 & 0 & 0 & N_iC_\ell & 0 \\ 0 & 0 & 0 & 0 & N_iS_\ell \end{bmatrix} \qquad (81a)$$

$$\bar{\underset{\sim}{N}}_i^{\ell} = \begin{bmatrix} N_i S_\ell & 0 & 0 & 0 & 0 \\ 0 & N_i C_\ell & 0 & 0 & 0 \\ 0 & 0 & N_i S_\ell & 0 & 0 \\ 0 & 0 & 0 & N_i S_\ell & 0 \\ 0 & 0 & 0 & 0 & N_i C_\ell \end{bmatrix} \qquad (81b)$$

are the shape function matrices corresponding to symmetrical and non-symmetrical displacement fields resp. Furthermore, $S_\ell = \sin \ell\theta$ and $C_\ell = \cos \ell\theta$.

Note that in (80) the harmonic 'zero' has been included. This term has a clear physical meaning and it corresponds to an axisymmetrical deformation.

To simplify the computation it is usual to evaluate the response of the shell to an arbitrary loading as an independent sum of the symmetric and non-symmetric part of the deformation. Thus, for the case in which all loads are symmetric with respect to a plane (which we will take here as that of $\theta = 0$ for simplicity) only matrix $\underset{\sim}{N}_i^{\ell}$ of (80) will be used.

The study of the non-symmetric part will be identical taking $\bar{\underset{\sim}{N}}_i^{\ell}$ instead of $\underset{\sim}{N}_i^{\ell}$.

Following precisely the same steps as those explained in Section 4.2 for the curved folded plate case, the decoupled system of stiffness equations for each harmonic term with the stiffness matrix may be obtained and written as

$$[\underset{\sim}{K}_{ij}^{\ell\ell}]^e = 2\pi \int_0^{a^e} (\bar{\underset{\sim}{B}}_i^{\ell})^T \underset{\sim}{D}\, \bar{\underset{\sim}{B}}_j^{\ell}\, r\, ds \quad \text{for } \ell = 0$$

$$[\underset{\sim}{K}_{ij}^{\ell\ell}]^e = \pi \int_0^{a^e} (\underset{\sim}{B}_i^{\ell})^T \underset{\sim}{D}\, \underset{\sim}{B}_j^{\ell}\, r\, ds \quad \text{for } \ell \neq 0 \qquad (82)$$

where matrix $\underset{\sim}{D}$ is given by (49) and matrix $\underset{\sim}{B}_i^{\ell}$ can be written in a general form for the axisymmetric and non-symmetric cases as

$$\bar{\underset{\sim}{B}}_i^{\ell} = \left[[\bar{\underset{\sim}{B}}_{mi}^{\ell}]^T, [\bar{\underset{\sim}{B}}_{bi}^{\ell}]^T, [\underset{\sim}{B}_{si}^{\ell}]^T \right]^T \qquad (83)$$

where

$$
\underset{\sim}{B}_{mi}^{\ell} = \begin{bmatrix} \frac{\partial N_i}{\partial s} & 0 & 0 & 0 & 0 \\ N_i \frac{\sin\phi}{r} & \frac{N_i}{r}\bar{\ell} & -\frac{N_i}{r}\cos\phi & 0 & 0 \\ -\frac{N_i}{r}\bar{\ell} & \frac{\partial N_i}{\partial s} - \frac{N_i}{r}\sin\phi & 0 & 0 & 0 \end{bmatrix}
$$

$$
\underset{\sim}{B}_{bi}^{\ell} = \begin{bmatrix} 0 & 0 & 0 & \frac{\partial N_i}{\partial s} & 0 \\ 0 & 0 & 0 & \frac{N_i}{r}\sin\phi & \frac{N_i}{r}\bar{\ell} \\ 0 & -\frac{\partial N_i}{\partial r}\frac{\cos\phi}{r} & 0 & -\frac{N_i}{r}\bar{\ell} & \frac{\partial N_i}{\partial s} - \frac{N_i}{r}\sin\phi \end{bmatrix} \tag{84}
$$

$$
\underset{\sim}{B}_{si}^{\ell} = \begin{bmatrix} 0 & 0 & \frac{\partial N_i}{\partial s} & N_i & 0 \\ 0 & \frac{N_i}{r}\cos\phi & -\frac{N_i}{r}\bar{\ell} & 0 & N_i \end{bmatrix}
$$

are the membrane bending and shear strain matrices for the ℓth harmonic with

$\bar{\ell} = \ell$ for the symmetric case

$\bar{\ell} = -\ell$ for the non-symmetric case

It is worth pointing out that matrix $\underset{\sim}{\bar{B}}_i^{\ell}$ can be directly obtained from its analogous expression for curved folded plates by simply substituting in (57) the value of $\frac{\ell\pi}{\alpha}$ by $\bar{\ell}$ and making $S_\ell = C_\ell = 1$.

This shows again the versatility of the general formulation of Section 4 and how it allows folded plates, plates and axisymmetric shells to be treated in a unified manner.

The transformation of the strip stiffness matrix into a global coordinate system follows precisely the steps presented in Section 4.3 for curved folded plates with the transformation matrix being identical to that of (67). The transformation will not be repeated here.

The loads are obtained using the method described in Section 4.2 with the following expansions

$$[\underset{\sim}{b}, \underset{\sim}{t}, \underset{\sim}{p}] = \sum_{\ell=0}^{n} [\underset{\sim}{S}_{\ell}\underset{\sim}{b}^{\ell}, \underset{\sim}{S}_{\ell}\underset{\sim}{t}^{\ell}, \underset{\sim}{S}_{\ell}\underset{\sim}{p}^{\ell}] \tag{85}$$

where

$$\underset{\sim}{S}_{\ell} = \begin{bmatrix} C_{\ell} & 0 & 0 & 0 & 0 \\ 0 & S_{\ell} & 0 & 0 & 0 \\ 0 & 0 & C_{\ell} & 0 & 0 \\ 0 & 0 & 0 & C_{\ell} & 0 \\ 0 & 0 & 0 & 0 & S_{\ell} \end{bmatrix} \tag{86}$$

for the symmetric loading case. For the non-symmetric case

$$\underset{\sim}{S}_{\ell} = \begin{bmatrix} S_{\ell} & 0 & 0 & 0 & 0 \\ 0 & C_{\ell} & 0 & 0 & 0 \\ 0 & 0 & S_{\ell} & 0 & 0 \\ 0 & 0 & 0 & S_{\ell} & 0 \\ 0 & 0 & 0 & 0 & C_{\ell} \end{bmatrix} \tag{87}$$

Details of the load vector for different loading cases are given in the next section.

If the loading is also axisymmetric the same formulation is directly applicable by simply evaluating the contribution of the zero harmonic term only (i.e. $\bar{\ell} = 0$ in (84)). However, a simpler formulation can be automatically derived taking into account the contribution of the non-zero displacements u, w and θ_s only in matrix $\bar{\underset{\sim}{B}}_i^{o}$. Details of this formulation can be found in reference [24].

8. COMPUTATION OF THE EQUIVALENT NODAL FORCE VECTOR

As already mentioned in Section 4.2, the external loads acting over the structure are expanded in the same way as the displacements, that is as the sum of the harmonic series along the longitudinal/circumferential direction of the structure. Moreover, the form of the Fourier expansions for the loads is the same as that chosen for the corresponding displacements, i.e.

Displacements	Loads
$\underset{\sim}{u} = \sum_{\ell=1}^{n} \underset{\sim}{S}^{\ell} \underset{\sim}{u}^{\ell}$	$\underset{\sim}{f} = \sum_{\ell=1}^{n} \underset{\sim}{S}^{\ell} \underset{\sim}{f}^{\ell}$

(88)

where $\underset{\sim}{f}^{\ell}$ and $\underset{\sim}{S}^{\ell}$ are the load amplitudes vector and the matrix of the harmonic functions for the ℓth harmonic term resp.

Thus, for the case of a rectangular plate

$$\underset{\sim}{P} = [P_w, M_{\theta x}, M_{\theta y}]^T = \sum_{\ell=1}^{n} \underset{\sim}{S}^{\ell} [P_w^{\ell}, M_{\theta x}^{\ell}, M_{\theta y}^{\ell}]^T = \sum_{\ell=1}^{n} \underset{\sim}{S}^{\ell} \underset{\sim}{P}^{\ell}$$

$$\underset{\sim}{t} = [t_w, M_{\theta x}, M_{\theta y}]^T = \sum_{\ell=1}^{n} \underset{\sim}{S}^{\ell} [t_w^{\ell}, M_{\theta x}^{\ell}, M_{\theta y}^{\ell}]^T = \sum_{\ell=1}^{n} \underset{\sim}{S}^{\ell} \underset{\sim}{t}^{\ell} \qquad (89)$$

$$\underset{\sim}{b} = [b_w, 0, 0]^T = \sum_{\ell=1}^{n} \underset{\sim}{S}^{\ell} [b_w^{\ell}, 0, 0]^T = \sum_{\ell=1}^{n_i} \underset{\sim}{S}^{\ell} \underset{\sim}{b}^{\ell}$$

Matrix $\underset{\sim}{S}^{\ell}$ can be easily deduced from (59) as

$$\underset{\sim}{S}^{\ell} = \begin{bmatrix} S_{\ell} & 0 & 0 \\ 0 & S_{\ell} & 0 \\ 0 & 0 & C_{\ell} \end{bmatrix} \qquad (90)$$

In (89) $\underset{\sim}{P}$, $\underset{\sim}{t}$ and $\underset{\sim}{b}$ are the vectors of point loads, surface loads and body forces loads, resp. The three components of such vectors correspond to the loads associated with the vertical deflection and the two rotations, resp. For the body forces case only the vertical component, b_w, corresponding to the self weight of the structure per unit area, has been considered.

The load amplitudes are calculated individually using

Euler's formula. For instance for a uniformly distributed vertical load we can write q_w^{ℓ} as

$$q_w^{\ell} = \frac{\int_0^b \sin \frac{\ell\pi}{b} y \, dy}{\int_0^b \sin^2 \frac{\ell\pi}{b} y \, dy} \simeq \frac{2q_w}{\ell\pi} (\ell - \cos \ell\pi) \tag{91}$$

An identical process will be followed for the evaluation of the rest of the loading amplitude terms.

The vector of nodal forces for each strip for the ℓth harmonic term can be written in a general form for the different structures studied in this chapter using (61) and the property of the orthogonal function S_ℓ and C_ℓ as

$$\underset{\sim}{f}_i^{\ell} = C \int_0^{a^e} N_i \underset{\sim}{b}^{\ell} rds + C \int_0^{a^e} N_i \underset{\sim}{t}^{\ell} rds + C\underset{\sim}{q}_i^{\ell} \tag{92}$$

For plates and folded plates $C = \frac{L}{2}$ where L is the length or angle of the structure for the rectangular of curved case, resp. For axisymmetric shells, $C = 2\pi$ and π for $\ell = 0$ and $\ell \neq 0$, resp. For right structures $r = 1$ in (92).

In Table 2, the expressions of $\underset{\sim}{f}_i^{\ell}$ for three typical loading cases of point load, uniform load and self weight, for plates, folded plates and axisymmetric shells (under symmetric loading) are given. The evaluation of the formulae of Table 2 for different strip elements mentioned in this chapter is simple and only involves the appropriate evaluation of the corresponding integrals for the strip shape functions over the strip length. This can be easily performed using the shape functions expressions of Figure 4, and it is left as an exercise for the reader. More details can be found in references [17] and [25].

LOADING CASE	PLATE	FOLDED PLATE	AXISYMMETRIC SHELL
POINT LOAD acting at node i at y = c	$f_i^\ell = \begin{Bmatrix} P_i \sin\frac{\ell\pi c}{L} \\ M_{txi}\sin\frac{\ell\pi c}{L} \\ M_{tyi}\cos\frac{\ell\pi c}{L} \end{Bmatrix}$	$f_i^\ell = \begin{Bmatrix} P_{xi}\sin\frac{\ell\pi c}{L} \\ P_{yi}\cos\frac{\ell\pi c}{L} \\ P_{zi}\sin\frac{\ell\pi c}{L} \\ M_{txi}\sin\frac{\ell\pi c}{L} \\ M_{tyi}\cos\frac{\ell\pi c}{L} \end{Bmatrix}$	$\ell = 0$: $f_i^\ell = \begin{Bmatrix} P_{ri} \\ 0 \\ P_{zi} \\ M_{tsi} \\ 0 \end{Bmatrix}$ $\ell \neq 0$: $f_i^\ell = \begin{Bmatrix} P_{ri}\cos\ell\pi \\ 0 \\ P_{zi}\cos\ell\pi \\ M_{tsi}\cos\ell\pi \\ 0 \end{Bmatrix}$
UNIFORM LOAD acting over the whole element between limits $b_0<y<b_1$ ($-\alpha_0<\alpha<+\alpha_0$ for axisymmetric shells)	$f_i^\ell = C_i\begin{Bmatrix} q_z(\cos\frac{\ell\pi b_0}{L} - \cos\frac{\ell\pi b_1}{L} \\ m_{tx}(\cos\frac{\ell\pi b_0}{L} - \cos\frac{\ell\pi b_1}{L}) \\ m_{ty}(\sin\frac{\ell\pi b_1}{L} - \sin\frac{\ell\pi b_0}{L}) \end{Bmatrix}$ $C_i = \frac{L}{\ell\pi}\int_0^a rN_i(x)dx$	$f_i^\ell = C_i\begin{Bmatrix} q_x(\cos\frac{\ell\pi b_0}{L} - \cos\frac{\ell\pi b_1}{L}) \\ q_y(\sin\frac{\ell\pi b_1}{L} \\ q_z(\cos\frac{\ell\pi b_0}{L} \\ m_{tx}(\cos\frac{\ell\pi b_0}{L} - \cos\frac{\ell\pi b_1}{L}) \\ m_{ty}(\sin\frac{\ell\pi b_1}{L} - \sin\frac{\ell\pi b_0}{L}) \end{Bmatrix}$ $C_i = \frac{L}{\ell\pi}\int_0^a rN_i(x)dx$	$\ell = 0$: $f_i^\ell = C_i\begin{Bmatrix} q_r\ell\alpha_0 \\ 0 \\ q_z\ell\alpha_0 \\ m_{ts}\ell\alpha_0 \\ 0 \end{Bmatrix}$ $\ell \neq 0$: $f_i^\ell = C_i\begin{Bmatrix} q_r\sin\ell\alpha_0\cos\ell- \\ 0 \\ q_z\sin\ell\alpha_0\cos\ell\pi \\ m_{ts}\sin\ell\alpha_0\cos\ell\pi \\ 0 \end{Bmatrix}$ $C_i = \frac{2}{\ell}\int_0^a N_i r d_r$
SELF WEIGHT ρ = density g = gravity	ℓ odd $f_i^\ell = \frac{2L\rho gt}{\ell\pi}\int_0^a rN_i(x)dx\begin{Bmatrix} 1 \\ 0 \\ 0 \end{Bmatrix}$ ℓ even $f_i^\ell = 0$	ℓ odd $f_i^\ell = \frac{2L\rho gt}{\ell\pi}\int_0^a rN_i(x)dx\begin{Bmatrix} 0 \\ 0 \\ 1 \\ 0 \\ 0 \end{Bmatrix}$ ℓ even f_i^ℓ	$\ell = 0$: $f_i^\ell = 2\rho gt\pi\int_0^a N_i r d_r\begin{Bmatrix} 0 \\ 0 \\ 1 \\ 0 \\ 0 \end{Bmatrix}$ $\ell \neq 0$: $f_i^\ell = 0$
	CURVED CASE L = α STRAIGHT CASE L = b, r = 1		Loading assumed symmetric with respect to α = 0

TABLE 2 EXPRESSIONS FOR THE LOAD FORCE VECTOR FOR NODE i FOR THE ℓTH HARMONIC TERM

9. THE REDUCED INTEGRATION FAMILY OF MINDLIN STRIP ELEMENTS FOR FOLDED PLATE AND AXISYMMETRIC SHELL ANALYSIS

The behaviour of Mindlin strip elements for folded plate or axisymmetric shell analysis is analogous to that explained in Section 3.4 for the plate case, i.e. the optimum performance of all strip elements occurs when selective or reduced integration is used. The Gaussian quadratures for the different integration rules for the linear, quadratic and cubic elements for the folded plate formulation presented in the last section are identical to those presented in Table 1. Moreover, it has been shown by Suarez[25] and Oñate and Suarez[17,18] that the linear strip with reduced integration (one single Gaussian point for all integrals) has an excellent performance in practical folded plate and axisymmetric shell problems. Indeed, useful 'explicit' expressions for all element matrices can be obtained by simply evaluating all integrals at the strip mid-point[17]. The accuracy of the reduced integration linear strip will be shown in the examples which are now presented.

10. EXAMPLES

10.1 Example 1: Curved simply supported plate

In this example a curved thin plate simply supported at the two ends is subjected to a point load. Both experimental and numerical results are available for the plate.

The geometry of the plate, material properties, loading position and finite strip mesh used in the analysis are given in Figure 15. Results for the mid-span deflections obtained with the reduced integration linear element using six non-zero harmonic terms can be seen in the same figure where experimental and theoretical results obtained by Coull and Das[26], finite strip solutions based on Kirchhoff's theory obtained by Thorpe[27] and Cheung[12], and finite element solutions reported by Sawko and Meriman[28] and Fam and Turkstra[29] for the same problem are also shown for comparison. The solution obtained using the linear strip element is accurate.

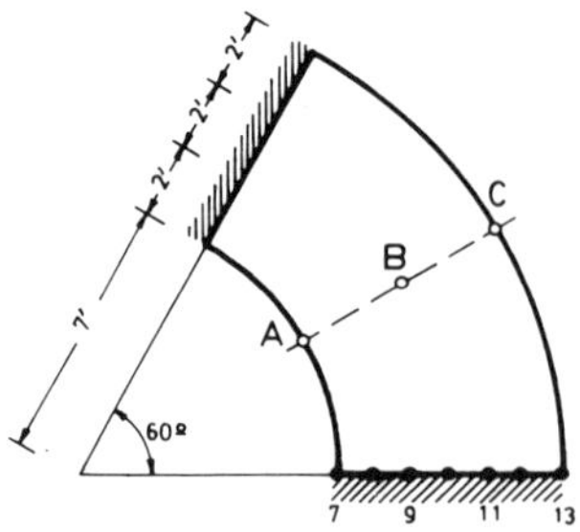

12 Elements

6 Non zero harmonics

$E = 42.460$ lb/inch^2

$\nu = 0.3$

$t = 0.172'$

Loading : Point load of 1 lb acting in A, B y C

		VERTICAL DEFLECTION (IN.)							
Loading Position	Rad. inc.	COULL & DAS		FINITE STRIP			FINITE ELEMENT		
		Experi.	Theor.	Thorpe	Cheung	*	Sawko	Fam & Turkstra	
A	13	0.876	0.752	0.882	0.995	0.874	0.851	0.880	0.881
	11	0.578	0.500	0.582	0.624	0.581	0.559	0.577	0.578
	9	0.353	0.300	0.356	0.388	0.357	0.344	0.353	0.354
	7	0.194	0.180	0.195	0.206	0.194	0.192	0.193	0.194
B	13	0.457	0.470	0.459	0.441	0.460	0.445	0.446	0.457
	11	0.342	0.370	0.343	0.329	0.348	0.333	0.342	0.343
	9	0.241	0.250	0.242	0.222	0.247	0.236	0.241	0.241
	7	0.155	0.170	0.156	0.147	0.158	0.154	0.154	0.158
C	13	0.195	0.150	0.195	0.180	0.195	0.192	0.193	0.194
	11	0.163	0.135	0.165	0.152	0.167	0.162	0.163	0.164
	9	0.157	0.125	0.153	0.149	0.155	0.151	0.151	0.151
	7	0.169	0.145	0.170	0.173	0.170	0.170	0.169	0.169
Reference		26		27	12		28	29	

* LINEAR MINDLIN STRIP ELEMENT   REDUCED INTEGRATION

Fig 15 Slab model of Coull and Das : Results for the deflection along ABC obtained by several authors.

10.2 Example 2: Right bridge - Simple supported concrete slab/beam bridge over the motorway - Nueve de Julio (Buenos Aires)

This example shows the adequacy of the linear strip element for practical bridge deck analysis. The example chosen is one of the bridges of the urban motorway, Nueve de Julio actually under construction in the city of Buenos Aires (Argentina). The geometry of the structure, loading position, material properties and finite strip discretisation used in the analysis can be seen in Figure 16. Numerical results for the vertical deflection, transverse bending moment and longitudinal resultant stress in the slab at the mid section obtained with 15 non-zero harmonic terms are plotted in Figure 17. The corresponding diagrams shown in the same figure are extrapolated from the finite strip results which are marked with a circle. To assess the validity of the numerical solution, an equilibrium check was performed comparing the total longitudinal bending moment in the mid-span section with the value obtained using simple beam theory. The percentage of error obtained is less than 5%, which can be considered as good for practical design purposes.

10.3 Example 3: Simply supported curved box girder bridge

The geometry of the structure, material properties and finite strip mesh of 18 linear strip elements (with reduced integration) used in the analysis can be seen in Figure 18. This problem has also been analysed by Cheung[16] using a Kirchhoff strip formulation. Results for the horizontal and vertical displacements of the mid-span section for three different positions of the point load are shown in Figure 19. In Figure 20 the axial circumferential stress resultant and the radial and circumferential bending moments are plotted together with some of Cheung's results which are shown for comparison. A total of 15 non-zero harmonic terms were used in the analysis.

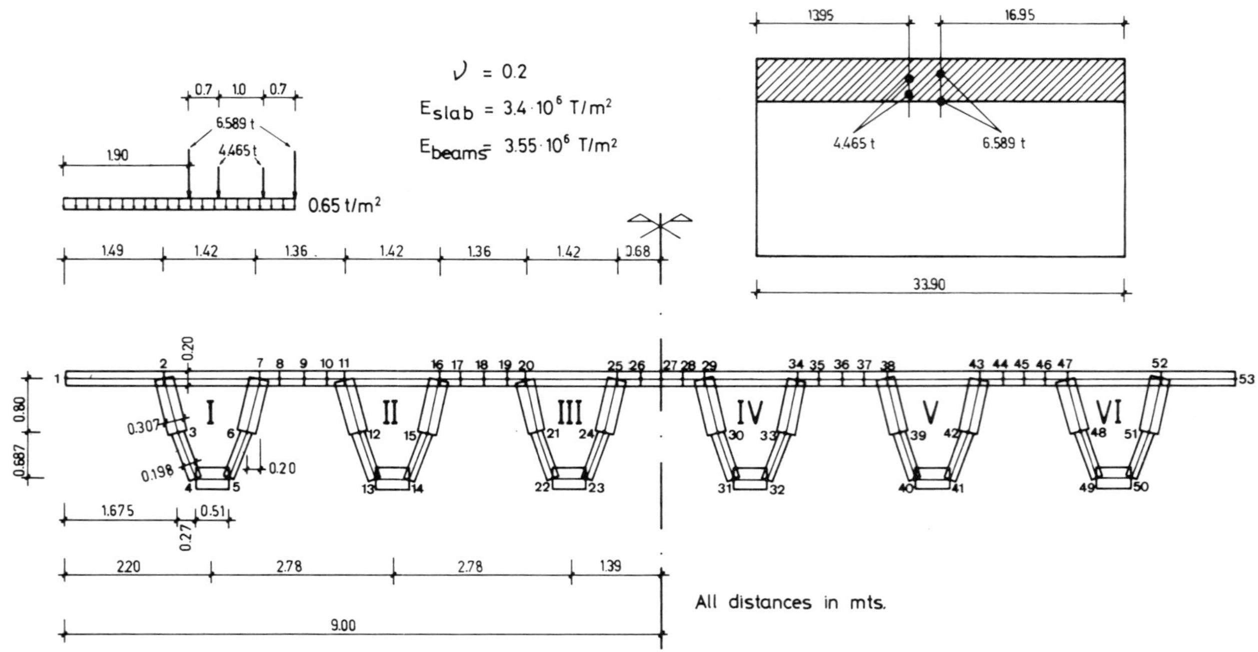

FIG. 16: Simple supported concrete bridge: Geometry of structure, loading position and idealisation into two noded finite strips.

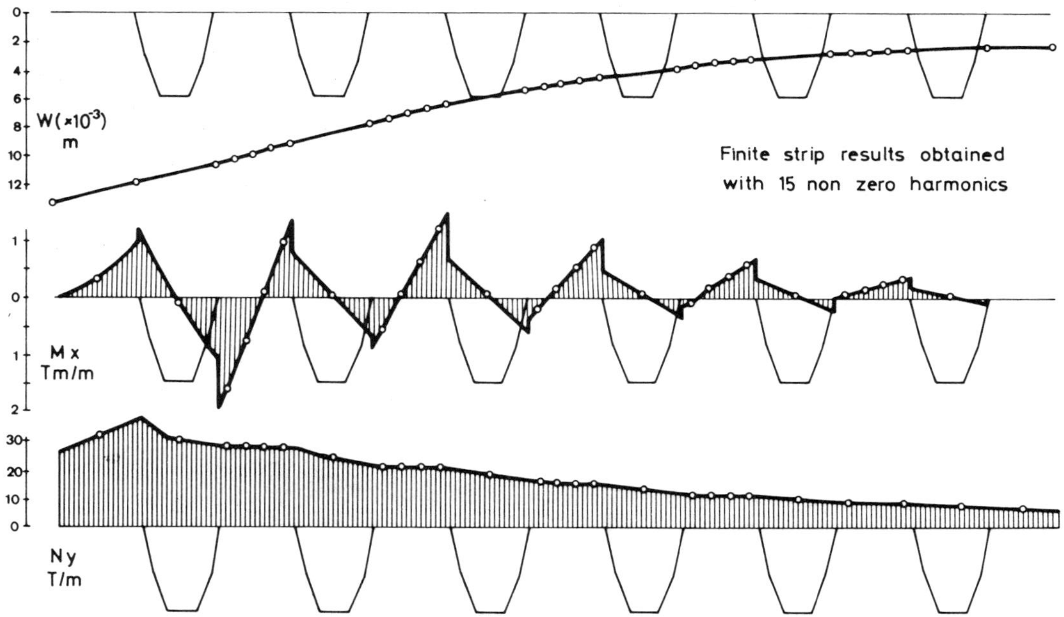

Fig. 17 Curves for the deflection, W, transverse bending moment, M_x, and longitudinal resultant stress, N_y, in the slab at the bridge mid section.

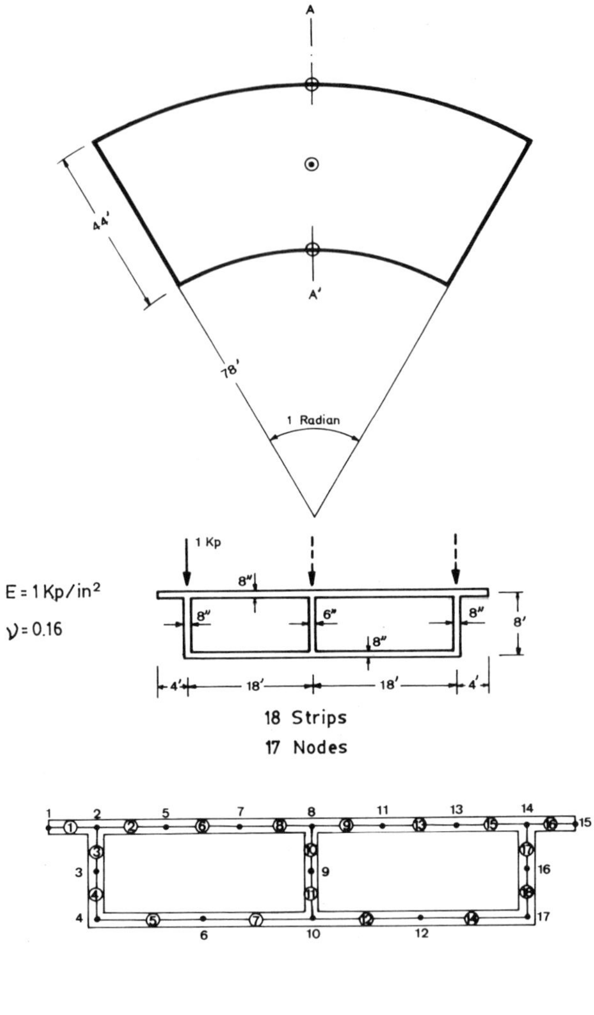

Fig 18 Curved box girder bridge. Geometry of the structure and finite strip idealisation into 18 strips.

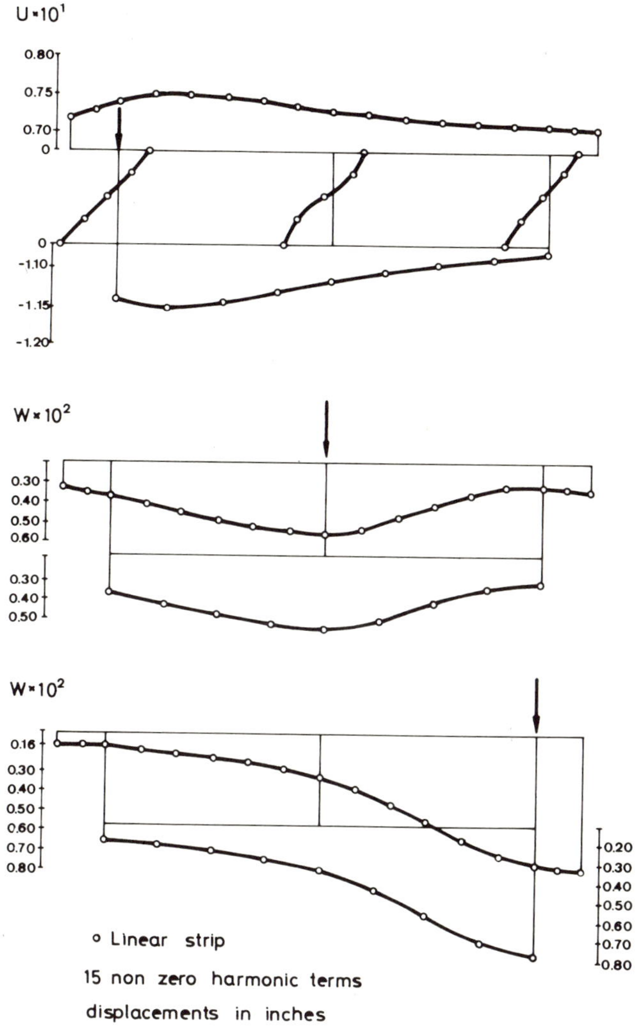

Fig. 19 Curved box girder bridge : Displacements at the mid section for several loading positions.

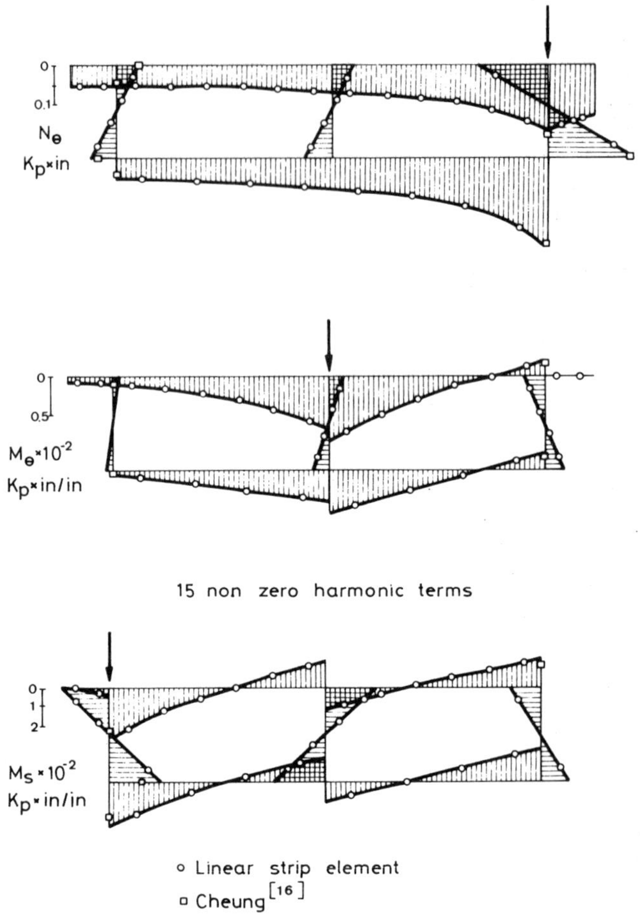

Fig. 20 Curved box girder bridge : Circunferencial resultant stress, N_θ, and bending moments M_s and M_θ at the central section.

10.4 Example 4: Circular plate under an eccentric point load

The fourth example is a thin circular plate subjected to a point load acting at a certain distance from the centre of the plate. The example is fully described in Figure 21 where results for the deflection and radial bending moment along several sections obtained with the linear axisymmetric element and with a mesh of 8-noded isoparametric reduced integrated Mindlin plate finite elements are shown for comparison. The Timoshenko[21] "exact" thin plate solution for the deflection under the load is also plotted. The accuracy of the linear element is again good.

10.5 Example 5: Pinched cylindrical shell

The last example is the classical thin cylindrical shell under two diametrically opposed point loads. The cylinder has rigid diaphragms at the two end sections (see Figure 22). This example, well-known in the shell literature, has been solved by several authors. Amongst others, there is an "exact" analytical double series solution due to Flugge[32]. Finite element solutions have been reported by Oñate et. al.[31], Lindberg et. al.[32], Ahmad et. al.[33] and many others. The solution presented here is probably the simplest one using only 20 linear axisymmetric shell elements. Nevertheless, it is highly accurate as demonstrated in Figure 22 where results obtained with the linear element for the displacements and axial forces along several sections using 15 non-zero harmonic terms compare well with the more sophisticated analytical and finite element solutions.

11. COMPUTER IMPLEMENTATION OF THE FINITE STRIP METHOD

In this section the general lay-out of a computer program for analysis of prismatic structures by the finite strip method will be presented. Also, a detailed computer listing of a finite strip program for the analysis of right or curved Mindlin

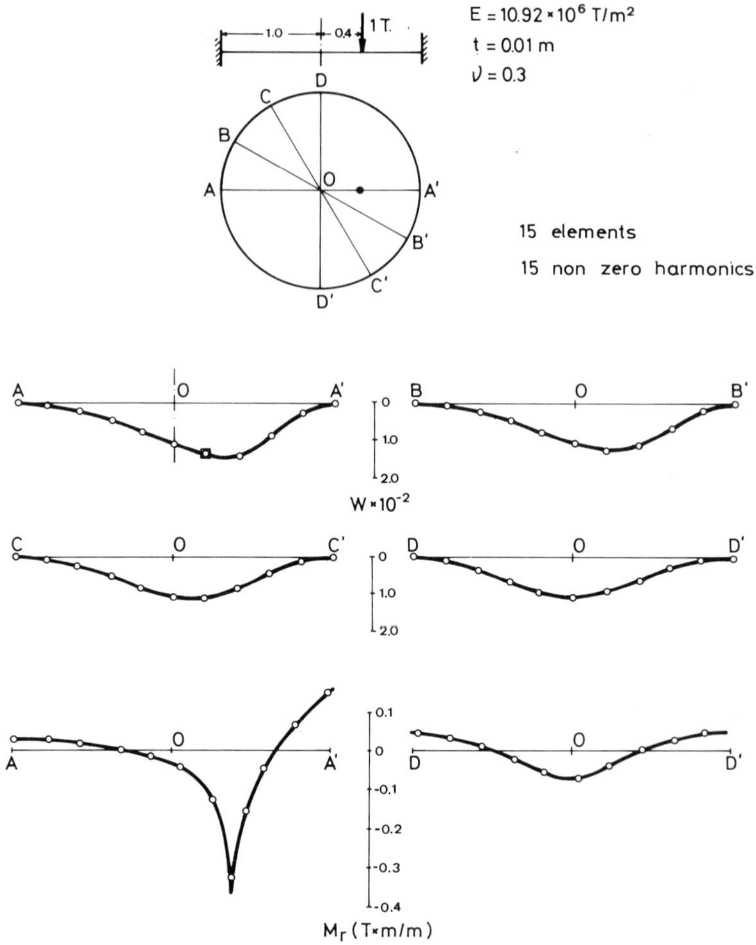

□ $w_{0.4} = 14 \times 10^{-2}$ m. (Timoshenko[21])

○ Linear strip element

— Finite element solution (8 noded Midlin plate element)[31]

Fig. 21 Circular thin plate under point load. Vertical deflections w, and radial bending moment, M_r, distributions along several sections.

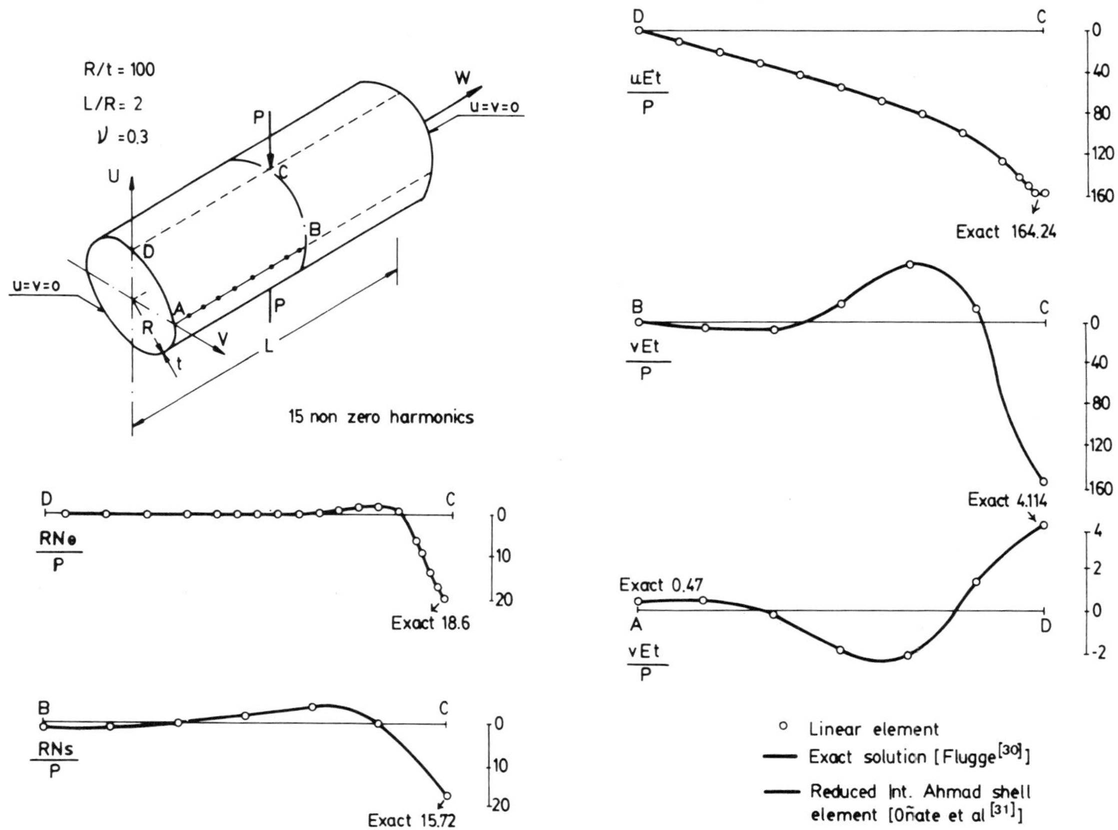

Fig. 22 **Pinched** cylindrical shell Displacements and axial forces along several sections.

plates will be provided together with some examples and user's instructions.

We will adopt here the notation used by Hinton and Owen [34] for the definition of the program variables and subroutines. Details of such a notation can be found in reference [34] and will not be given here. Also, details of standard subroutines, like that of solution of the system of linear equations by the frontal method which can also be found in references [34] and [35], will be omitted.

Program lay-out

Figure 23 shows the flow chart of a standard finite strip program. It can be seen that the construction of a finite strip program falls into four well-defined phases.

Phase 1 - Definition of data input

One of the advantages of the finite strip method is that relatively little data is needed compared with that required with the finite element method. Input data for a finite strip program is equivalent to that needed for a standard computer program for the analysis of framed structures. The subroutine controlling the input data is named INPUT.

Phase 2 - Stiffness and stress matrices and applied load vector generation for each harmonic term

The strip stiffness and stress matrices are calculated in subroutine STIFFS. The load vector is evaluated in subroutine LOADFS. Both subroutines are within a loop which implies that the calculations are performed for each harmonic term of the series used in the definition for the displacement field. [Note that after subsequent solution for the nodal displacement amplitudes for each harmonic term, the strip stress matrices are employed in the evaluation of the stress resultants for that harmonic. This task is performed in subroutine STREFS.]

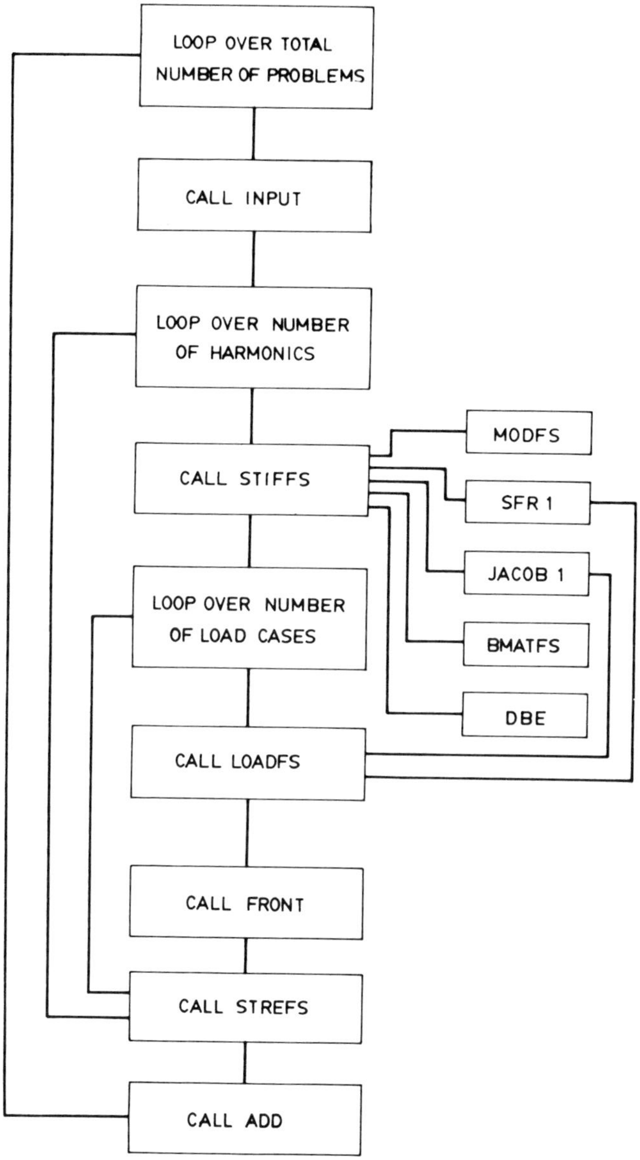

FIG. 23 FLOW CHART OF PROGRAM PBSTRIP

Phase 3 - Solution of the stiffness equations for each harmonic term

Subroutine FRONT is the equation solution subroutine which assembles the strip stiffness equations for each harmonic term and solves the unknown displacement amplitudes for the current harmonic using the frontal elimination technique [34,35].

Phase 4 - Summation of displacement and stresses for all harmonics

Subroutine ADD sums the contribution of all harmonic terms to evaluate the nodal displacements and stress resultants at any point in the structure.

The different phases of the program will be detailed in the following sections. As an example, and for the sake of clarity, all phases will be described for a plate bending finite strip program. The FORTRAN computer listing is also given. However, most steps in such a program are completely general and applicable to the different structures studied in this chapter.

12. PROGRAM PBSTRIP FOR THE ANALYSIS OF RIGHT OR CURVED PLATES BY THE FINITE STRIP METHOD

In this section the different phases of the finite strip program introduced earlier will be described for the particular case of a program for analysis of right and curved plates using any of the reduced integration family of Mindlin plate strip elements studied in Section 3. Most subroutines of the program are equally valid for direct use in more general finite strip programs for analysis of folded plates or axisymmetric shell structures. This point will be stressed throughout the following sections.

12.1 Main segment

A listing of the master segment of the program is now given. This is followed in Table 2 with a Nassi-Schneiderman (NS)-chart which describes the main segment.

```
C     MASTER PBSTRIP
      DIMENSION ADISP(3,40),AFORC(5,4,20),ASDIS(120),COORD(40,1),
     .DISPL(120),DIZPL(120),EFORC(400),ELOAD(20,12),FORCE(400),
     .IFPRE(2,3),LNODS(20,4),MATNO(20),NOFIX(2),POSGP(4),PRESC(2,3),
     .PROPS(20,5),TITLE(18),WEIGP(4)
C
C     DIMENSION STATEMENTS ASSOCIATED WITH FRONTAL SOLUTION
C
      DIMENSION EQRHS(1),EQUAT(12,1),FIXED(120),GLOAD(12),
     .GSTIF(72),IFFIX(120),NACVA(12),NAMEV(1),NPIVO(1),VECRV(12)
      DATA MELEM/20/,MFACT/400/,MMATS/20/,MPOIN/40/,MTOTV/120/
C
C     DATA STATEMENT ASSOCIATED WITH FRONTAL SOLUTION ARRAYS
C
      DATA MBUFA/1/,MFRON/12/,MSTIF/72/,MVFIX/2/
C  ***********************************************************
C                                                            *
C   PROGRAM FOR THE ANALYSIS OF RIGHT AND CURVED PLATES
C   BY THE MINDLIN FINITE STRIP METHOD USING LINEAR,         *
C   QUADRATIC OR CUBIC STRIP ELEMENTS WITH FULL, REDUCED OR  *
C   SELECTIVE INTEGRATION                                    *
C                                                            *
C  ***********************************************************
      READ(5,900) NPROB
  900 FORMAT(I5)
      WRITE(6,905) NPROB
  905 FORMAT(// 5X,'NUMBER OF PROBLEMS TO BE SOLVED=',I3)
      DO 40 IPROB=1,NPROB
      REWIND 7
      REWIND 8
      READ(5,910) TITLE
  910 FORMAT(18A4)
      WRITE(6,915) IPROB,TITLE
  915 FORMAT(/////,6X,12HPROBLEM NO. ,I3,10X,18A4)
      READ(5,920) NHARM,NSYME
  920 FORMAT(2I5)
      WRITE(6,925) NHARM,NSYME
  925 FORMAT(//, ' NUMBER OF HARMONIC TERMS TO BE USED =',I5,
     ./,' INDICATOR FOR SYMMETRY OF LOADING ( 1 FOR NON SYMMETRY,
     . 2 FOR SYMMETRY) =',I5)
C
C*** CALL THE SUBROUTINE WHICH READS MOST OF THE PROBLEM DATA.
C
      CALL INPUT(COORD,IFPRE,LNODS,MATNO,MELEM,MMATS,MPOIN,MVFIX,
     .NCASE,NDIME,NDOFN,NELEM,NEVAB,NGAUB,NGAUS,NMATS,NNODE,NOFIX,
     .NPOIN,NPROP,NSTRE,NTYPE,NVFIX,PRESC,PROPS,TLENG)
C
C*** LOOP OVER ALL THE HARMONICS
C
      DO 30 IHARM=1,NHARM,NSYME
      REWIND 1
      REWIND 2
      REWIND 3
      REWIND 4
      REWIND 9
C
C*** NEXT CREATE THE STRIP ELEMENT STIFFNESS FILE FOR EACH
C*** HARMONIC
C
      CALL STIFFS(COORD,IHARM,LNODS,MATNO,MELEM,MMATS,MPOIN,NELEM,
     .NEVAB,NGAUB,NGAUS,NNODE,NSTRE,NTYPE,POSGP,PROPS,TLENG,WEIGP)
      DO 20 ICASE=1,NCASE
C
C*** COMPUTE LOADS,FOR EACH HARMONIC,AFTER READING THE RELEVANT
C*** EXTRA DATA
C
      CALL LOADFS(COORD,ELOAD,ICASE,IHARM,LNODS,MATNO,MELEM,MMATS,
     .MPOIN,NDOFN,NELEM,NEVAB,NGAUS,NNODE,NPOIN,NTYPE,POSGP,PROPS,
     .TLENG,WEIGP)
C
C*** MERGE AND SOLVE THE RESULTING EQUATIONS BY THE FRONTAL SOLVER
C*** FOR EACH HARMONIC
C
      CALL FRONT(ASDIS,ELOAD,EQRHS,EQUAT,FIXED,GLOAD,GSTIF,ICASE,
     .IFFIX,IFPRE,LNODS,MBUFA,MELEM,MFRON,MSTIF,MTOTV,MVFIX,NACVA,
     .NAMEV,NDOFN,NELEM,NEVAB,NNODE,NOFIX,NPIVO,NPOIN,NVFIX,PRESC,
     .VECRV)
      DO 10 IPOIN=1,NPOIN
      NGASH=IPOIN*NDOFN
      NGISH=NGASH-NDOFN+1
```

```
C                                                                    MAIN  81
C*** WRITE IN DISC DISPLACEMENTS FOR EACH HARMONIC                   MAIN  82
C                                                                    MAIN  83
      WRITE(7) (ASDIS(IGASH),IGASH=NGISH,NGASH)                      MAIN  84
   10 CONTINUE                                                       MAIN  85
C                                                                    MAIN  86
C***  COMPUTE THE STRESSES IN ALL THE STRIPS FOR EACH HARMONIC        MAIN  87
C                                                                    MAIN  88
      CALL STREFS(ASDIS,LNODS,MELEM,MTOTV,NDOFN,NELEM,NGAUS,NNODE,   MAIN  89
     .NSTRE)                                                         MAIN  90
   20 CONTINUE                                                       MAIN  91
   30 CONTINUE                                                       MAIN  92
C                                                                    MAIN  93
C*** SUM DISPLACEMENTS AND STRESSES FOR ALL HARMONICS                MAIN  94
C                                                                    MAIN  95
      CALL ADD(ADISP,AFORC,DISPL,DIZPL,EFORC,FORCE,MELEM,MFACT,MPOIN, MAIN  96
     .MTOTV,NCASE,NDOFN,NELEM,NGAUS,NHARM,NPOIN,NSTRE,NSYME,TLENG)   MAIN  97
   40 CONTINUE                                                       MAIN  98
      STOP                                                           MAIN  99
      END                                                            MAIN 100
```

Set of maximum dimensions and read and write number of problems. MAIN 11-27 — 1

Loop over number of problems. MAIN 28 — 2

Rewind tapes for displacements and stresses. MAIN 29-30 — 3

Read and write problem title and data. MAIN 31-40 — 4

Call INPUT to read input data. MAIN 44-46 — 5

Loop over number of harmonics. MAIN 50 — 6

Rewind tapes 1-4,9. MAIN 51-55 — 7

Call STIFFS to evaluate element stiffness. MAIN 60-61 — 8

A

Table 2 Nassi-Schneiderman (NS) chart for main segment of PBSTRIP

A

Loop over each load case. MAIN 62	9
Evaluate load vector by calling LOADFS. MAIN 67-69	10
Call FRONT to solve for unknowns. MAIN 74-77	11
Write displacements to peripheral storage device 7. MAIN 78-85	12
Call STREFS to evaluate stress resultants for current harmonic. MAIN 89-90	13
Call ADD to sum displacements and stress resultants for all harmonics and then write our results. MAIN 96-97	14

Table 2 Nassi-Schneiderman (NS) chart for main segment of PBSTRIP

12.2 Input data subroutine INPUT

The input data for a finite strip program can be subdivided into three main classifications. Firstly, the data required to define the geometry of the structure and the support conditions must be supplied. Secondly, the properties of the constituent materials must be specified and finally, the applied loading must be defined.

The input data subroutine presented in this section is concerned with the geometrical and consitutive properties only. All the loading data is supplied in subroutine LOADFS. Subroutine INPUT, as presented here, can be used for any of the structual applications of the finite strip method studied in this chapter. As already mentioned, all variables have essentially the same meaning as in reference [35] and full details will not be given here. However, for clarity, a definition of the control parameters and main variables is presented at the end of the chapter.

The geometry of the structure needed for the analysis is completely defined by the prescription of the nodal coordinates of the strips which discretise the transverse cross-section of the structure. The coordinate of each node must be defined with reference to an xz global coordinate system which may have an arbitrarily located origin.

The prescribed degrees of freedom are constrained along the entire length of the structure. Simply supported end boundary conditions at $y = 0$ and $y = b$ require no definition of prescribed nodes since such a condition is automatically satisfied by the Fourier expansions adopted.

Subroutine INPUT is now presented. Descriptive comments in the form of a NS chart in Table 3 are provided after the FORTRAN listing of the subroutine.

```
      SUBROUTINE INPUT(COORD,IFPRE,LNODS,MATNO,MELEM,MMATS,MPOIN,
     .MVFIX,NCASE,NDIME,NDOFN,NELEM,NEVAB,NGAUB,NGAUS,NMATS,NNODE,
     .NOFIX,NPOIN,NPROP,NSTRE,NTYPE,NVFIX,PRESC,PROPS,TLENG)
      DIMENSION COORD(MPOIN,1),IFPRE(MVFIX,3),LNODS(MELEM,4),
     .MATNO(MELEM),NOFIX(MVFIX),PRESC(MVFIX,3),PROPS(MMATS,5)
C
C*** DATA INPUT SUBROUTINE
C
C
C*** READ THE FIRST DATA CARD, AND ECHO IT IMMEDIATELY.
C
      READ(5,900) NPOIN,NELEM,NVFIX,NCASE,NTYPE,NNODE,NMATS,
     .NGAUB,NGAUS
      NDOFN=3
      NPROP=5
      NSTRE=5
      NDIME=1
  900 FORMAT(13I5)
      READ(5,905) TLENG
  905 FORMAT(F10.5)
      IF(NTYPE.EQ.1) GO TO 5
      WRITE(6,910) TLENG
  910 FORMAT(/5X,'ANGLE OF THE PLATE =',F10.5)
      GO TO 10
    5 WRITE(6,915) TLENG
  915 FORMAT(/ 5X,'LENGTH OF THE PLATE=',F10.5)
   10 CONTINUE
      NEVAB=NDOFN*NNODE
      WRITE(6,920) NPOIN,NELEM,NVFIX,NCASE,NTYPE,NNODE,NDOFN,NMATS,
     .NPROP,NGAUB,NGAUS,NDIME,NSTRE,NEVAB
  920 FORMAT(//8H NPOIN =,I4,4X,8H NELEM =,I4,4X,8H NVFIX =,I4,4X,
     .8H NCASE =,I4,4X,8H NTYPE =,I4,4X,8H NNODE =,I4,4X,8H NDOFN =,I4//
```

```
     . 8H NMATS =,I4,4X,8H NPROP =,I4,4X,8H NGAUB =,I4,4X,8H NGAUS =,I4,
     . 4X,8H NDIME =,I4,4X,8H NSTRE =,I4,4X,8H NEVAB =,I4)
C
C*** READ THE ELEMENT NODAL CONNECTIONS, AND THE PROPERTY NUMBERS.
C
      WRITE(6,925)
  925 FORMAT(//8H ELEMENT,3X,8HPROPERTY,6X,12HNODE NUMBERS)
      DO 15 IELEM=1,NELEM
      READ(5,900) NUMEL,MATNO(NUMEL),(LNODS(NUMEL,INODE),INODE=1,NNODE)
   15 WRITE(6,930) NUMEL,MATNO(NUMEL),(LNODS(NUMEL,INODE),INODE=1,
     .NNODE)
  930 FORMAT(1X,I5,I9,6X,8I5)
C
C*** ZERO ALL THE NODAL COORDINATES, PRIOR TO READING SOME OF THEM.
C
      DO 20 IPOIN=1,NPOIN
      DO 20 IDIME=1,NDIME
   20 COORD(IPOIN,IDIME)=0.0
C
C*** READ SOME NODAL COORDINATES, FINISHING WITH THE LAST NODE OF ALL.
C
      WRITE(6,935)
  935 FORMAT(//25H  NODAL POINT COORDINATES)
      WRITE(6,940)
  940 FORMAT(6H  NODE,7X,1HX,9X)
   25 READ(5,945) IPOIN,(COORD(IPOIN,IDIME),IDIME=1,NDIME)
  945 FORMAT(I5,2F10.5)
      IF(IPOIN.NE.NPOIN) GO TO 25
      DO 30 IPOIN=1,NPOIN
   30 WRITE(6,950) IPOIN,(COORD(IPOIN,IDIME),IDIME=1,NDIME)
  950 FORMAT(1X,I5,3F10.5)
      IF(NVFIX.EQ.0) GO TO 40
C
C*** READ THE FIXED VALUES.
C
      WRITE(6,955)
  955 FORMAT(// ,' RESTRAINED NODES')
      WRITE(6.960)
  960 FORMAT(2X,'NODE',1X,'CODE',6X,'FIXED VALUES')
      DO 35 IVFIX=1,NVFIX
      READ(5,965) NOFIX(IVFIX),(IFPRE(IVFIX,IDOFN),IDOFN=1,NDOFN),
     .(PRESC(IVFIX,IDOFN),IDOFN=1,NDOFN)
   35 WRITE(6,965) NOFIX(IVFIX),(IFPRE(IVFIX,IDOFN),IDOFN=1,NDOFN),
     .(PRESC(IVFIX,IDOFN),IDOFN=1,NDOFN)
  965 FORMAT(I5,2X,3I1,3F10.5)
      CONTINUE
   40 CONTINUE
C
C*** READ THE AVAILABLE SELECTION OF ELEMENT PROPERTIES.
C
      WRITE(6,970)
  970 FORMAT(//21H  MATERIAL PROPERTIES)
      WRITE(6,975)
  975 FORMAT(8H  NUMBER,25X,10HPROPERTIES)
      DO 45 IMATS=1,NMATS
      READ(5,980) NUMAT,(PROPS(NUMAT,IPROP),IPROP=1,NPROP)
  980 FORMAT(I5,5F10.5)
   45 WRITE(6,985) NUMAT,(PROPS(NUMAT,IPROP),IPROP=1,NPROP)
  985 FORMAT(1X,I5,7X,5E14.6)
      RETURN
      END
```

Read and write the control parameters and define some further parameters. INPU 12-34	1.
Read and write the strip number, NUMEL, nodal connection numbers LNODS (NUMEL,INODE), and material identification number MATNO (NUMEL) for each strip. INPU 38-44	2.
Initialise nodal coordinates array COORD and then read and write node number IPOIN and coordinates COORD (IPOIN,1) for each node. INPU 48-63	3.
For prescribed nodes, read and write the prescribed node number, NVFIX (IVFIX), the prescribed degree of freedom indicator IFPRE (IVFIX, IDOFN) and the prescribed values PRESC (IVFIX, IDOFN) for each degree of freedom IDOFN. INPU 64-79	4.
Read and write the material properties PROPS (NUMAT, IPROP) for each individual material NUMAT. INPU 83-91	5.

Table 3 NS chart for subroutine INPUT

12.3 Stiffness matrix subroutine STIFFS

The purpose of this subroutine is to evaluate the stiffness and stress matrices for each strip and for each harmonic terms of the series. Both matrices are stored on disc file for subsequent use in the equation solution subroutine FRONT, and in the subroutine STREFS to compure the stress resultants resp. A NS chart describing STIFFS is provided in Table 4.

```
      SUBROUTINE STIFFS(COORD,IHARM,LNODS,MATNO,MELEM,MMATS,MPOIN,
     .NELEM,NEVAB,NGAUB,NGAUS,NNODE,NSTRE,NTYPE,POSGP,PROPS,TLENG,
     .WEIGP)
      DIMENSION BMATX(5,12),CARTD(1,4),COORD(MPOIN,1),DBMAT(5,12),
     .DERIV(1,4),DMATX(5,5),ELCOD(1,4),ESTIF(12,12),GPCOD(1,4),
     .LNODS(MELEM,4),MATNO(MELEM),POSGP(4),PROPS(MMATS,5),SHAPE(4),
     .SMATX(5,12,4),WEIGP(4)
C
C*** EVALUATION OF THE STIFFNESS MATRIX FOR THE
C*** LINEAR,QUADRATIC OR CUBIC STRIP ELEMENT
C
C
C*** LOOP OVER EACH ELEMENT
C
      DO 55 IELEM=1,NELEM
      LPROP=MATNO(IELEM)
      DO 5 INODE=1,NNODE
      LNODE=LNODS(IELEM,INODE)
      ELCOD(1,INODE)=COORD(LNODE,1)
    5 CONTINUE
C
C*** EVALUATE THE D-MATRIX
C
      CALL MODPB(DMATX,LPROP,MMATS,NSTRE,PROPS)
C
C*** INITIALIZE THE ELEMENT STIFFNESS MATRIX
C
      DO 10 IEVAB=1,NEVAB
      DO 10 JEVAB=1,NEVAB
   10 ESTIF(IEVAB,JEVAB)=0.0
      KGASP=0
      CALL GAUSSQ(NGAUB,POSGP,WEIGP)
C
C*** FULL OR SELECTIVE INTEGRATION FOR BENDING (NO. GAUSS POINTS=NGAUB)
C
      DO 25 IGAUS=1,NGAUB
      KGASP=KGASP+1
      EXISP=POSGP(IGAUS)
      CALL SFR1(DERIV,EXISP,NNODE,SHAPE)
      CALL JACOB1(CARTD,DERIV,DJACB,ELCOD,GPCOD,IELEM,KGASP,
     .NNODE,SHAPE)
      DLENG=DJACB*WEIGP(IGAUS)
      CALL BMATFS(BMATX,CARTD,GPCOD,IGAUS,IHARM,NEVAB,NNODE,NSTRE,
     .NTYPE,SHAPE,TLENG)
      CALL DBE(BMATX,DBMAT,DMATX,NEVAB,NSTRE)
      KSTRE=NSTRE-2
      DO 20 IEVAB=1,NEVAB
      DO 20 JEVAB=IEVAB,NEVAB
      DO 20 ISTRE=1,KSTRE
      IF(NTYPE.EQ.2) GO TO 15
      ESTIF(IEVAB,JEVAB)=ESTIF(IEVAB,JEVAB)+BMATX(ISTRE,IEVAB)*
     .DBMAT(ISTRE,JEVAB)*DLENG*TLENG/2.0
      GO TO 20
   15 ESTIF(IEVAB,JEVAB)=ESTIF(IEVAB,JEVAB)+GPCOD(1,IGAUS)
     .*BMATX(ISTRE,IEVAB)*DBMAT(ISTRE,JEVAB)*DLENG*TLENG/2.0
   20 CONTINUE
   25 CONTINUE
C
C*** REDUCED INTEGRATION FOR SHEAR TERMS (NO. GAUSS POINTS=NGAUS)
C
      CALL GAUSSQ(NGAUS,POSGP,WEIGP)
C
C*** ENTER LOOPS FOR AREA NUMERICAL INTEGRATION
C
      KGASP=0
      DO 45 IGAUS=1,NGAUS
      KGASP=KGASP+1
      EXISP=POSGP(IGAUS)
      CALL SFR1(DERIV,EXISP,NNODE,SHAPE)
      CALL JACOB1(CARTD,DERIV,DJACB,ELCOD,GPCOD,IELEM,KGASP,
     .NNODE,SHAPE)
      DLENG=DJACB*WEIGP(IGAUS)
      CALL BMATFS(BMATX,CARTD,GPCOD,IGAUS,IHARM,NEVAB,NNODE,NSTRE,
     .NTYPE,SHAPE,TLENG)
      CALL DBE(BMATX,DBMAT,DMATX,NEVAB,NSTRE)
      LSTRE=NSTRE-1
      DO 35 IEVAB=1,NEVAB
      DO 35 JEVAB=IEVAB,NEVAB
      DO 35 ISTRE=LSTRE,NSTRE
      IF(NTYPE.EQ.2) GO TO 30
```

```
C                                                                       STIF  81
C*** STRAIGHT PLATE                                                     STIF  82
C                                                                       STIF  83
      ESTIF(IEVAB,JEVAB)=ESTIF(IEVAB,JEVAB)+BMATX(ISTRE,IEVAB)*DBMAT     STIF  84
     .(ISTRE,JEVAB)*DLENG*TLENG/2.0                                     STIF  85
      GO TO 35                                                          STIF  86
C                                                                       STIF  87
C*** CURVED PLATE                                                       STIF  88
C                                                                       STIF  89
   30 ESTIF(IEVAB,JEVAB)=ESTIF(IEVAB,JEVAB)+GPCOD(1,IGAUS)*BMATX(       STIF  90
     .ISTRE,IEVAB)*DBMAT(ISTRE,JEVAB)*DLENG*TLENG/2.0                   STIF  91
   35 CONTINUE                                                          STIF  92
C                                                                       STIF  93
C*** STORE THE COMPONENTS OF THE DB MATRIX FOR THE ELEMENT              STIF  94
C                                                                       STIF  95
      DO 40 ISTRE=1,NSTRE                                               STIF  96
      DO 40 IEVAB=1,NEVAB                                               STIF  97
   40 SMATX(ISTRE,IEVAB,KGASP)=DBMAT(ISTRE,IEVAB)                       STIF  98
   45 CONTINUE                                                          STIF  99
C                                                                       STIF 100
C*** CONSTRUCT THE LOWER TRIANGLE OF THE STIFFNESS MATRIX               STIF 101
C                                                                       STIF 102
      DO 50 IEVAB=1,NEVAB                                               STIF 103
      DO 50 JEVAB=1,NEVAB                                               STIF 104
   50 ESTIF(JEVAB,IEVAB)=ESTIF(IEVAB,JEVAB)                             STIF 105
C                                                                       STIF 106
C*** STORE THE STIFFNESS MATRIX,STRESS MATRIX AND SAMPLING              STIF 107
C*** POINT COORDINATES FOR EACH ELEMENT ON DISC FILE                    STIF 108
C                                                                       STIF 109
      WRITE(1) ESTIF                                                    STIF 110
      WRITE(3) SMATX,GPCOD                                              STIF 111
   55 CONTINUE                                                          STIF 112
      RETURN                                                            STIF 113
      END                                                               STIF 114
```

Loop over each strip IELEM. STIF 15	1
Identify the material set number for current strip. STIF 16	2
Extract nodal coordinates for current strip. STIF 17-20	3
Call MODPB to evaluate matrix $\underline{D}$ for the strip. STIF 24	4
Initialise (zero) stiffness matrix array and Gauss point counter. STIF 28-31	5
Call GAUSSQ to set the Gauss-Legendre quadrature parameters for bending. STIF 32	6

A

Table 4 NS chart for subroutine STIFFS

A

Loop over bending Gauss points. STIF 36 7

Increment Gauss point counter, extract sampling position and call SFR1 to evaluate shape function N_i and local derivatives $\partial N_i/\partial\xi$. STIF 37-39 8

Call JACOB1 to evaluate $\partial N_i/\partial x$ and then dx. STIF 40-42 9

Call BMATFS to evaluate strain matrix. STIF 43-44 10

Call DBE to evaluate stress matrix. STIF 45 11

Accumulate contributions to the strip stiffness matrix for either right or curved plates. STIF 47-57 12

Call GAUSSQ to set the Gauss-Legendre quadrature parameters for shear. STIF 61 13

Zero Gauss point counter. STIF 65 14

Loop over shear Gauss points. STIF 66 15

Increment Gauss point counter, extract sampling coordinate and call SFR1 to evaluate shape function N_i and local derivatives $\partial N_i/\partial\xi$. STIF 67-69 16

B

Table 4 NS chart for subroutine STIFFS

B

Call JACOB1 to evaluate $\partial N_i/\partial x$ and then dx. STIF 70-72	17
Call BMATFS to evaluate strain matrix. STIF 73-74	18
Call DBE to evaluate stress matrix. STIF 75	19
Accumulate contributions to the strip stiffness matrix for either right or curved plates. STIF 76-92	20
Extract components of the stress matrix for later use. STIF 96-99	21
Construct lower triangular part of strip stiffness matrix. STIF 103-105	22
Store on peripheral storage devices stiffness matrix, stress matrix and sampling point coordinates for each strip. STIF 110-112	23

Table 4 NS chart for subroutine STIFFS

The computer listing of STIFFS, provided above for the plate case, can be of direct use for folded plate of axisymmetric shell programs with the only (conceptual) difference being that for the axisymmetric shell case matrix $\tilde{B}_i^{*\ell}$ of (57), relating local strains with global displacements, should be used instead of matrix $\tilde{B}_i^{\ell}$ of (26). Also, the matrix $\tilde{D}$ of (49) would substitute $\tilde{D}$ of (18), used in the plate program.

12.4 Subroutines used by STIFFS

12.4.1 Subroutine MODFS

This subroutine calculates the matrix of plate rigidities $\underline{D}$ given in (18) and (19)

```
      SUBROUTINE MODPB(DMATX,LPROP,MMATS,NSTRE,PROPS)                   MODP   1
      DIMENSION DMATX(5,5),PROPS(MMATS,5)                               MODP   2
C                                                                       MODP   3
C*** EVALUATE D MATRIX                                                  MODP   4
C                                                                       MODP   5
      DO 5 ISTRE=1,NSTRE                                                MODP   6
      DO 5 JSTRE=1,NSTRE                                                MODP   7
      DMATX(ISTRE,JSTRE)=0.0                                            MODP   8
    5 CONTINUE                                                          MODP   9
      YOUNG=PROPS(LPROP,1)                                              MODP  10
      POISS=PROPS(LPROP,2)                                              MODP  11
      THICK=PROPS(LPROP,3)                                              MODP  12
      DMATX(1,1)=YOUNG*THICK*THICK*THICK/(12.0*(1.0-POISS*POISS))       MODP  13
      DMATX(1,2)=POISS*DMATX(1,1)                                       MODP  14
      DMATX(2,2)=DMATX(1,1)                                             MODP  15
      DMATX(2,1)=DMATX(1,2)                                             MODP  16
      DMATX(3,3)=(1.0-POISS)*DMATX(1,1)/2.0                             MODP  17
      DMATX(4,4)=YOUNG*THICK/(2.4*(1.0+POISS))                          MODP  18
      DMATX(5,5)=DMATX(4,4)                                             MODP  19
      RETURN                                                            MODP  20
      END                                                               MODP  21
```

The NS chart is now given in Table 5.

Initialise (zero) $\underset{\sim}{D}$ matrix. MODP 5-8	1.
Extract Young's modulus, E, Poisson's ratio ν and strip thickness t. MODPB 9-12	2.
Compute $\underset{\sim}{D}$ matrix. MODPB 13-19	3.

Table 5 NS chart for MODFS

For folded plate and axisymmetric shell programs, matrix $\underset{\sim}{D}$ should be extended to include membrane effects according to (49).

12.4.2 Subroutine GAUSSQ

This subroutine sets up the local Gauss point coordinates with a strip and their resp. weights.

```
      SUBROUTINE GAUSSQ(NGAUT,POSGP,WEIGP)
      DIMENSION POSGP(4),WEIGP(4)
C
C*** SET UP GAUSS POINT COORDINATES AND WEIGHTS
C
      IF(NGAUT.GT.3) GO TO 30
      IF(NGAUT.GT.2) GO TO 20
      IF(NGAUT.GT.1) GO TO 10
      POSGP(1)=0.0
      WEIGP(1)=2.0
      GO TO 50
   10 POSGP(1)=-0.577350269189626
      WEIGP(1)=1.0
      GO TO 40
   20 POSGP(1)=-0.774596669241483
      POSGP(2)=0.0
      WEIGP(1)=0.555555555555556
      WEIGP(2)=0.888888888888889
      GO TO 40
   30 POSGP(1)=-0.8611363115
      POSGP(2)=-0.3399810435
      WEIGP(1)=0.3478548452
      WEIGP(2)=0.6521451548
   40 KGAUT=NGAUT/2
      DO 50 IGASH=1,KGAUT
      JGASH=NGAUT+1-IGASH
      POSGP(JGASH)=-POSGP(IGASH)
      WEIGP(JGASH)=WEIGP(IGASH)
   50 CONTINUE
      RETURN
      END
```

Note that NGAUT may take values from 1 to 4 for the resp. Gauss-Legendre rule. This routine is valid for folded plate and axisymmetric shell programs.

12.4.3 Subroutine SFR1

This subroutine calculates the shape functions SHAPE and their derivatives DERIV of a specified sampling point with coordinate S.

```
      SUBROUTINE SFR1(DERIV,S,NNODE,SHAPE)
      DIMENSION DERIV(1,4),SHAPE(4)
C
C***  CALCULATES SHAPE FUNCTIONS AND THEIR DERIVATIVES
C
C
C***     SHAPE FUNCTIONS
C
      IF(NNODE.EQ.4) GO TO 10
      IF(NNODE.EQ.3) GO TO 5
C
C*** TWO NODED STRIP
C
      SHAPE(1)=(1.0-S)/2.0
      SHAPE(2)=(1.0+S)/2.0
      DERIV(1,1)=-1.0/2.0
      DERIV(1,2)=1.0/2.0
      GO TO 15
C
C*** THREE NODED STRIP
C
```

```
      5 SHAPE(1)=(S*S-S)/2.0                                  SFR1  22
        SHAPE(2)=1.0-S*S                                      SFR1  23
        SHAPE(3)=(S+S*S)/2.0                                  SFR1  24
        DERIV(1,1)=(2.0*S-1.0)/2.0                            SFR1  25
        DERIV(1,2)=-2.0*S                                     SFR1  26
        DERIV(1,3)=(1.0+2.0*S)/2.0                            SFR1  27
        GO TO 15                                              SFR1  28
C                                                             SFR1  29
C*** FOUR NODED STRIP                                         SFR1  30
C                                                             SFR1  31
     10 SHAPE(1)=-9.0/16.0*(S*S*S-S/9.0-S*S+1.0/9.0)          SFR1  32
        SHAPE(2)=27.0/16.0*(S*S*S-S-S*S/3.0+1.0/3.0)          SFR1  33
        SHAPE(3)=-27.0/16.0*(S*S*S-S+S*S/3.0-1.0/3.0)         SFR1  34
        SHAPE(4)=9.0/16.0*(S*S*S-S/9.0+S*S-1.0/9.0)           SFR1  35
        DERIV(1,1)=-9.0/16.0*(3.0*S*S-1.0/9.0-2.0*S)          SFR1  36
        DERIV(1,2)=27.0/16.0*(3.0*S*S-1.0-2.0/3.0*S)          SFR1  37
        DERIV(1,3)=-27.0/16.0*(3.0*S*S-1.0+2.0/3.0*S)         SFR1  38
        DERIV(1,4)=9.0/16.0*(3.0*S*S-1.0/9.0+2.0*S)           SFR1  39
     15 CONTINUE                                              SFR1  40
        RETURN                                                SFR1  41
        END                                                   SFR1  42
```

The NS chart is now given in Table 6

Evaluate shape functions and their local derivatives for a linear, 2-node strip. SFR1 14-17	1.
Evaluate shape functions and their local derivatives for a quadratic, 3-node strip. SFR1 22-27	2.
Evaluate shape functions and their local derivatives for a cubic, 4-node strip. SFR1 32-39	3.

Table 6 NS chart for subroutine SFR1

12.4.4 Subroutine JACOB1

This subroutine calculates the Gauss point coordinates, the Cartesian shape function derivatives dN_i/dx and an elemental length dx.

```
      SUBROUTINE JACOB1(CARTD,DERIV,DJACB,ELCOD,GPCOD,IELEM,KGASP,       JACO   1
     .NNODE,SHAPE)                                                       JACO   2
      DIMENSION CARTD(1,4),DERIV(1,4),ELCOD(1,4),GPCOD(1,4),SHAPE(4)     JACO   3
C                                                                        JACO   4
C***  CALCULATES COORDINATES OF GAUSS POINTS AND THE JACOBIAN            JACO   5
C***  MATRIX AND ITS DETERMINANT AND THE INVERSE                         JACO   6
C                                                                        JACO   7
      DJACB=0.0                                                          JACO   8
      GPCOD(1,KGASP)=0.0                                                 JACO   9
C                                                                        JACO  10
C***     CALCULATE COORDINATES OF SAMPLING POINT                         JACO  11
C                                                                        JACO  12
      DO 5 INODE=1,NNODE                                                 JACO  13
      GPCOD(1,KGASP)=GPCOD(1,KGASP)+SHAPE(INODE)*ELCOD(1,INODE)          JACO  14
    5 CONTINUE                                                           JACO  15
C                                                                        JACO  16
C***   CALCULATE DETERMINANT OF JACOBIAN MATRIX                          JACO  17
C                                                                        JACO  18
      DO 10 INODE=1,NNODE                                                JACO  19
      DJACB=DJACB+DERIV(1,INODE)*ELCOD(1,INODE)                          JACO  20
   10 CONTINUE                                                           JACO  21
      IF(DJACB) 15,15,20                                                 JACO  22
   15 WRITE(6,900) IELEM                                                 JACO  23
      STOP                                                               JACO  24
C                                                                        JACO  25
C***    CALCULATE CARTESIAN DERIVATIVES                                  JACO  26
C                                                                        JACO  27
   20 DO 25 INODE=1,NNODE                                                JACO  28
      CARTD(1,INODE)=DERIV(1,INODE)/DJACB                                JACO  29
   25 CONTINUE                                                           JACO  30
  900 FORMAT(//,10X,36H PROGRAM HALTED IN SUBROUTINE JACOB1,/ 11X,       JACO  31
     .22H ZERO OR NEGATIVE AREA,/ 10X,16H ELEMENT NUMBER ,I5)            JACO  32
      RETURN                                                             JACO  33
      END                                                                JACO  34
```

The NS chart is now given in Table 6. Note that JACOB1 is valid for folded plate and axisymmetric shell programs.

Initialise (zero) Jacobian determinant and the Gauss point coordinate vector. JACO 8-9	1.
Calculate the coordinates of the sampling point. JACO 13-15	2.
Evaluate the Jacobian determinant and if it is zero or negative write a message and terminate run. JACO 19-24	3.
Evaluate Cartesian shape function derivatives dN_i/dx. JACO 28-30	4.

Table 6 NS chart for subroutine JACOB1

12.4.5 Subroutine BMATFS

This subroutine calculates the strain matrix given by (27) at a sampling point.

```
      SUBROUTINE BMATFS(BMATX,CARTD,GPCOD,IGAUS,IHARM,NEVAB,NNODE,
     .NSTRE,NTYPE,SHAPE,TLENG)
      DIMENSION BMATX(5,12),CARTD(1,4),GPCOD(1,4),SHAPE(4)
C
C*** EVALUATE B MATRIX
C
      DO 5 ISTRE=1,NSTRE
      DO 5 IEVAB=1,NEVAB
      BMATX(ISTRE,IEVAB)=0.0
    5 CONTINUE
      JGASH=0
      DO 15 INODE=1,NNODE
      IGASH=JGASH+1
      IF(NTYPE.EQ.2) GO TO 10
C
C***  STRAIGHT PLATE
C
      BMATX(4,IGASH)=CARTD(1,INODE)
      BMATX(5,IGASH)=SHAPE(INODE)*FLOAT(IHARM)*3.14159265/TLENG
      IGASH=IGASH+1
      JGASH=IGASH+1
      BMATX(1,IGASH)=-CARTD(1,INODE)
      BMATX(3,IGASH)=-SHAPE(INODE)*FLOAT(IHARM)*3.14159265/TLENG
      BMATX(4,IGASH)=-SHAPE(INODE)
      BMATX(2,JGASH)=SHAPE(INODE)*FLOAT(IHARM)*3.14159265/TLENG
      BMATX(3,JGASH)=-CARTD(1,INODE)
      BMATX(5,JGASH)=-SHAPE(INODE)
      GO TO 15
C
C***  CURVED PLATE
C
   10 BMATX(4,IGASH)=CARTD(1,INODE)
      BMATX(5,IGASH)=(SHAPE(INODE)/GPCOD(1,IGAUS))*FLOAT(IHARM)*
     .3.14159265/TLENG
      IGASH=IGASH+1
      JGASH=IGASH+1
      BMATX(1,IGASH)=-CARTD(1,INODE)
      BMATX(2,IGASH)=-SHAPE(INODE)/GPCOD(1,IGAUS)
      BMATX(3,IGASH)=-(SHAPE(INODE)/GPCOD(1,IGAUS))*FLOAT(IHARM)*
     .3.14159265/TLENG
      BMATX(4,IGASH)=-SHAPE(INODE)
      BMATX(2,JGASH)=(SHAPE(INODE)/GPCOD(1,IGAUS))*FLOAT(IHARM)*
     .3.14159265/TLENG
      BMATX(3,JGASH)=-CARTD(1,INODE)+SHAPE(INODE)/GPCOD(1,IGAUS)
      BMATX(5,JGASH)=-SHAPE(INODE)
   15 CONTINUE
      RETURN
      END
```

For folded plate and axisymmetric shell programs, subroutine BMATFS should evaluate the strain matrix according to (56) - (57). The NS chart for NMATFS is now given in Table 8.

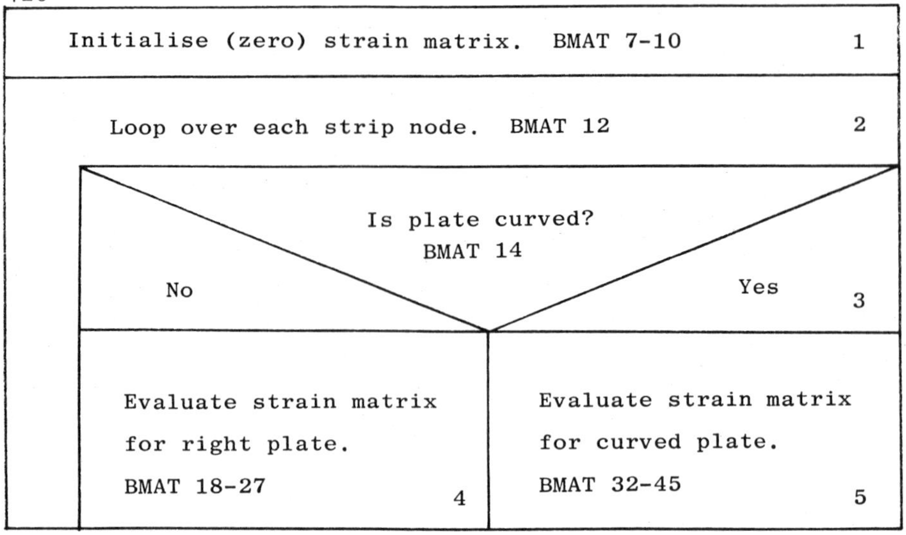

Table 8 NS chart for subroutine BMATFS

12.4.6 Subroutine DBE

This subroutine simply calculates the stress matrix $\underline{\underline{DB}}^{\ell}$ at any sampling point.

```
      SUBROUTINE DBE(BMATX,DBMAT,DMATX,NEVAB,NSTRE)
      DIMENSION BMATX(5,12),DBMAT(5,12),DMATX(5,5)
C
C***  CALCULATE D X B
C
      DO 5 ISTRE=1,NSTRE
      DO 5 IEVAB=1,NEVAB
      DBMAT(ISTRE,IEVAB)=0.0
      DO 5 JSTRE=1,NSTRE
      DBMAT(ISTRE,IEVAB)=DBMAT(ISTRE,IEVAB)+
     .DMATX(ISTRE,JSTRE)*BMATX(JSTRE,IEVAB)
    5 CONTINUE
      RETURN
      END
```

12.5 Evaluation of equivalent nodal load vector for each harmonic term: Subroutine LOADFS

Only two loading cases will be considered in the plate program presented here: a) a vertical point load acting at a node at a certain distance from the simply supported end and b) a uniformly distributed load acting over a certain number of strips. Extensions of the program for the consideration of other loading situations is easy once the basic steps for the two cases presented here are fully understood, and can be attempted by the reader.

A FORTRAN listing of the subroutine is now presented.

```
      SUBROUTINE LOADFS(COORD,ELOAD,ICASE,IHARM,LNODS,MATNO,MELEM,
     .MMATS,MPOIN,NDOFN,NELEM,NEVAB,NGAUS,NNODE,NPOIN,NTYPE,POSGP,
     .PROPS,TLENG,WEIGP)
      DIMENSION CARTD(1,4),COORD(MPOIN,1),DERIV(1,4),ELCOD(1,4),
     .ELOAD(MELEM,12),GPCOD(1,4),LNODS(MELEM,4),MATNO(MELEM),POINT(3),
     .POSGP(4),PROPS(MMATS,5),SHAPE(4),TITLE(18),WEIGP(4)
C
C*** EVALUATE NODAL FORCE VECTOR
C
      IF(IHARM.GT.1) GO TO 5
C
C*** READ AND WRITE TITLE OF LOAD CASE
C
      READ(5,900) TITLE
  900 FORMAT(18A4)
C
C*** READ AND WRITE INDICATORS FOR TYPE OF LOAD
C
      READ(5,905) IPLOD,IUNIF
  905 FORMAT(2I5)
      WRITE(9) IPLOD,IUNIF
      WRITE(6,910) ICASE
  910 FORMAT(/ 10X,' LOAD CASE NUMBER ',1X,I2)
      WRITE(6,900) TITLE
      WRITE(6,905) IPLOD,IUNIF
      GO TO 10
    5 READ(9) IPLOD,IUNIF
   10 CONTINUE
      DO 15 IELEM=1,NELEM
      DO 15 IEVAB=1,NEVAB
   15 ELOAD(IELEM,IEVAB)=0.0
C
C***  POINT LOAD
C
      IF(IPLOD.EQ.0) GO TO 50
      IF(IHARM.EQ.1) WRITE(6,915)
   20 CONTINUE
      IF(IHARM.GT.1) GO TO 25
      READ(5,920)  LODPT,POINT(1),YLOAD
      WRITE(9) LODPT,(POINT(IDOFN),IDOFN=1,NDOFN),YLOAD
  915 FORMAT(/10X,'POINT LOAD')
      WRITE(6,920) LODPT,(POINT(IDOFN),IDOFN=1,NDOFN),YLOAD
  920 FORMAT(I5,4F10.5)
      GO TO 30
   25 READ(9) LODPT,(POINT(IDOFN),IDOFN=1,NDOFN),YLOAD
   30 CONTINUE
C
C***  CALCULATE LOADS AND ASSOCIATE WITH NODAL POINTS
C
      FPLOA=FLOAT(IHARM)*3.14159265*YLOAD/TLENG
      DO 35 IELEM=1,NELEM
      DO 35 INODE=1,NNODE
      NLOCA=LNODS(IELEM,INODE)
      IF(LODPT.EQ.NLOCA) GO TO 40
   35 CONTINUE
   40 DO 45 IDOFN=1,NDOFN
      IF(IDOFN.EQ.1.OR.IDOFN.EQ.2) SINCO=SIN(FPLOA)
      IF(IDOFN.EQ.3) SINCO=COS(FPLOA)
      NGASH=(INODE-1)*NDOFN+IDOFN
      ELOAD(IELEM,NGASH)=ELOAD(IELEM,NGASH)+POINT(IDOFN)*SINCO
   45 CONTINUE
```

```
      IF(LODPT.LT.NPOIN) GO TO 20
   50 CONTINUE
C
C***  UNIFORM DISTRIBUTED LOAD
C
      VLEPI=3.14159265*FLOAT(IHARM)
      COEFL=2.0*TLENG/VLEPI
C
C***       LOOP OVER EACH STRIP
C
      CALL GAUSSQ(NGAUS,POSGP,WEIGP)
      DO 75 IELEM=1,NELEM
      LPROP=MATNO(IELEM)
      VDLOD=COEFL*PROPS(LPROP,4)
      IF(VDLOD.EQ.0.0) GO TO 75
      KGASP=0
C
C***    EXTRACT ELEMENT NODAL COORDINATES
C
      DO 55 INODE=1,NNODE
      LNODE=LNODS(IELEM,INODE)
      ELCOD(1,INODE)=COORD(LNODE,1)
   55 CONTINUE
C
C***ENTER LOOPS FOR NUMERICAL INTEGRATION
C
      DO 70 IGAUS=1,NGAUS
      KGASP=KGASP+1
      EXISP=POSGP(IGAUS)
C
C*** EVALUATE THE SHAPE FUNCTIONS AT THE SAMPLING POINTS AND
C*** ELEMENTAL LENGTH
C
      CALL SFR1(DERIV,EXISP,NNODE,SHAPE)
      CALL JACOB1(CARTD,DERIV,DJACB,ELCOD,GPCOD,IELEM,KGASP,NNODE,
     .SHAPE)
C
C***    CALCULATE LOADS AND ASSOCIATE WITH ELEMENT NODAL POINTS
C
      DO 65 INODE=1,NNODE
      NPOSN=(INODE-1)*NDOFN+1
      IF(NTYPE.EQ.2) GO TO 60
C
C***    STRAIGHT PLATE
C
      ELOAD(IELEM,NPOSN)=ELOAD(IELEM,NPOSN)+
     .SHAPE(INODE)*VDLOD*DJACB*WEIGP(IGAUS)
      GO TO 65
C
C***    CURVED PLATE
C
   60 ELOAD(IELEM,NPOSN)=ELOAD(IELEM,NPOSN)+GPCOD(1,IGAUS)*
     .SHAPE(INODE)*VDLOD*DJACB*WEIGP(IGAUS)
   65 CONTINUE
   70 CONTINUE
   75 CONTINUE
      RETURN
      END
```

The NS chart for LOADFS is presented in Table 9

For the first harmonic only read and write the load case title and the load type code. LOAD 10-28		1
Initialise (zero) the load vector. LOAD 29-31		2
If there are point loads continue, otherwise jump to 10. LOAD 35		3
Is this the first harmonic? LOAD 38 — Yes	No	4
Read and write nodal load data and then write on file 9. LOAD 39-43 (5)	Read load data from file 9. LOAD 45	6
Identify loaded node with a strip. LOAD 52-55		7
Compute nodal forces for either a right or curved strip. LOAD 56-61		8
If all nodes have been considered continue; otherwise jump back to 4. LOAD 62		9
Set loading parameters for right and curved strip plate cases. LOAD 67-68		10
Call GAUSSQ to obtain Gauss point weights and sampling positions. LOAD 72		11
Loop over each strip. LOAD 73		12

A

Table 9 NS chart for subroutine LOADFS

A

Identify strip material set number and extract strip load intensity parameter PROPS (LPROP,4).
LOAD 74-75 13

If load intensity is zero, jump to next strip.
LOAD 76 14

Zero Gauss point counter and extract element nodal coordinates. LOAD 77-84 15

Loop over each Gauss point. LOAD 88 16

Increment Gauss point counter and identify local coordinate of Gauss point. LOAD 89-90 17

Call SFR1 to evaluate shape functions and their local derivatives. LOAD 95 18

Call JACOB1 to evaluate elemental length dx, etc. LOAD 96-97 19

Accumulate contributions to the consistent load vector for either right or curved strips.
LOAD 101-115 20

Table 9 NS chart for subroutine LOADFS

12.6 Computation of the strip stress resultants for each harmonic term: Subroutine STREFS

This subroutine evaluates the stress resultants at the shear Gauss points for each strip and for each harmonic term accordingly to (25) making use of the stress matrix (see Section 12.3) and the strip nodal displacements computed in subroutine FRONT. The strip stress resultants for each harmonic are stored on a disc file for subsequent use in subroutine ADD

```
      SUBROUTINE STREFS(ASDIS,LNODS,MELEM,MTOTV,NDOFN,NELEM,NGAUS,      STRE   1
     .NNODE,NSTRE)                                                      STRE   2
      DIMENSION ASDIS(MTOTV),ELDIS(3,4),LNODS(MELEM,4),GPCOD(1,4),      STRE   3
     .SMATX(5,12,4),STRSG(5)                                            STRE   4
C                                                                       STRE   5
C*** EVALUATE STRESSES AT THE SHEAR GAUSS POINTS FOR                    STRE   6
C*** EACH STRIP                                                         STRE   7
C                                                                       STRE   8
C                                                                       STRE   9
C***     LOOP OVER EACH STRIP ELEMENT                                   STRE  10
C                                                                       STRE  11
      DO 20 IELEM=1,NELEM                                               STRE  12
C                                                                       STRE  13
C***   READ THE STRESS MATRIX ,SAMPLING POINT COORDINATES               STRE  14
C***   FOR THE STRIP                                                    STRE  15
C                                                                       STRE  16
      READ(3) SMATX,GPCOD                                               STRE  17
C                                                                       STRE  18
C***    IDENTIFY THE DISPLACEMENTS OF THE ELEMENT NODAL POINTS          STRE  19
C                                                                       STRE  20
      DO 5 INODE=1,NNODE                                                STRE  21
      LNODE=LNODS(IELEM,INODE)                                          STRE  22
      NPOSN=(LNODE-1)*NDOFN                                             STRE  23
      DO 5 IDOFN=1,NDOFN                                                STRE  24
      NPOSN=NPOSN+1                                                     STRE  25
      ELDIS(IDOFN,INODE)=ASDIS(NPOSN)                                   STRE  26
    5 CONTINUE                                                          STRE  27
      KGASP=0                                                           STRE  28
C                                                                       STRE  29
C***    ENTER LOOPS OVER EACH SHEAR SAMPLING POINT                      STRE  30
C                                                                       STRE  31
      DO 15 IGAUS=1,NGAUS                                               STRE  32
      KGASP=KGASP+1                                                     STRE  33
      DO 10 ISTRE=1,NSTRE                                               STRE  34
      STRSG(ISTRE)=0.0                                                  STRE  35
      KGASH=0                                                           STRE  36
C                                                                       STRE  37
C***       COMPUTE THE STRESS RESULTANTS                                STRE  38
C                                                                       STRE  39
      DO 10 INODE=1,NNODE                                               STRE  40
      DO 10 IDOFN=1,NDOFN                                               STRE  41
      KGASH=KGASH+1                                                     STRE  42
      STRSG(ISTRE)=STRSG(ISTRE)+SMATX(ISTRE,KGASH,KGASP)*ELDIS(IDOFN,   STRE  43
     .INODE)                                                            STRE  44
   10 CONTINUE                                                          STRE  45
C                                                                       STRE  46
C***         WRITE STRESSES IN DISC FOR EACH HARMONIC                   STRE  47
C                                                                       STRE  48
      WRITE(8) (STRSG(ISTRE),ISTRE=1,NSTRE)                             STRE  49
   15 CONTINUE                                                          STRE  50
   20 CONTINUE                                                          STRE  51
      RETURN                                                            STRE  52
      END                                                               STRE  53
```

The NS chart for STREFS is now given in Table 10

Loop over each strip. STRE 12	1
Read from file 3 the stress resultant matrix and the sampling point coordinates. STRE 17	2
Extract the displacements of the strip nodal points from the global displacement vector ASDIS. STRE 21-27	3

A

Table 10 NS chart for Subroutine STREFS

A

Zero Gauss point counter. STRE 28	4
Loop over Gauss points. STRE 32	5
Increment Gauss point counter. STRE 33	6
Compute stress resultants at each Gauss point. STRE 34-45	7
Write in disc file 8 the Gauss point stress resultants. STRE 49	8

Table 10 NS chart for subroutine STREFS

This routine is identical for all of the different structural applications given in this chapter.

12.7 Subroutine for summing the contributions from all harmonics: Subroutine ADD

This subroutine evaluates the displacements and stress resultants at a given longitudinal section of the structure by summing the nodal displacements and Gauss point stress resultants obtained for all harmonic terms chosen for the analysis.

```
      SUBROUTINE ADD(ADISP,AFORC,DISPL,DIZPL,EFORC,FORCE,MELEM,MFACT,
     .MPOIN,MTOTV,NCASE,NDOFN,NELEM,NGAUS,NHARM,NPOIN,NSTRE,NSYME,
     .TLENG)
      DIMENSION ADISP(3,MPOIN),AFORC(5,4,MELEM),DISPL(MTOTV),
     .DIZPL(MTOTV),EFORC(MFACT),FORCE(MFACT),GPCOD(1,4),SMATX(5,12,4),
     .YSECT(5)
C
C*** SUM DISPLACEMENTS AND STRESSES FOR ALL HARMONICS
C
C*** READ AND WRITE SECTION ANALYSIS DATA
C
      READ(5,900) NSECT
  900 FORMAT(I5)
      WRITE(6,950) NSECT
      READ(5,905) (YSECT(ISECT),ISECT=1,NSECT)
  905 FORMAT(8F10.3)
      VARAD=3.14159265/TLENG
      DO 105 ICASE=1,NCASE
      DO 100 ISECT=1,NSECT
      WRITE(6,915) YSECT(ISECT)
      RADIA=VARAD*YSECT(ISECT)
      WRITE(6,910) ICASE
  910 FORMAT(//,5X,14H LOAD CASE NO=,I3,//)
      REWIND 7
      REWIND 8
```

```
      NUMDR=NPOIN
      DO 5 IDOFN=1,NDOFN
      DO 5 IUMDR=1,NUMDR
    5 ADISP(IDOFN,IUMDR)=0.0
      DO 10 IELEM=1,NELEM
      DO 10 ISTRE=1,NSTRE
      DO 10 IGAUS=1,NGAUS
   10 AFORC(ISTRE,IGAUS,IELEM)=0.0
      WRITE(6,920)
      WRITE(6,925)
      LDDIS=NUMDR*NDOFN
C
C***   READ BACK DISPLACEMENTS
C
      DO 95 IHARM=1,NHARM,NSYME
      DO 30 ICASO=1,NCASE
      DO 15 IPOIN=1,NPOIN
      NGASH=IPOIN*NDOFN
      NGISH=NGASH-NDOFN+1
   15 READ(7) (DIZPL(IDDIS),IDDIS=NGISH,NGASH)
      IF(ICASO.NE.ICASE) GO TO 25
      DO 20 IDDIS=1,LDDIS
   20 DISPL(IDDIS)=DIZPL(IDDIS)
   25 CONTINUE
   30 CONTINUE
      HARMO=FLOAT(IHARM)
      FCTOR=HARMO*RADIA
C
C***   SUM DISPLACEMENTS FOR ALL HARMONICS
C
      DO  35 IUMDR=1,NUMDR
      IWDIS=(IUMDR-1)*3+1
      IROTX=IWDIS+1
      IROTY=IWDIS+2
      ADISP(1,IUMDR)=ADISP(1,IUMDR)+DISPL(IWDIS)*SIN(FCTOR)
      ADISP(2,IUMDR)=ADISP(2,IUMDR)+DISPL(IROTX)*SIN(FCTOR)
      ADISP(3,IUMDR)=ADISP(3,IUMDR)+DISPL(IROTY)*COS(FCTOR)
   35 CONTINUE
C
C***   PRINT DISPLACEMENTS
C
      IF(IHARM.LT.NHARM) GO TO 45
      DO 40 IPOIN=1,NPOIN
      WRITE(6,930) IPOIN,(ADISP(IDOFN,IPOIN),IDOFN=1,NDOFN)
   40 CONTINUE
   45 CONTINUE
C
C***   STRESSES AT SHEAR GAUSS POINTS
C
      LDSTR=NELEM*NSTRE*NGAUS
C
C***   READ BACK STRESSES AT SHEAR GAUSS POINTS
C
      DO 70 ICASO=1,NCASE
      DO 55 IELEM=1,NELEM
      KSTEG=NSTRE*NGAUS
      LSTEG=(IELEM-1)*KSTEG
      DO 50 IGAUS=1,NGAUS
      NGEST=IGAUS*NSTRE+LSTEG
      NGIST=NGEST-NSTRE+1
   50 READ(8) (EFORC(IDSTR),IDSTR=NGIST,NGEST)
   55 CONTINUE
      IF(ICASO.NE.ICASE) GO TO 65
      DO 60 IDSTR=1,LDSTR
   60 FORCE(IDSTR)=EFORC(IDSTR)
   65 CONTINUE
   70 CONTINUE
      REWIND 3
C
C*** SUM SHEAR GAUSS POINT STRESSES FOR ALL HARMONICS
C
      DO 90 IELEM=1,NELEM
      MSTEG=NSTRE*NGAUS
      NSTEG=(IELEM-1)*MSTEG
      READ(3) SMATX,GPCOD
      DO 85 IGAUS=1,NGAUS
      ISXDI=(IGAUS-1)*NSTRE+1+NSTEG
      ISYDI=ISXDI+1
      ISXYD=ISXDI+2
      ISXZD=ISXDI+3
```

```
      ISYZD=ISXDI+4                                                     ADD 106
      AFORC(1,IGAUS,IELEM)=AFORC(1,IGAUS,IELEM)+FORCE(ISXDI)*SIN(FCTOR) ADD 107
      AFORC(2,IGAUS,IELEM)=AFORC(2,IGAUS,IELEM)+FORCE(ISYDI)*SIN(FCTOR) ADD 108
      AFORC(3,IGAUS,IELEM)=AFORC(3,IGAUS,IELEM)+FORCE(ISXYD)*COS(FCTOR) ADD 109
      AFORC(4,IGAUS,IELEM)=AFORC(4,IGAUS,IELEM)+FORCE(ISXZD)*SIN(FCTOR) ADD 110
      AFORC(5,IGAUS,IELEM)=AFORC(5,IGAUS,IELEM)+FORCE(ISYZD)*COS(FCTOR) ADD 111
      IF(IHARM.LT.NHARM) GO TO 80                                       ADD 112
      IF(IGAUS.GT.1) GO TO 75                                           ADD 113
      WRITE(6,935)                                                      ADD 114
      WRITE(6,940)                                                      ADD 115
      WRITE(6,945)                                                      ADD 116
   75 CONTINUE                                                          ADD 117
C                                                                       ADD 118
C***   PRINT FINAL SHEAR GAUSS POINT STRESSES                           ADD 119
C                                                                       ADD 120
      WRITE(6,935) IELEM,IGAUS,(AFORC(ISTRE,IGAUS,IELEM),ISTRE=1,       ADD 121
     .NSTRE),GPCOD(1,IGAUS)                                             ADD 122
   80 CONTINUE                                                          ADD 123
   85 CONTINUE                                                          ADD 124
   90 CONTINUE                                                          ADD 125
   95 CONTINUE                                                          ADD 126
  100 CONTINUE                                                          ADD 127
  105 CONTINUE                                                          ADD 128
  915 FORMAT(1H0,'TOTAL DISPLACEMENTS AND STRESSES AT Z=',F8.3,//)      ADD 129
  920 FORMAT(1H0,5X,13HDISPLACEMENTS)                                   ADD 130
  925 FORMAT(1H0,5X,4HNODE,6X,5HDISP.,8X,9H  XZ-ROT.,7X,8H YZ-ROT.)     ADD 131
  930 FORMAT(I10,3E16.6)                                                ADD 132
  935 FORMAT(2I5,6E15.6,F10.4)                                          ADD 133
  940 FORMAT(// 1X,'STRESSES AT THE SHEAR GAUSS POINTS',/)              ADD 134
  945 FORMAT(/ 4X,'EL',3X,'GP.',13H XX-MOMENT   ,                       ADD 135
     .15H   YY-MOMENT   ,15H   XY-MOMENT   ,15H   XZ-FORCE    ,         ADD 136
     .15H   YZ-FORCE    ,1X,'X-CORD GAUS P.')                           ADD 137
  950 FORMAT(// 1X,'NUMBER OF SECTIONS TO BE ANALYSED=',I3)             ADD 138
      RETURN                                                            ADD 139
      END                                                               ADD 140
```

The NS chart for subroutine ADD is now given in Table 11.

Read and write the numbers of longitudinal sections at which displacements and stress resultants are to be calculated. Also read and write the y coordinate of the sections. ADD 11-16

Preset π/ℓ. ADD 17

Loop over each load case. ADD 18

> Loop over the number of longitudinal sections. ADD 19
>
> > Write section coordinates and evaluate $\ell\pi/a$ and $\ell\pi/\alpha$ for right or curved plate case, resp. ADD 20 - 23

A

Table 11 NS chart for subroutine ADD

A

Rewind displacement and stress resultant discs and initialise displacement and stress resultant vectors. ADD 24-33

Write titles. ADD 34-35

Loop over all harmonic terms. ADD 40

Read back nodal displacements from disc file 7. Identify displacements for current load case and accumulate displacements for all harmonics at the current longitudinal section. ADD 41-63

If this is the last harmonic, then write nodal displacements at the current longitudinal section. ADD 67-71

Read back stress resultants at the Gauss points for current strip from disc 8. Identify stress resultants for the current load case and accumulate stress resultants for all harmonics for current strip at the current longitudinal section. ADD 79-111

If this is the last harmonic, then write stress resultants at the Gauss point for current strip at current longitudinal section. ADD 112-122

Table 11 NS chart for subroutine ADD

This subroutine is almost identical for all the different structures studied in this chapter. The only difference lays in the number of displacement and stress components and the type of expansions chosen for each particular problem.

12.9 Subroutine to solve the stiffness equations for each harmonic term: Subroutine FRONT

This subroutine solves the uncoupled system of stiffness equations for each harmonic term (see (35)) by the FRONTAL elimination technique [35]. Full descriptive comments of this subroutine, together with the corresponding comments can be found in reference [34] and they will not be given here.

```
      SUBROUTINE FRONT(ASDIS,ELOAD,EQRHS,EQUAT,FIXED,GLOAD,GSTIF,
     .ICASE,IFFIX,IFPRE,LNODS,MBUFA,MELEM,MFRON,MSTIF,MTOTV,MVFIX,
     .NACVA,NAMEV,NDOFN,NELEM,NEVAB,NNODE,NOFIX,NPIVO,NPOIN,NVFIX,
     .PRESC,VECRV)
      DIMENSION ASDIS(MTOTV),ELOAD(MELEM,12),EQRHS(MBUFA),
     .EQUAT(MFRON,MBUFA),ESTIF(12,12),FIXED(MTOTV),GLOAD(MFRON),
     .GSTIF(MSTIF),IFFIX(MTOTV),IFPRE(MVFIX,3),LNODS(MELEM,4),
     .LOCEL(12),NACVA(MFRON),NAMEV(MBUFA),NDEST(12),NOFIX(MVFIX),
     .NPIVO(MBUFA),PRESC(MVFIX,3),VECRV(MFRON)
C
C*** MERGE AND SOLVE THE RESULTING EQUATIONS
C*** BY THE FRONTAL METHOD FOR EACH HARMONIC
C
C*** FOR A COMPLETE DESCRIPTION OF THIS SUBROUTINE
C*** SEE THE BOOK "FINITE ELEMENTS IN PLASTICITY:
C*** THEORY AND PRACTICE" BY OWEN AND HINTON
C*** PINERIDGE PRESS, SWANSEA 1980
C
C     RETURN
C     END
```

13. EXAMPLES

13.1 Example 1: Simply supported square thick plate under uniformly distributed load

The geometry of the plate and the material properties are indicated in Figure 24. A mesh of five, three-noded strips with full integration has been used. This example has been discussed earlier in the chapter in Section 3.5. Figure 24 shows a plot of the numerical results obtained for the bending moment M_x along the central transverse section of the plate versus the theoretically exact values [32]. Accuracy of the numerical solution is noticeable. In section 14 computer listings of the data input and output results for this example are given.

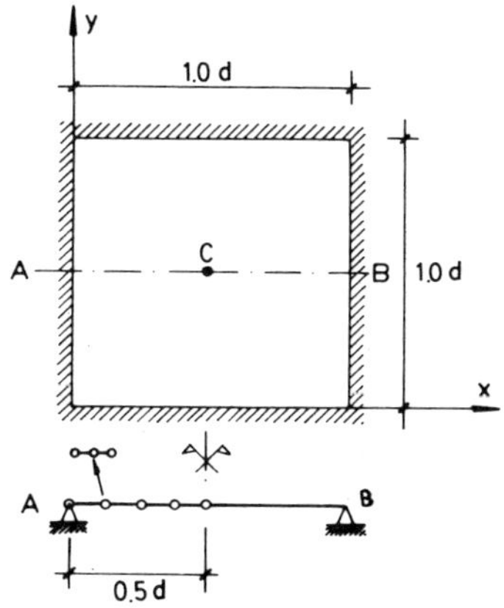

Uniformly distributed load q

$q = 1.0\ (F/d^2)$

$E = 10.920\ (F/d^2)$

$\nu = 0.3$

thickness = 0.1 d

5 three noded strips

Full integration

6 Non zero harmonics

W_C (obtained) = 0.4272×10^{-2} (d)

W_C (exact) [21] = 0.4270×10^{-2} (d)

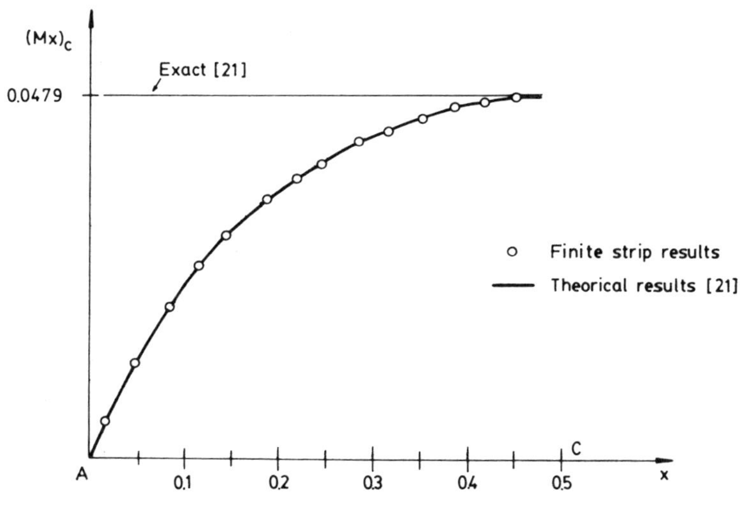

Fig. 24 Example 1. Thick square plate under uniformly distributed loading.

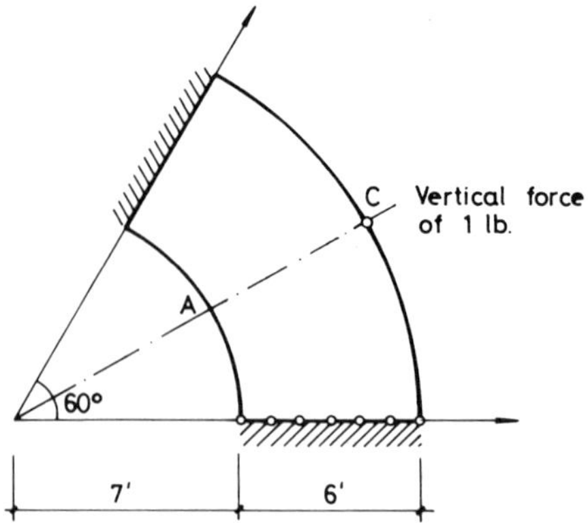

E = 42.460 lb/in²

ν = 0.3

thickness = 0.172 in

12 two noded strips

Reduced integration

6 non zero harmonics

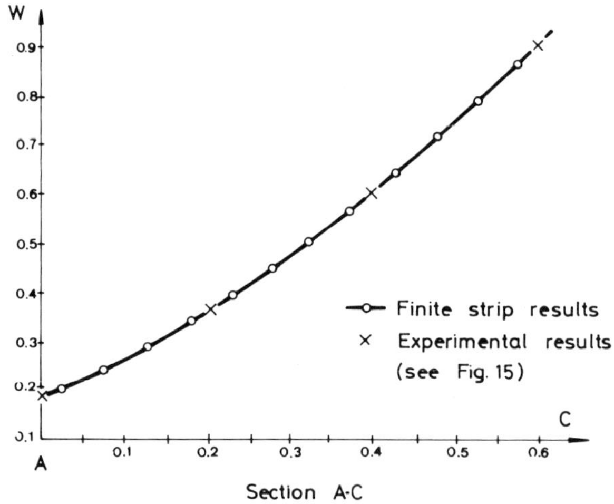

Fig. 25 Example 2 : Circular slab simply supported at two ends under point load acting at the center of a free edge.

13.2 Example 2: Circular slab simply supported at two ends under point load acting at the centre of a free edge

This example was discussed earlier in Section 10.1. The geometry and material properties are shown in Figure 25. A mesh of twelve, two-noded strips with reduced integration has been used for the analysis. In the same figure a plot of the numerical solution obtained by Coull and Das for the same problem [26] is also shown in the figure for comparison. Accuracy of the numerical results is again good. Data input for this problem is given in Section 15. A listing of the numerical results obtained is also presented in Section 15.

14. INPUT AND OUTPUT DATA FOR EXAMPLE 1

Input data for Example 1 is now listed.

```
   1
SIMPLY SUPPORTED SQUARE PLATE. QUADRATIC STRIP WITH FULL INTEGRATION
  11    2
  11    5    2    1    1    3    1    3    3
   1.
   1    1    1    2    3
   2    1    3    4    5
   3    1    5    6    7
   4    1    7    8    9
   5    1    9   10   11
   1   .0
   2   .05
   3   .1
   4   .15
   5   .2
   6   .25
   7   .30
   8   .35
   9   .40
  10   .45
  11   .5
   1  101
  11  010
   110920.      .3          .1      1.
   UNIFORM LOAD
   0    1
   1
   0.5
```

Typical output data for Example 1 is now listed.

```
     PROBLEM NO.   1          SIMPLY SUPPORTED SQUARE PLATE. QUADRATIC STRIP WITH FULL INTEGRATION

 NUMBER OF HARMONIC TERMS TO BE USED =   11
 INDICATOR FOR SYMMETRY OF LOADING ( 1 FOR NON SYMMETRY,        2 FOR SYMMETRY) =    2

    LENGHT OF THE PLATE=   1.00000

 NPOIN =  11     NELEM =   5     NVFIX =   2     NCASE =   1     NTYPE =   1     NNODE =   3     NDOFN =   3

 NMATS =   1     NPROP =   5     NGAUB =   3     NGAUS =   3     NDIME =   1     NSTRE =   5     NEVAB =   9

 ELEMENT   PROPERTY      NODE NUMBERS
     1         1          1    2    3
     2         1          3    4    5
     3         1          5    6    7
     4         1          7    8    9
     5         1          9   10   11

 NODAL POINT COORDINATES
 NODE       X
    1   0.00000
    2   0.05000
    3   0.10000
    4   0.15000
    5   0.20000
    6   0.25000
    7   0.30000
    8   0.35000
    9   0.40000
   10   0.45000
   11   0.50000

 RESTRAINED NODES
 NODE CODE       FIXED VALUES
   1  101    0.00000    0.00000    0.00000
  11  010    0.00000    0.00000    0.00000

 MATERIAL PROPERTIES
 NUMBER                     PROPERTIES
    1        0.109200E+05  0.300000E+00  0.100000E+00  0.100000E+01  0.000000E+00
```

```
        LOAD CASE NUMBER   1
   UNIFORM LOAD
   0    1

 NUMBER OF SECTIONS TO BE ANALYSED=  1

TOTAL DISPLACEMENTS AND STRESSES AT Z=   0.500

   LOAD CASE NO=  1

   DISPLACEMENTS
```

NODE	DISP.	XZ-ROT.	YZ-ROT.
1	0.000000E+00	0.134854E-01	0.000000E+00
2	0.714032E-03	0.132200E-01	-0.923487E-10
3	0.139803E-02	0.125430E-01	-0.182944E-09
4	0.203262E-02	0.115109E-01	-0.267722E-09
5	0.260408E-02	0.102355E-01	-0.344004E-09
6	0.310131E-02	0.875862E-02	-0.410277E-09
7	0.351682E-02	0.714906E-02	-0.465529E-09
8	0.384477E-02	0.543559E-02	-0.509111E-09
9	0.408153E-02	0.365863E-02	-0.540538E-09
10	0.422445E-02	0.183859E-02	-0.559517E-09
11	0.427228E-02	0.000000E+00	-0.565858E-09

STRESSES AT THE SHEAR GAUSS POINTS

EL	GP.	XX-MOMENT	YY-MOMENT	XY-MOMENT	XZ-FORCE	YZ-FORCE	X-CORD GAUS P.
1	1	0.344624E-02	0.224752E-02	0.129786E-08	0.344417E+00	0.777929E-09	0.112702E-01
1	2	0.111826E-01	0.868759E-02	0.128385E-08	0.266092E+00	0.207174E-08	0.500000E-01
1	3	0.188927E-01	0.150399E-01	0.125730E-08	0.274231E+00	0.122838E-08	0.887298E-01

STRESSES AT THE SHEAR GAUSS POINTS

EL	GP.	XX-MOMENT	YY-MOMENT	XY-MOMENT	XZ-FORCE	YZ-FORCE	X-CORD GAUS P.
2	1	0.231534E-01	0.186229E-01	0.121308E-08	0.247447E+00	0.208295E-09	0.111270E+00
2	2	0.281676E-01	0.238984E-01	0.112975E-08	0.192361E+00	-0.100963E-08	0.150000E+00
2	3	0.330994E-01	0.288991E-01	0.103397E-08	0.188392E+00	-0.212736E-08	0.188730E+00

STRESSES AT THE SHEAR GAUSS POINTS

EL	GP.	XX-MOMENT	YY-MOMENT	XY-MOMENT	XZ-FORCE	YZ-FORCE	X-CORD GAUS P.
3	1	0.356971E-01	0.316057E-01	0.969291E-09	0.167687E+00	-0.270671E-08	0.211270E+00
3	2	0.387503E-01	0.355468E-01	0.851558E-09	0.129077E+00	-0.355219E-08	0.250000E+00
3	3	0.416819E-01	0.390826E-01	0.730621E-09	0.118339E+00	-0.425440E-08	0.288730E+00

STRESSES AT THE SHEAR GAUSS POINTS

EL	GP.	XX-MOMENT	YY-MOMENT	XY-MOMENT	XZ-FORCE	YZ-FORCE	X-CORD GAUS P.
4	1	0.431349E-01	0.408900E-01	0.656436E-09	0.102008E+00	-0.459466E-08	0.311270E+00
4	2	0.447815E-01	0.433952E-01	0.525483E-09	0.740215E-01	-0.507922E-08	0.350000E+00
4	3	0.462832E-01	0.454172E-01	0.393004E-09	0.593722E-01	-0.544764E-08	0.388730E+00

STRESSES AT THE SHEAR GAUSS POINTS

EL	GP.	XX-MOMENT	YY-MOMENT	XY-MOMENT	XZ-FORCE	YZ-FORCE	X-CORD GAUS P.
5	1	0.469282E-01	0.463202E-01	0.314104E-09	0.456623E-01	-0.560721E-08	0.411270E+00
5	2	0.474945E-01	0.473364E-01	0.177364E-09	0.240985E-01	-0.579543E-08	0.450000E+00
5	3	0.479053E-01	0.478345E-01	0.401654E-10	0.643182E-02	-0.588011E-08	0.488730E+00

15. INPUT AND OUTPUT DATA FOR EXAMPLE 2

Input data for Example 2 is now listed.

```
    1
CIRCULAR PLATE SUPPORTED AT TWO ENDS.LINEAR ELEMENT WITH REDUCED INTEGRATION
   11    2
   13   12    0    1    2    2    1    1    1
 1.04720
    1    1    1    2
    2    1    2    3
    3    1    3    4
    4    1    4    5
    5    1    5    6
    6    1    6    7
    7    1    7    8
    8    1    8    9
    9    1    9   10
   10    1   10   11
   11    1   11   12
   12    1   12   13
    1    7.
    2    7.5
    3    8.
    4    8.5
    5    9.
    6    9.5
    7   10.
    8   10.5
    9   11.
   10   11.5
   11   12.
   12   12.5
   13   13.
    142460.3      .3        .172
  CIRCULAR PLATE UNDER EDGE POINT LOAD
    1
   13   1.       .5236
    1
    .5236
```

Typical output data for Example 2 is now listed.

```
      PROBLEM NO.   1         CIRCULAR PLATE SUPPORTED AT TWO ENDS.LINEAR ELEMENT

    NUMBER OF HARMONIC TERMS TO BE USED =   11
    INDICATOR FOR SYMMETRY OF LOADING ( 1 FOR NON SYMMETRY,        2 FOR SYMMETRY) =    2

       ANGLE OF THE PLATE =   1.04720


    NPOIN =  13     NELEM =  12     NVFIX =   0     NCASE =   1     NTYPE =   2     NNODE =   2     NDOFN =   3

    NMATS =   1     NPROP =   5     NGAUB =   1     NGAUS =   1     NDIME =   1     NSTRE =   5     NEVAB =   6


    ELEMENT   PROPERTY      NODE NUMBERS
        1         1           1    2
        2         1           2    3
        3         1           3    4
        4         1           4    5
        5         1           5    6
        6         1           6    7
        7         1           7    8
        8         1           8    9
        9         1           9   10
       10         1          10   11
       11         1          11   12
       12         1          12   13


    NODAL POINT COORDINATES
    NODE        X
       1    7.00000
       2    7.50000
       3    8.00000
       4    8.50000
       5    9.00000
       6    9.50000
       7   10.00000
       8   10.50000
       9   11.00000
      10   11.50000
      11   12.00000
      12   12.50000
      13   13.00000
```

```
MATERIAL PROPERTIES
NUMBER                         PROPERTIES
   1        0.424603E+05  0.300000E+00  0.172000E+00  0.000000E+00  0.000000E+00

         LOAD CASE NUMBER    1
CIRCULAR PLATE UNDER EDGE POINT LOAD
  1    0

         POINT LOAD
  13   1.00000   0.00000   0.00000   0.52360

NUMBER OF SECTIONS TO BE ANALYSED=  1

TOTAL DISPLACEMENTS AND STRESSES AT Z=   0.524



     LOAD CASE NO=  1


     DISPLACEMENTS

     NODE      DISP.          XZ-ROT.        YZ-ROT.
        1   0.194096E+00   0.691507E-01  -0.370931E-08
        2   0.229811E+00   0.737695E-01  -0.398639E-08
        3   0.268167E+00   0.797073E-01  -0.441637E-08
        4   0.309768E+00   0.867174E-01  -0.479085E-08
        5   0.355035E+00   0.943580E-01  -0.517592E-08
        6   0.404263E+00   0.102548E+00  -0.565801E-08
        7   0.457668E+00   0.111042E+00  -0.602984E-08
        8   0.515410E+00   0.119893E+00  -0.670909E-08
        9   0.577619E+00   0.128876E+00  -0.709822E-08
       10   0.644414E+00   0.138241E+00  -0.826672E-08
       11   0.715932E+00   0.147683E+00  -0.885258E-08
       12   0.792337E+00   0.157792E+00  -0.114704E-07
       13   0.873764E+00   0.167534E+00  -0.130092E-07
```

```
STRESSES AT THE SHEAR GAUSS POINTS

  EL   GP. XX-MOMENT      YY-MOMENT     XY-MOMENT     XZ-FORCE      YZ-FORCE     X-CORD GAUS P.
  1    1  -0.230874E-01  0.477447E+00  0.905515E-08 -0.692050E-01  0.286227E-07  0.725000E+01


STRESSES AT THE SHEAR GAUSS POINTS

  EL   GP. XX-MOMENT      YY-MOMENT     XY-MOMENT     XZ-FORCE      YZ-FORCE     X-CORD GAUS P.
  2    1  -0.700285E-01  0.479293E+00  0.112162E-07 -0.631946E-01 -0.294223E-07  0.775000E+01


STRESSES AT THE SHEAR GAUSS POINTS

  EL   GP. XX-MOMENT      YY-MOMENT     XY-MOMENT     XZ-FORCE      YZ-FORCE     X-CORD GAUS P.
  3    1  -0.105912E+00  0.488393E+00  0.105541E-07 -0.249140E-01  0.145402E-07  0.825000E+01


STRESSES AT THE SHEAR GAUSS POINTS

  EL   GP. XX-MOMENT      YY-MOMENT     XY-MOMENT     XZ-FORCE      YZ-FORCE     X-CORD GAUS P.
  4    1  -0.126036E+00  0.496996E+00  0.109210E-07 -0.803686E-02 -0.227869E-07  0.875000E+01

STRESSES AT THE SHEAR GAUSS POINTS

  EL   GP. XX-MOMENT      YY-MOMENT     XY-MOMENT     XZ-FORCE      YZ-FORCE     X-CORD GAUS P.
  5    1  -0.140982E+00  0.513161E+00  0.126238E-07  0.798940E-02  0.169662E-07  0.925000E+01


STRESSES AT THE SHEAR GAUSS POINTS

  EL   GP. XX-MOMENT      YY-MOMENT     XY-MOMENT     XZ-FORCE      YZ-FORCE     X-CORD GAUS P.
  6    1  -0.146456E+00  0.531339E+00  0.115947E-07  0.336699E-01 -0.325593E-07  0.975000E+01
```

STRESSES AT THE SHEAR GAUSS POINTS

EL	GP.	XX-MOMENT	YY-MOMENT	XY-MOMENT	XZ-FORCE	YZ-FORCE	X-CORD GAUS P.
7	1	-0.150729E+00	0.559987E+00	0.167009E-07	0.370290E-01	0.346755E-07	0.102500E+02

STRESSES AT THE SHEAR GAUSS POINTS

EL	GP.	XX-MOMENT	YY-MOMENT	XY-MOMENT	XZ-FORCE	YZ-FORCE	X-CORD GAUS P.
8	1	-0.145828E+00	0.592085E+00	0.140238E-07	0.815567E-01	-0.661010E-07	0.107500E+02

STRESSES AT THE SHEAR GAUSS POINTS

EL	GP.	XX-MOMENT	YY-MOMENT	XY-MOMENT	XZ-FORCE	YZ-FORCE	X-CORD GAUS P.
9	1	-0.144149E+00	0.643693E+00	0.273290E-07	0.738511E-01	0.775397E-07	0.112500E+02

STRESSES AT THE SHEAR GAUSS POINTS

EL	GP.	XX-MOMENT	YY-MOMENT	XY-MOMENT	XZ-FORCE	YZ-FORCE	X-CORD GAUS P.
10	1	-0.128782E+00	0.704029E+00	0.237931E-07	0.169320E+00	-0.155664E-06	0.117500E+02

STRESSES AT THE SHEAR GAUSS POINTS

EL	GP.	XX-MOMENT	YY-MOMENT	XY-MOMENT	XZ-FORCE	YZ-FORCE	X-CORD GAUS P.
11	1	-0.120586E+00	0.811450E+00	0.607846E-07	0.168711E+00	0.155147E-06	0.122500E+02

STRESSES AT THE SHEAR GAUSS POINTS

EL	GP.	XX-MOMENT	YY-MOMENT	XY-MOMENT	XZ-FORCE	YZ-FORCE	X-CORD GAUS P.
12	1	-0.654474E-01	0.951187E+00	0.601473E-07	0.450070E+00	-0.445427E-06	0.127500E+02

16. DATA INPUT INSTRUCTIONS FOR PROGRAM PBSTRIP

DATA SET 1	PROBLEM DATA (I5). One record
Cols. 1-5 NPROB	Total number of problems to be solved in one run.
DATA SET 2	TITLE (12A6). One record
Cols. 1-72 TITLE	Title of the problem - limited to 72 alphanumeric characters.
DATA SET 3	NUMBER OF HARMONICS (2I5). One record
Cols. 1-5 NHARM	Total number of harmonics to be used for each problem.
Cols. 6-10 NSYME	Symmetry parameter = 0 non-symmetric loading = 1 symmetric loading (with respect to the axis $y = \frac{b}{2}$).
DATA SET 4	CONTROL DATA (9I5). One record
Cols. 1-5 NPOIN	Total number of nodal points.
" 6-10 NELEM	Total number of strips.
" 11-15 NVFIX	Total number of restrained points where one or more degrees of freedom are restrained along the longitudinal direction.
" 16-20 NCASE	Total number of load cases to be analysed.
" 21-25 NTYPE	Indicator for right or curved plate. = 1 right plate = 2 curved plate.
" 26-30 NNODE	Number of nodes per strip. = 2 linear strip = 3 quadratic strip = 4 cubic strip.
" 31-35 NMATS	Total number of different materials.

Cols. 36-40	NGAUB	Order of integration formulae for numerical integration of bending stiffness (see Table 1).
" 41-45	NGAUS	Order of integration formulae for numerical integration of shear stiffness (see Table 1).

<u>DATA SET 5</u> PLATE LENGTH/ANGLE (F10.5). One record

Cols. 1-10	TLENG	Plate length (right plate) or angle (curved plate).

<u>DATA SET 6</u> ELEMENT DATA (6I5). One record for each strip element. Total of NELEM cards.

Cols. 1-5	NUMEL	Strip element number.
" 6-10	MATNO(NUMEL)	Material property number.
" 11-15	LNODS(NUMEL,1)	1st nodal connection number.
" 16-20	LNODS(NUMEL,2)	2nd nodal connection number.
" 21-25	LNODS(NUMEL,3)	3rd nodal connection number (for quadratic or cubic strips only).
" 26-30	LNODS(NUMEL,4)	4th nodal connection number (for cubic strip only).

<u>DATA SET 7</u> NODE CARDS (I5,2F10.5). One record for each nodal point. Total of NPOIN records.

Cols. 1-5	IPOIN	Nodal point number.
" 6-15	COORD(IPOIN,1)	The x coordinate of the node.
" 16-25	COORD(IPOIN,2)	The z coordinate of the node.

<u>DATA SET 8</u> RESTRAINED NODE DATA (I5,2x,3I1, 3F10.5). One record for each restrained node. Total of NVFIX records.

Cols. 1-5	NVFIX(IVFIX)	Restrained node number.
" 8	IFPRE(IVFIX,1)	Condition of longitudinal restraint on nodal displacement w.
" 9	IFPRE(IVFIX,2)	Condition of longitudinal restraint on nodal rotation θ_x.

Cols.	10	IFPRE(IVFIX,3)	Condition of longitudinal restraint on nodal rotation θ_y. In Cols. 8-10: IFPRE = 0 No displacement = 1 Nodal displacement restrained.
"	11-20	PRESC(IVFIX,1)	The prescribed value of nodal displacement w.
"	21-30	PRESC(IVFIX,2)	The prescribed value of nodal rotation θ_x.
"	31-40	PRESC(IVFIX,3)	The prescribed value of nodal rotation θ_y.

Note: The program is able to deal with only zero prescribed values. Other prescribed values, different from zero, should be input separately for each harmonic term accordingly to the contribution of the prescribed movement to each harmonic, which should be properly evaluated beforehand.

<u>DATA SET 9</u> — MATERIAL DATA (I5,4F10.5). One record for each different material. Total of NMATS records.

Cols.	1-5	NUMAT	Material identification number.
"	6-15	PROPS(NUMAT,1)	Young's modulus, E.
"	16-25	PROPS(NUMAT,2)	Poisson's ratio, ν.
"	26-35	PROPS(NUMAT,3)	Element thickness.
"	36-45	PROPS(NUMAT,4)	Distributed load intensity.

<u>DATA SET 10</u> — LOAD CASE TITLE DATA (12A6). One record.

Cols.	1-72	TITLE	Title of the load case - limited to 72 alphanumeric characters.

<u>DATA SET 11</u> — LOAD CONTROL DATA (2I5). One record.

Cols.	1-5	IPLOD	Applied vertical point load control parameter = 0 no applied vertical nodal loads to be input.

= 1 applied vertical nodal loads to be input.

" 6-10 IUNIF — Uniformly distributed load control parameter.

= 0 No distributed load case to be considered.

= 1 Distributed load case to be considered.

DATA SET 12 — APPLIED VERTICAL POINT LOAD DATA (I5,2F10.5). One record for each loaded nodal point.

Cols. 1-5	LODPT	Node number.
" 6-15	POINT(i)	Value of the vertical point load acting at the node.
" 16-25	YLOAD	The y coordinate of the point at which the load acts.

Notes: 1) The last record should be that for the highest numbered node whether it is loaded or note.

2) If IPLOD = 0 in Data Set 9, omit this set.

DATA SETS 10 TO 12 TO BE REPEATED FOR EACH LOAD CASE IN ACCORDANCE WITH NCASE IN DATA SET 4.

DATA SETS 2 TO 12 TO BE REPEATED FOR EACH PROBLEM IN ACCORDANCE WITH NPROB IN DATA SET 1.

DATA SET 13 — NUMBER OF OUTPUT SECTIONS DATA (I5). One record.

Cols. 1-5	NSECT	Number of output section (≤ 5).

DATA SET 14 — SECTION COORDINATES DATA (8F10.5). One record.

Cols. 1-10	YSECT(1)	The y (right plate) or θ (curved plate) coordinate of the first output section.
" 11-20	YSECT(2)	ditto for the second section
⋮		⋮
" 41-50	YSECT(5)	ditto for the fifth section.

17. DICTIONARY OF MAIN VARIABLE NAMES

ASDIS (NTOTV) - Vector of nodal displacement amplitudes for each harmonic.

BMATX (NSTRE,NEVAB) - Matrix $\underset{\sim}{B}^{\ell}$ for each element.

CARTD (1,INODE) - Cartesian shape function derivative $\frac{\partial N_i^{(e)}}{\partial x}$.

COORD (NPOIN,1) - Coordinates of nodal points.

DBMAT (NSTRE,NEVAR) - Matrix $\underset{\sim}{D}\,\underset{\sim}{B}^{\ell}$ for each element.

DERIV (1 NODE) - Shape function derivative $\frac{\partial N_i^{(e)}}{\partial \xi}$.

DMATX (NSTRE,NSTRE) - Matrix $\underset{\sim}{D}$ for each element.

ELOAD (NELEM,NEVAB) - Nodal forces for each element.

ESTIF (NEVAB,NEVAB) - The element stiffness matrix $[\underset{\sim}{K}^{e}]^{\ell}$.

GPCOD (1,NGAUS) - Coordinate x of the shear integration points.

IFPRE (2,NDOFN) - Integer code to specify prescribed degrees of freedom.

LNODS (10,NNODE) - Element node number listed for each element.

MATNO (NELEM) - Material set number for each element.

NDIME - Number of coordinate components required to define each node (=1).

NDOFN - Number of degrees of freedom per node.

NELEM - Number of strip elements.

NEVAB - Number of variables per element = NNODE * NDOFN.

NGAUB - Number of Gaussian integration points for the flexure terms.

NGAUS - Number of Gaussian integration points for the shear terms.

NHARM - Number of harmonic terms to be used in the analysis.

NNODE - Number of nodes per strip element.

NOFIX (2) - List of prescribed node numbers (maximum of two nodes to be prescribed).

NPOIN - Total number of nodal points.

NPROP - Number of material parameters required to define the characteristics of a material completely.

NSTRE - Number of stress components per element.

NSYME - Code for symmetric loading.

NTYPE - Code for right (= 1), or curved (= 2) plate.

POINT - Value of the vertical point load.

POSGP (4) - Coordinates of integration points.

PRESC (2,NODFN) - Values of the prescribed degrees of freedom.

PROPS (1,5) - Values of the material parameters.

TLENG - Plate length (right) or angle (curved).

YLOAD - y coordinate of the transverse section where the point load acts.

WEIGP (4) - Weights of the integrating points.

18. REFERENCES

1. GRAFTON, P.E. and STROME, D.R. - Analysis of axisymmetric shells by the direct stiffness method. J.A.I.A.A., 2342-7, 1963.

2. AHMAD, S., IRONS, B.M. and ZIENKIEWICZ, O.C. - Curved thick shell and membrane elements with particular reference to axisymmetric shell problems. Proc. 2nd Conf. on Matrix Methods in Structural Mechanics. Wright-Patterson A.F. Base, Ohio, AFFDL-TR-68-150, 1968.

3. WILSON, E.L. - Structural analysis of axisymmetric solids. J.A.I.A.A., 3, 2269-74, 1965.

4. CHEUNG, Y.K. - Finite strip method analysis of elastic slabs. Proc. A.S.C.E., 94, 1365-1378, 1968.

5. CHEUNG, Y.K. - The finite strip method in the analysis of elastic plates with two opposite simply supported ends. Proc. I.C.E., 40, 1-7, 1968.

6. CHEUNG, Y.K. - Analysis of box girder bridges by finite strip method. A.C.I. Publications, SP 26, 357-378, 1969.

7. CHEUNG, Y.K. - Folded plate structures by finite strip method. Proc. A.S.C.E., 96, 2963-2979, 1969.

8. CHEUNG, Y.K. - The analysis of cylindrical orthotropic curved bridge decks. Pub. Int. Ass. Struct. Engrs. 29, 41-52, 1969.

9. WILLAM, K.J. and SCORDELIS, A.C. - Analysis of orthotropic folded plates with eccentric stiffeners. Report No. SESM 70-2. Dept. of Civil Engng., Univ. of California, Berkeley, 1970.

10. LOO, Y.C. and CUSENS, A.R. - A refined finite strip method for the analysis of orthotropic plates. Proc. I.C.E., 48, 85-91, 1970.

11. CUSENS, A.R. and LOO, Y.C. - Application of the finite strip method in the analysis of concrete box girders. Proc. I.C.E., 57-II, 251-273, 1974.

12. CHEUNG, Y.K. - The Finite Strip Method in Structural Analysis, p. 232, Pergamon Press, 1976.

13. BENSON, P.R. and HINTON, E. - A thick finite strip solution for static free vibration and stability problems. Int. J. Num. Meth. Engng., 10, 665-678, 1976.

14. OÑATE, E. - Comparisons of finite strip method for the analysis of box girder bridges. M.Sc. Thesis Dept. of Civil Engng., Univ. College of Swansea, 1976.

15. LOO, Y.C. and CUSENS, A.R. - The Finite Strip Method in Bridge Engineering, Viewpoint Public, 1978.

16. CHEUNG, M.S. and CHEUNG, Y.K. - Analysis of curved bridges by the finite strip method. Research Report Dept. of Civil Engng., Univ. of Calgary, Canada, 1970.

17. OÑATE, E. and SUAREZ, B. - A unified approach for the analysis of bridges, plates and axisymmetric shells using the linear Mindlin strip elements. Computers and Structures, 17, 407-426, 1983.

18. OÑATE, E. and SUAREZ, B. - A comparison of the linear quadratic and cubic Mindlin strip elements for the analysis of thick and thin plates. Computers and Structures, 17, 427-439, 1983.

19. TIMOSHENKO, S. - Strength of Materials, Van Nostrand, New York, 1955.

20. ZIENKIEWICZ, O.C. - The Finite Element Method, McGraw-Hill, New York, 1979.

21. TIMOSHENKO, S. and WOINOWSKY-KRIEGER, S. - Theory of Plates and Shells, 2nd Edition McGraw-Hill, New York, 1959.

22. ALWAR, R.S. and RAMACHANDRAN, K.N. - Theoretical and photoelastic analysis of thick slabs subject to highly localised loads. Buid. Sci., 7, 159-166, Pergamon Press, 1972.

23. CLOUGH, R.W. and WILSON, E.L. - Dynamic finite element analysis of arbitrary thin shells. Computers and Structures, 1, 35, 1971.

24. ZIENKIEWICZ, O.C., BAUER, O.C., MORGAN, K. and ONATE, E. - A simple and efficient element for axisymmetric shells, Int. J. Num. Meth. Engng., 11, 1545-1558, 1977.

25. SUAREZ, B. - La formulación de Bandas finitas de Reissner Mindlin para análisis de placas, puentes y láminas de revolución, Ph.D. Thesis E.T.S. Ing. Caminos, Barcelona, 1982.

26. COULL, A. and DAS, Y.P.C. - Analysis of curved bridge decks. Proc. I.C.E., 37, 75-85, 1967.

27. THORPE, J. - Ph.D. Thesis, University of Dundee, 1976.

28. SAWKO, F. and MERRIMAN, Y.P.A. - An annular segment finite element for plate bending. Int. J. Num. Meth. Engng., 3, 119-129, 1971.

29. FAM, A. and TURKSTRA, Y.C. - A finite element method for box bridge analysis. Computers and Structures, 5, 179-186, 1975.

30. FLUGGE, W. - Stresses in Shells, p.525, Springer-Verlag, Berlin, 1973.

31. OÑATE, E., HINTON, E. and GLOVER, Y.N. - Techniques for improving the performance of Ahmad shell elements. Int. Conf. Appl. Num. Modelling, Edited by Pentech Press, Madrid, 1979.

32. LINDBERG, G.M., OLSEN, M.D. and COWPER, Y.G.R. - New Developments in the Finite Element Analysis of Shells. Struc. Material Laboratory, 1-38, 1969.

33. AHMAD, S., IRONS, B.M. and ZIENKIEWICZ, O.C. - Curved thick shell and membrane elements with particular reference to axisymmetric problems. Proc. 2nd Conference Matrix Methods in Struct. Mech., Wright-Patterson A.F. Base, Ohio, 1968.

34. HINTON, E. and OWEN, D.R.J. - Finite Element Programming. Academic Press, London, 1979.

35. IRONS, B.M. - A frontal solution program for finite element analysis. Int. J. Num. Meth. Engng., 2, 5-32, 1970.

CHAPTER 3

MINDLIN PLATE FINITE ELEMENTS

E. HINTON* and H.H. ABDEL RAHMAN**

* University College of Swansea, U.K.
** University of Petroleum and Minerals, Dhahran, Saudi Arabia

1. INTRODUCTION

The main objectives of this chapter are as follows:

a) to briefly review various plate elements based on Mindlin plate theory,

b) to describe a hierarchical version of the heterosis plate element,

c) to present a documented program called MINDLIN in which the heterosis element is implemented, and

d) to provide a set of user instructions for MINDLIN and also examples demonstrating the use of the program and the performance of the heterosis plate element.

The main equations of Mindlin plate theory have already been presented in the chapter dealing with closed form solutions. However, for completeness, a brief resume is now provided.

2. REVIEW OF MINDLIN PLATE THEORY

2.1 Mindlin plate formulation - displacement approach

Mindlin plate theory [1] allows for transverse shear deformation effects and thus offers an attractive alternative to classical Kirchhoff thin plate theory. The main assumptions are that

a) displacements are small compared to the plate thickness,

b) the stress normal to the plate mid-surface is negligible,

and

c) normals to the mid-surface before deformation remain straight but not necessarily normal to the mid-surface after deformation.

A typical Mindlin plate is shown in Figure 2.1.1(a) and if the xy-plane is taken as the reference plane then the plate displacements can be expressed as

$$U(x,y,z) = z\theta_x(x,y)$$
$$V(x,y,z) = z\theta_y(x,y) \qquad (2.1.1)$$
$$W(x,y,z) = w(x,y)$$

where $w(x,y)$, $\theta_x(x,y)$ and $\theta_y(x,y)$ are weighted averages for the transverse deflection and rotations of the normal in the xz-and yz-planes respectively.

The bending and shear strains have the form

$$z\underline{\varepsilon}_b = z(\theta_{x,x}, \theta_{y,y}, (\theta_{x,y}+\theta_{y,x}))^T$$
$$z(K_x, K_y, K_{xy})^T \qquad (2.1.2)$$

and

$$\underline{\varepsilon}_s = (w_{,x} + \theta_x, w_{,y} + \theta_y)$$
$$(\gamma_{xz}, \gamma_{yz}) \qquad (2.1.3)$$

where $\theta_{x,x}$ denotes $\partial\theta_x/\partial x$ etc., K_x and K_y are the bending curvatures, K_{xy} is the twisting curvature, γ_{xz} and γ_{yz} represent the average transverse shear strains (or shear angles).

For the general anisotropic case under the generalised plane stress assumptions, the bending moments may be written as

$$\begin{bmatrix} M_x \\ M_y \\ M_{xy} \end{bmatrix} = \begin{bmatrix} D_{11} & D_{12} & D_{16} \\ D_{12} & D_{22} & D_{26} \\ D_{16} & D_{26} & D_{66} \end{bmatrix} \begin{bmatrix} K_x \\ K_y \\ K_{xy} \end{bmatrix} \qquad (2.1.4)$$

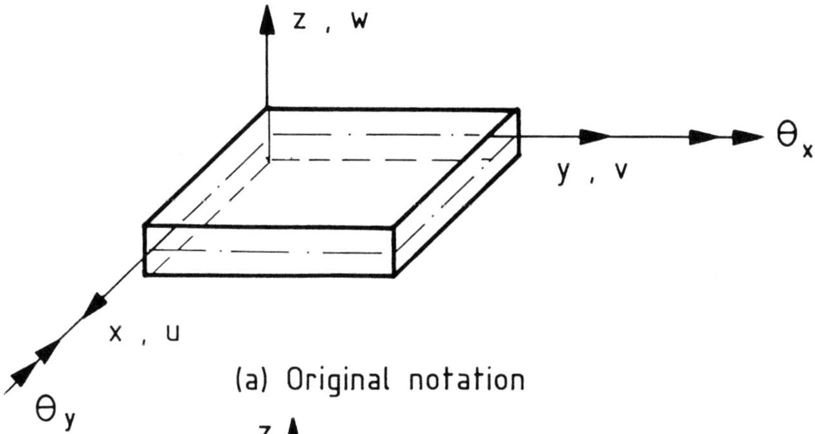

(a) Original notation

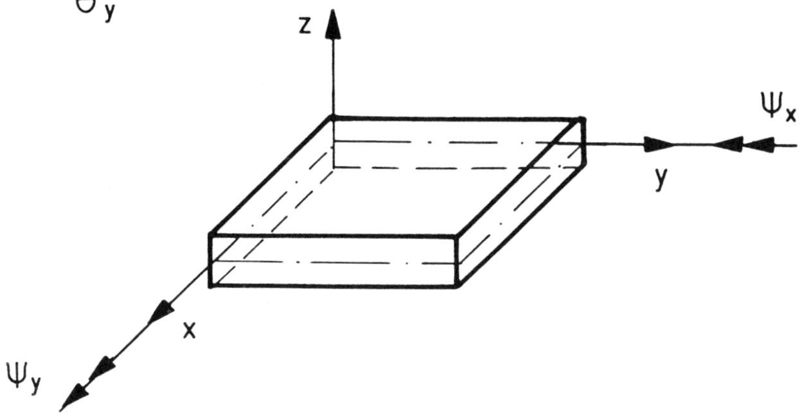

(b) Modified notation used in finite element formulation

Figure 2.1.1 Definitions of rotations

or $\underline{\sigma}_b = \underline{D}_b \, \underline{\varepsilon}_b$

and the shear forces may be expressed as

$$\begin{bmatrix} Q_x \\ Q_y \end{bmatrix} = \begin{bmatrix} D_{55} & D_{45} \\ D_{45} & D_{44} \end{bmatrix} \begin{bmatrix} \gamma_{xz} \\ \gamma_{yz} \end{bmatrix} \qquad (2.1.5)$$

or $\underline{\sigma}_s = \underline{D}_s \, \underline{\varepsilon}_s$

in which

$$\begin{aligned} D_{ij} &= C_{ij} t^3/12 \qquad (1,j=1,2,6) \\ &= k_{ij}^2 \, C_{ij} t \qquad (i,j=4,5) \end{aligned} \qquad (2.1.6)$$

where t is the plate thickness, C_{ij} represent the elastic moduli under generalised plane stress assumptions and k_{ij}^2 are shear correction factors. For non-homogeneous cases, approximate values of the plate rigidities D_{ij} may be evaluated using the expressions given in Chapter 1.

The total potential energy for a typical Mindlin plate is then given as

$$\begin{aligned} \pi(w,\theta_x,\theta_y) = &\frac{1}{2} \iint_A \underline{\varepsilon}_b^T \, \underline{D}_b \, \underline{\varepsilon}_b \, dxdy + \frac{1}{2} \iint_A \underline{\varepsilon}_s^T \, \underline{D}_s \, \underline{\varepsilon}_s \, dxdy \\ &- \iint_A wq \, dxdy \\ &- \int_{S_\sigma} (\bar{M}_n \, \theta_n + \bar{M}_{ns} \, \theta_s + \bar{Q}_n w) ds \end{aligned} \qquad (2.1.7)$$

in which q is the normal distributed load per unit area, $\bar{M}_n$, $\bar{M}_{ns}$ and $\bar{Q}_n$ are the moments and transverse shear forces per unit length of the portion S_σ of the plate boundary S. The terms in (2.1.7) are respectively the bending and transverse shear strain energies and the potential energy associated with the lateral loads and boundary forces.

For the case where $D_{45} = 0$, the shear strain energy expression may be written as

$$\frac{1}{2}\iint_A [D_{55}(\theta_x + w_{,x})^2 + D_{44}(\theta_y + w_{,y})^2]\,dxdy$$

$$= \frac{1}{2}\iint_A \underline{\varepsilon}_s^T\, \underline{D}_s\, \underline{\varepsilon}_s\, dxdy \qquad (2.1.8)$$

which may be regarded as a penalty term which gradually enforces the thin plate conditions

$$\gamma_{xz} = \theta_x + w_{,x} = 0$$

$$\gamma_{yz} = \theta_y + w_{,y} = 0 \qquad (2.1.9)$$

as the plate thickness-to-span ratio is reduced.

2.2 Alternative Formulation

Another popular functional which is used as a basis for constructing finite elements is the modified Hellinger-Reissner functional which has the form

$$\pi_{HR}(w,\theta_x,\theta_y,Q_x,Q_y) = \frac{1}{2}\iint_A \underline{\varepsilon}_b^T\, \underline{D}_b\, \underline{\varepsilon}_b\, dxdy$$

$$+ \iint_A \underline{\sigma}_s^T\, \underline{\varepsilon}_s\, dxdy - \frac{1}{2}\iint_A \sigma_s\, \underline{D}_s^{-1}\, \underline{\sigma}_s\, dxdy$$

$$- \iint_A wq\,dxdy + \int_{S_\sigma} (\bar{M}_n\, \theta_n + \bar{M}_{ns}\, \theta_s + \bar{Q}_n w)\,ds \qquad (2.2.1)$$

2.3 Modified notation

In the following section, in which a finite element formulation is presented, a modified definition of the rotations is adopted in which

$$\begin{bmatrix}\theta_x\\ \theta_y\end{bmatrix} = \begin{bmatrix}-1 & 0\\ 0 & -1\end{bmatrix}\begin{bmatrix}\psi_x\\ \psi_y\end{bmatrix} \qquad (2.3.1)$$

See Figure 2.1.1(b).

3. MINDLIN PLATE ELEMENTS

3.1 Finite element formulation

Finite elements based on Mindlin's assumptions have one important advantage over elements based on classical thin plate theory. Mindlin plate elements require only C(0) continuity of the lateral displacement w and the independent normal rotations ψ_x and ψ_y. However, elements based on classical thin plate Kirchhoff theory require C(1) continuity. In other words, $w_{,x}$ and $w_{,y}$ as well as w should ideally be continuous across element interfaces although this condition is relaxed in non-conforming plate elements. Thus, it would appear that Mindlin plate elements are simpler to formulate and they have the added advantage of being able to model shear-weak as well as shear-stiff plates - if transverse shear effects are present in the plate they are automatically modelled with Mindlin elements.

The displacements and normal rotations at any point (ξ,η) in a typical n-noded element e may be expressed as

$$\begin{bmatrix} w \\ \theta_x \\ \theta_y \end{bmatrix} = \sum_{i=1}^{n} \begin{bmatrix} N_i^e & 0 & 0 \\ 0 & -N_i^e & 0 \\ 0 & 0 & -N_i^e \end{bmatrix} \begin{bmatrix} w_i^e \\ \psi_{xi}^e \\ \psi_{yi}^e \end{bmatrix} \tag{3.1.1}$$

or

$$\underline{u} = \sum_{i=1}^{n} \underline{N}_i^e \, \underline{a}_i^e$$

where the shape functions N_i^e are used to interpolate both the nodal lateral displacements w and the normal rotations ψ_x and ψ_y. The shape functions N_i^e are expressed in terms of the natural element coordinate system (ξ,η).

The curvatures may now be written as

$$\begin{bmatrix} \theta_{x,x} \\ \theta_{y,y} \\ (\theta_{x,y} + \theta_{y,x}) \end{bmatrix} = \sum_{i=1}^{n} \begin{bmatrix} 0 & -N^e_{i,x} & 0 \\ 0 & 0 & -N^e_{i,y} \\ 0 & -N^e_{i,y} & -N^e_{i,x} \end{bmatrix} \begin{bmatrix} w^e_i \\ \psi^e_{xi} \\ \psi^e_{yi} \end{bmatrix} \qquad (3.1.2)$$

or

$$\underline{\varepsilon}_b = \sum_{i=1}^{n} \underline{B}^e_{bi} \, \underline{a}^e_i$$

where $\underline{B}^e_{bi}$ is the curvature-displacement matrix associated with element e.

The shear strains can be written as

$$\begin{bmatrix} w_{,x} + \theta_x \\ w_{,y} + \theta_y \end{bmatrix} = \begin{bmatrix} N^e_{i,x} & -N^e_i & 0 \\ N^e_{i,y} & 0 & -N_i^e \end{bmatrix} \begin{bmatrix} w^e_i \\ \psi^e_{xi} \\ \psi^e_{yi} \end{bmatrix} \qquad (3.1.3)$$

or

$$\underline{\varepsilon}_s = \sum_{i=1}^{n} \underline{B}^e_{si} \, \underline{a}^e_i$$

where $\underline{B}^e_{si}$ is the shear strain-displacement matrix.

Thus, neglecting point loads or couples, the contribution to the Total Potential Energy from element e may be expressed as

$$I^e = \sum_{i=1}^{n} \sum_{j=1}^{n} \Big(\frac{1}{2} \iint [\underline{a}^e_i]^T ([\underline{B}^e_{bi}]^T \underline{D}_b \underline{B}^e_{bj} + [\underline{B}^e_{si}]^T \underline{D}_s \underline{B}^e_{sj}) dxdy) \, \underline{a}^e_j\Big) - \sum_{i=1}^{n} \iint [\underline{a}^e_i]^T [\underline{N}^e_i]^T \underline{q} \, dxdy \qquad (3.1.4)^*$$

---

* $\underline{q} = [q,0,0]^T$.

$$= \sum_{i=1}^{n} \sum_{j=1}^{n} [\underline{a}_i^e]^T (\underline{K}_{bij}^e + \underline{K}_{sij}^e) \underline{a}_j^e$$

$$- \sum_{i=1}^{n} [\underline{a}_i^e]^T \underline{f}_i^e$$

where $\underline{K}_{bij}^e$ and $\underline{K}_{sij}^e$ are the contributions to the submatrix of the element stiffness matrix linking nodes i and j and respectively associated with the bending and shear strain energies.

3.2 Isoparametric representation

In an isoparametric element the coordinates are expressed as

$$\begin{bmatrix} x \\ y \end{bmatrix} = \sum_{i=1}^{n} \begin{bmatrix} N_i^e & 0 \\ 0 & N_i^e \end{bmatrix} \begin{bmatrix} x_i^e \\ y_i^e \end{bmatrix} \tag{3.2.1}$$

where x_i^e and y_i^e are the coordinates of node i.

The Jacobian matrix of the element is defined as

$$\underline{J} = \begin{bmatrix} x_{,\xi} & y_{,\xi} \\ x_{,\eta} & y_{,\eta} \end{bmatrix} \tag{3.2.2}$$

$$\begin{bmatrix} \sum_{i=1}^{n} N_{i,\xi}^e x_i^e & \sum_{i=1}^{n} N_{i,\xi}^e y_i^e \\ \sum_{i=1}^{n} N_{i,\eta}^e x_i^e & \sum_{i=1}^{n} N_{i,\eta}^e y_i^e \end{bmatrix}$$

The inverse of $\underline{J}$ may be expressed as

$$\underline{J}^{-1} = \begin{bmatrix} \xi_{,x} & \eta_{,x} \\ \xi_{,y} & \eta_{,y} \end{bmatrix} \tag{3.2.3}$$

$$= \frac{1}{\det \underline{J}} \begin{bmatrix} y_{,\eta} & -y_{,\xi} \\ -x_{,\eta} & x_{,\xi} \end{bmatrix}$$

where $\det\underline{J}$ is the determinant of the Jacobian matrix.

To evaluate the Cartesian shape function derivatives, the chain rule of differentiation must be used so that

$$N_{i,x} = N_{i,\xi}\,\xi_{,x} + N_{i,\eta}\,\eta_{,x}$$

$$N_{i,y} = N_{i,\xi}\,\xi_{,y} + N_{i,\eta}\,\eta_{,y} \qquad (3.2.4)$$

An element of area may be expressed as

$$dxdy = \det\underline{J}\; d\xi d\eta$$

Thus the contributions to the stiffness matrices and load vector may be written as

$$\underline{K}^e_{bij} = \int_{-1}^{+1}\int_{-1}^{+1} [\underline{B}^e_{bi}]^T\,\underline{D}_b\,\underline{B}^e_{bj}\;\det\underline{J}\;d\xi d\eta$$

$$\underline{K}^e_{sij} = \int_{-1}^{+1}\int_{-1}^{+1} [\underline{B}^e_{si}]^T\,\underline{D}_s\,\underline{B}^e_{sj}\;\det\underline{J}\;d\xi d\eta$$

$$\underline{f}^e_i = \int_{-1}^{+1}\int_{-1}^{+1} [N^e_i]^T\,\underline{q}\;\det\underline{J}\;d\xi d\eta \qquad (3.2.5)$$

4. DESIRABLE CHARACTERISTICS OF A MINDLIN PLATE ELEMENT

4.1 Locking

The requirement of only C(0) continuity of displacements and independent rotations by the Mindlin formulation allows the use of a wide range of interpolation schemes. Serendipity and Lagrangian displacement elements have been widely used with such formulations. However, when exact numerical integration is used with standard Mindlin finite elements, very disappointing results are obtained in applications to thin plates. This phenomenon which is termed locking, is caused by the imposition of the constraints $\gamma_{xz} = \gamma_{yz} = 0$ by the shear strain energy terms in the Total Potential Energy when limiting thin plate situations are approached. The shear strain energy terms may be interpreted as penalty functions which force the shear strains to equal zero as the plate thickness-to-span ratio is reduced. The imposition of these constraints leads to the deterioration of the stiffness matrix and overstiff results are obtained.

4.2 Reduced and selective integration

When a reduced order of numerical integration of the stiffness terms is used on some isoparametric displacement-based Mindlin plate elements improved behaviour is often obtained. However, when this method was first adopted [2] a thorough understanding of the reduced integration concept was lacking and the technique was initially regarded as a 'trick' rather than a legitimate methodology. The use of reduced integration, though successful in some cases, produces the further problem of rank deficiency of the stiffness matrix [3-5].

Selective integration procedures were developed [6] in which a reduced integration rule is used to evaluate the stiffness matrix associated with the troublesome shear strain energy. This is done to alleviate the overconstraining effects of this portion of the stiffness matrix. Full integration is used on the remaining terms in an attempt to retain the required rank of the overall stiffness matrix. Figure 4.2.1 shows the exact, reduced and selective integration rules used for typical Serendipity and Lagrangian Mindlin plate elements.

An example [7] demonstrating locking behaviour is given in Figure 4.2.2. A thin, uniformly loaded, square, clamped plate is analysed for decreasing thickness using a sequence of 8x8 finite element meshes. Results from these Mindlin plate bending analyses using 4, 8, 9, 12 and 16 noded quadrilateral elements with full, reduced and selective integration are compared with the exact classical thin plate solution. Lagrangian elements with selective integration give optimal performance.

For reduced and selective integration schemes, the number of zero energy modes in excess of the three rigid body modes are given in Table 4.2.1. It can be seen that when selective integration is used Serendipity elements suffer no rank deficiency whilst linear Lagrangian elements possess two spurious zero energy modes and higher order elements possess only one spurious zero energy mode. In some cases the zero energy modes are not communicable from element to element. However, the existence of zero energy modes must always be viewed with suspicion.

In mixed methods, based on the modified Hellinger-Reissner functional described earlier, displacement, rotation and shear force representations are used to derive the element stiffness matrix. Because of the form of the functional the shear forces may be interpolated within each element independently. If discontinuous shear force representations are indeed used then their contribution to the element stiffness matrix may be eliminated by static condensation at the element level prior to assembly.

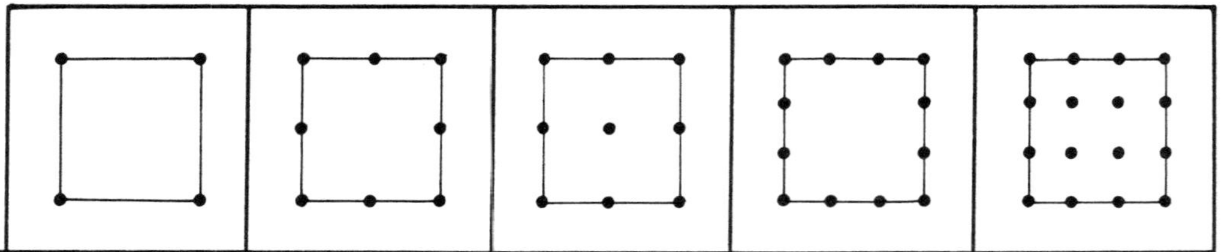

Shape functions Formulation	Linear Langrangian	Quadratic Serendipity	Quadratic Langrangian	Cubic Serendipity	Cubic Lagrangian
Full integration	2 x 2 LF1	3 x 3 SF2	3 x 3 LF2	4 x 4 SF3	4 x 4 LF3
Reduced integration	1 x 1 LR1	2 x 2 SR2	2 x 2 LR2	3 x 3 SR3	3 x 3 LR3
Selective integration	shear 1 x 1 bending 2 x 2 LS1	2 x 2 3 x 3 SS2	2 x 2 3 x 3 LS2	3 x 3 4 x 4 SS3	3 x 3 4 x 4 LS3

Figure 4.2.1 Integration rules for Serendipity and Lagrangian Mindlin plate elements

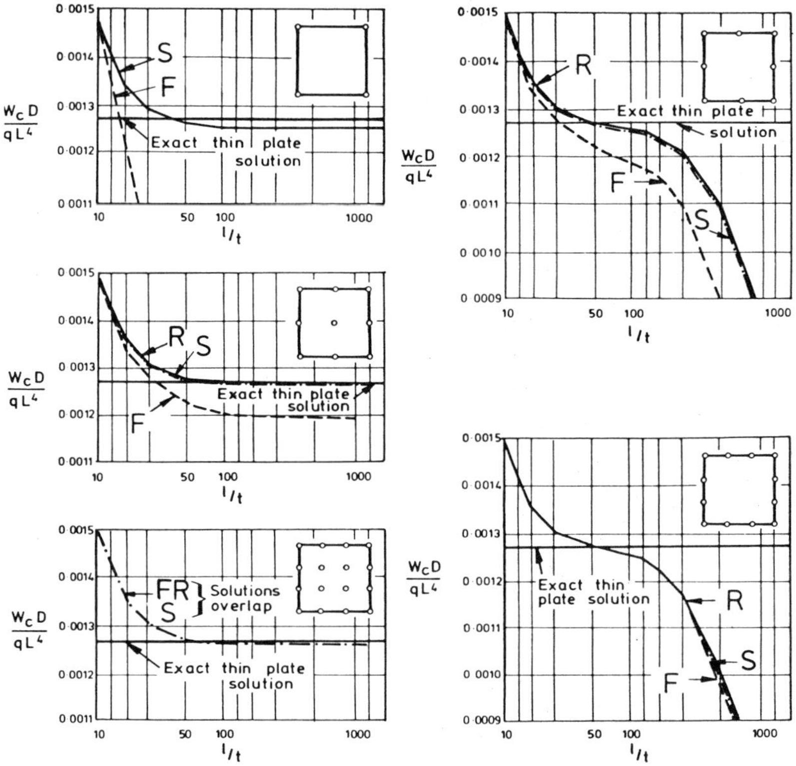

Figure 4.2.2 Locking tests for Lagrangian and Serendipity Mindlin plate elements

Equivalence has been established [8] between mixed finite elements based on the modified Hellinger-Reissner functional and displacement elements developed using selective integration. This elevates the selective integration approach from being regarded as a mere trick to be a legitimate methodology and enables convergence proofs and error estimates developed for the mixed methods to be applied to selectively integrated displacement-based elements.

TABLE 4.2.1 Number of zero eigenvalues for single unconstrained Mindlin plate element stiffness matrix using various integration rules

No. of nodes per element	Number of zero eigenvalues for stiffness matrix of a single unconstrained element		
	Full	Selective	Reduced
4	3	5	7
8	3	3	4
9	3	4	7
12	3	3	3
16	3	4	7

A very useful outcome of the equivalence between selectively integrated displacement-based Mindlin plate elements and those based on mixed formulations is that this equivalence justifies the use of local stress extrapolation techniques over individual elements. In these techniques the values of the shear forces at the integration points, which are considered to be accurate, are therefore extrapolated with each element separately. Nodal values which are of prime interest can then be obtained by averaging the extrapolated contributions from adjacent elements sharing the same node. It can be shown that Gauss point stresses in certain cases may be interpreted as being the unknown parameters in the mixed formulation and, furthermore, that the stress distribution obtained by extrapolation is the same distribution used to represent the shear force function in the mixed formulation.

Recently, displacement elements based on the Kirchhoff-mode criteria have been introduced [9,10]. In these elements, the interpolation functions have to be chosen to allow shear angles to obey the Kirchhoff constraints (2.1.9) throughout the element domain or at certain key positions. Different interpolation patterns have therefore to be used for the displacement and rotations. These elements, though considered effective, are not convenient for practical implementation. Special procedures have to be adopted to retain the simplicity of the resulting element. Based on these considerations, some highly effective quadrilateral elements have been developed [9-17].

Based on all of the preceding considerations it is possible to summarise the main features of the ideal Mindlin plate element,

a) The element should not lock in thin plate situations.

b) The element should not contain any spurious zero energy modes.

c) The element should satisfy the usual isotropy and convergence requirements [18].

d) The formulation should not be based on numerically adjusted factors.

e) The element should be capable of providing accurate displacements, bending moments and shear forces and be relatively insensitive to element distortions.

f) The element should be easy to implement and use.

In the next section the heterosis element developed by Hughes [9] is presented. This element has many of the features of the ideal Mindlin plate element. However, it should be mentioned that there has been much research activity in recent years directed towards the development of the ideal Mindlin plate and related degenerate shell elements. Some recent contenders are those developed by Hughes [9,12], Bathe [14], Crisfield [15], Belytschko [17], Park [16] and Macneal [11].

5. HETEROSIS MINDLIN PLATE ELEMENT

5.1 Introduction

The 9-node Heterosis element was developed by Hughes et al. [19]. The basic idea which led to its development is that the spurious zero energy mode possessed by LS2 (see Figure 4.2.1 for notation) is eliminated if the 8-node Serendipity shape functions are used for the transverse displacement w. The Heterosis element is thus formulated using 9-node Lagrangian shape functions for rotations (ψ_x and ψ_y) and 8-node Serendipity shape functions for lateral displacement w. The Heterosis element, which has improved characteristics, has given a consistently better performance than the 'parent' elements in a wide range of problems [9]. When used with selective integration, it does not 'lock' in very thin plate situations and unlike LS2, it has correct rank. Results of an aspect ratio study given in Figure 5.1.1 support the above conclusion.

5.2 Hierarchical formulation of the Heterosis element

Since it is useful for research purposes to implement the 8-node Serendipity and the 9-node Lagrangian elements as well as the Heterosis element in MINDLIN,it is more convenient to adopt a hierarchical formulation to represent all degrees of freedom. Thus, for a typical element e the following Serendipity shape functions are adopted for nodes 1 to 8

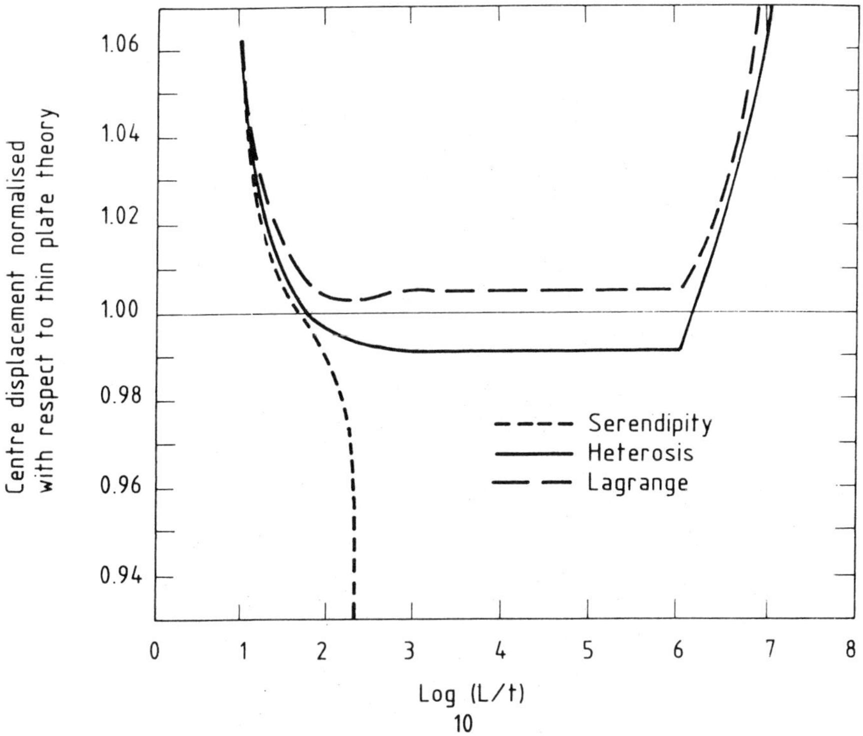

Figure 5.1.1 Aspect ratio study for a uniformly simply supported square plate

so that

for corner node i = 1,3,5,7

$$N_i^e = \frac{1}{4}(1+\xi\xi_i)(1+\eta\eta_i)(\xi\xi_i+\eta\eta_i-1) \qquad (5.2.1)$$

for midside nodes i=2,4,6,8

$$N_i^e = \frac{\xi_i^{\,2}}{2}(1+\xi\xi_i)(1-\eta^2) + \frac{\eta_i^{\,2}}{2}(1+\eta\eta_i)(1-\xi^2) \qquad (5.2.2)$$

and for node 9 the bubble function is used

$$N_9^e\,(\xi,\eta) = (1-\xi^2)(1-\eta^2) \qquad (5.2.3)$$

see Figure 5.2.1. Thus, the hierarchical degrees of freedom at node 9 are perturbations from the associated Serendipity

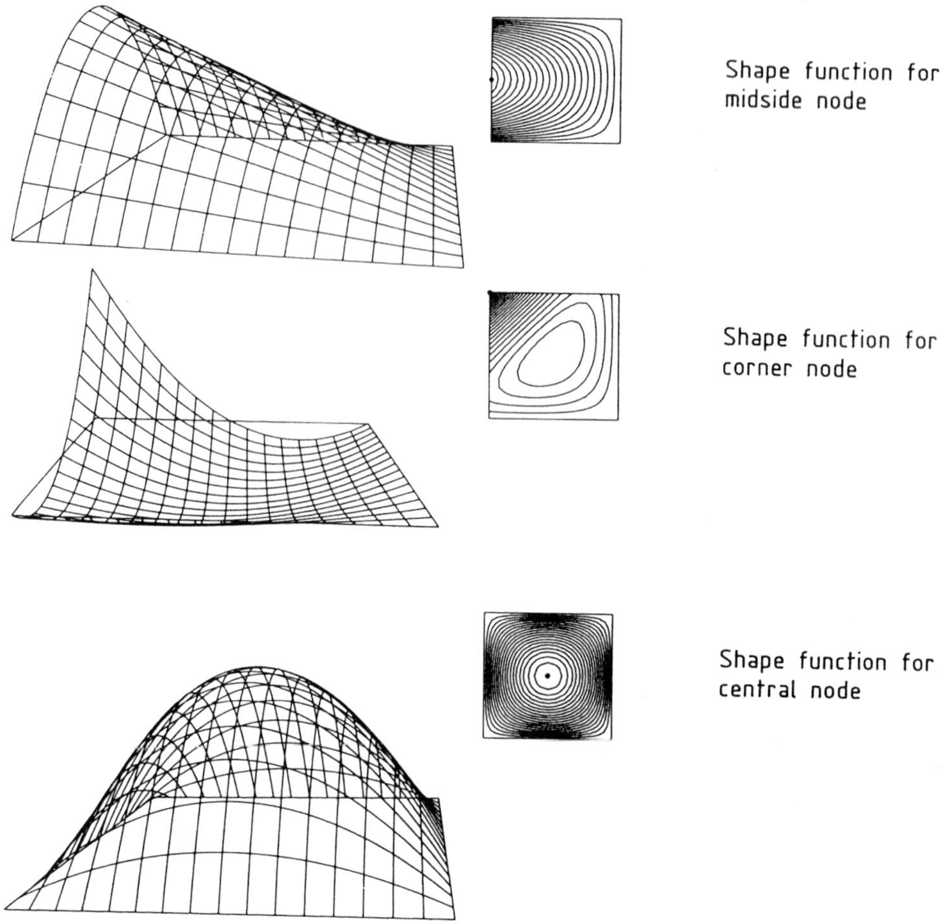

Figure 5.2.1 Hierarchical shape functions for the hetereosis element

interpolants. The hierarchical formulation can be used for geometrical representation as well as for interpolating displacements.

In MINDLIN, any of the 8-node, 9-node or Heterosis elements can be chosen where:

(a) To obtain the 8-node Serendipity representation, all degrees of freedom at node 9 are constrained to zero.

(b) To obtain an element equivalent to the 9-node Lagrangian representation, all degrees of freedom at node 9 are left free. The total displacements at node 9 are calculated using the expression

$$\underline{u} = \sum_{i=1}^{8} N_i^e\,(0,\,0)\,\underline{a}_i^e + \underline{a}_9^e \qquad (5.2.4)$$

(c) To obtain the Heterosis element representation, only the hierarchical lateral displacement at node 9 (w_9^e) is restrained to zero and again (5.2.4) is used to interpret the displacements at node 9.

5.3 Evaluation of stress resultants

In MINDLIN, the bending moments are evaluated directly using the bending moment-displacement relations at the 3x3 Gauss-Legendre sampling points and at the nodal points. The shear forces are also evaluated directly using the shear force-displacement relations at the 2x2 Gauss-Legendre sampling points and they are then extrapolated to the corner nodes (1,3,5,7) using the expression [20]

$$\begin{bmatrix} Q_1 \\ Q_3 \\ Q_5 \\ Q_7 \end{bmatrix} = \begin{bmatrix} (1+\frac{\sqrt{3}}{2}) & -\frac{1}{2} & (1-\frac{\sqrt{3}}{2}) & -\frac{1}{2} \\ -\frac{1}{2} & (1+\frac{\sqrt{3}}{2}) & -\frac{1}{2} & (1-\frac{\sqrt{3}}{2}) \\ (1-\frac{\sqrt{3}}{2}) & -\frac{1}{2} & (1+\frac{\sqrt{3}}{2}) & -\frac{1}{2} \\ -\frac{1}{2} & (1-\frac{\sqrt{3}}{2}) & -\frac{1}{2} & (1+\frac{\sqrt{3}}{2}) \end{bmatrix} \begin{bmatrix} Q_I \\ Q_{II} \\ Q_{III} \\ Q_{IV} \end{bmatrix} \qquad (5.3.1)$$

where, for example, Q_1 is the shear force at node 1 and Q_I is the shear force evaluated at (ξ_I, η_I) the first Gauss-Legendre point. See Figure 5.3.1.

Q takes the value Q_x or Q_y. Note that (5.3.1) is a bilinear extrapolation which can be justified, as mentioned earlier, on the grounds of the equivalence between elements obtained using a mixed method and elements obtained using selectively integrated displacement-based elements [8].

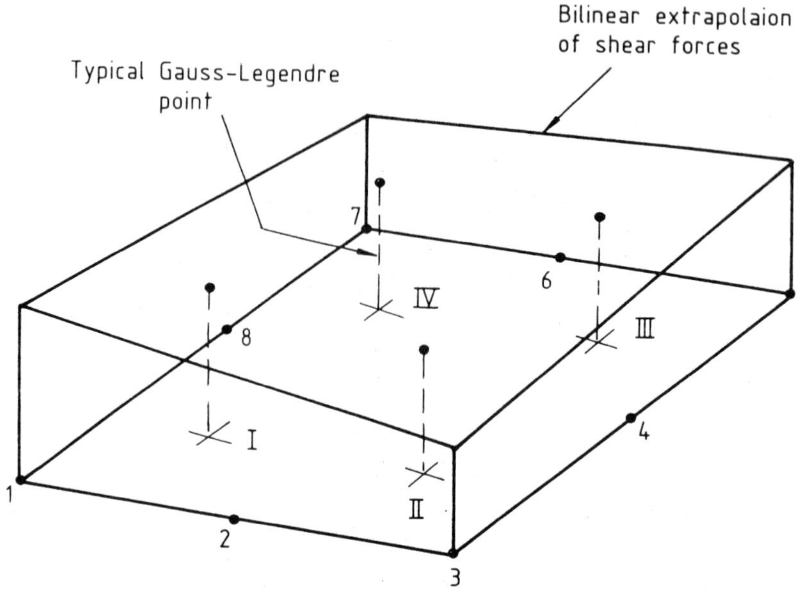

Figure 5.3.1 Bilinear extrapolation of plate shear forces at 2 x 2 Gauss-Legendre points to nodal points

In the present case, the Heterosis, selectively integrated Mindlin plate element has an equivalent mixed element with the same nodal displacement degrees of freedom but with bilinear Q_x and Q_y fields which are discontinuous over element interfaces. The unknown parameters defining Q_x and Q_y may be eliminated prior to assembly in the mixed element and the resulting stiffness matrix is identical to that obtained using the displacement approach.

6. PROGRAM MINDLIN

6.1 Introduction

This section describes MINDLIN which is a program for the linear elastic analysis of plates using Mindlin plate bending elements. For convenience, all element stiffness matrices are obtained using selective integration. Attention will be focussed on the heterosis element as it is the most successful and has no major defects unlike the Serendipity element which locks in thin plate situations or the Lagrangian element which has a transmittable mechanism.

Program MINDLIN consists of 5 main modules

(a) input module,

(b) stiffness module,

(c) load module,

(d) solution module, and

(e) output and stress resultant module.

The main variables used in MINDLIN will now be described in a glossary.

6.2 Glossary of variable names

Main arrays

Array name	Description
ASDIS(MTOTV)	nodal unknowns written in the sequence w, ψ_x, ψ_y
COORD(MPOIN,2)	x and y coordinates
ELOAD(MELEM.27)	element load vectors
ESTIF(27,27)	element stiffness matrix
EQRHS(MBUFA)	reduced RHS in FRONT
EQUAT(MFRON,MBUFA)	reduced equations in FRONT
FIXED(MTOTV)	prescribed variables
IFFIX(MTOTV)	fixity code for each variable 0 - free 1 - prescribed
GLOAD(MFRON)	global load vector stored in FRONT
GSTIF(MSTIF)	upper triangular part of global stiffness matrix stored in FRONT
KOUNT(MPOIN)	number of elements attached to a node
LNODS(MELEM,9)	element nodal connectivities
LOCEL(27)	vector which locates the global position of each element variable in FRONT

Array name	Description
MATNO(MELEM)	material set numbers for each element
NACVA(MFRON)	vector of active variables in FRONT
NAMEV(MBUFA)	variable used in FRONT
NDEST(27)	destination vector in FRONT
NOFIX(MVFIX)	nodes with at least one prescribed variable
NPIVO(MBUFA)	pivot positions in FRONT
POSGP(3)	Gauss-Legendre sampling positions
PRESC(MVFIX,3)	prescribed variables
PROPS(MMATS,8)	properties for each material set
STRND(MPOIN,5)	averaged nodal stress resultants
TREAC(MVFIX,3)	reactions at prescribed nodes
VECRV(MFRON)	vector of running variables in which the evaluated degrees of freedom are stored
VIXED(MTOTV)	reactions
WEIGP(3)	Gauss-Legendre weights

Main variables

Variable	Description
ICASE	current load case
MBUFA	maximum buffer size
MELEM	maximum number of elements
MEVAB	=27, maximum number of element variables
MFRON	maximum frontwidth
MMATS	maximum number of material sets
MPOIN	maximum number of nodal points
MTOTV	maximum total number of variables
MVFIX	maximum number of prescribed nodes
NCASE	number of load cases
NDIME	=2, number of dimensions
NDOFN	=3, number of degrees of freedom per node
NELEM	number of elements
NEVAB	=27, number of element variables
NGAUB	=3, Gauss-Legendre rule to evaluate the bending stiffness matrix
NGAUS	=2, Gauss-Legendre rule to evaluate the shear stiffness matrix
NMATS	number of material sets
NNODE	=9, number of nodes per element
NPOIN	number of nodal points
NSTRE	=5, number of stress resultants
NTOTV	total number of variables
NVFIX	number of prescribed nodes

6.3 Main Segment

The main segment which controls the calling of the major subroutines is now described. The Nassi-Schneiderman (NS) chart for this routine is shown in Figure 6.3.1.

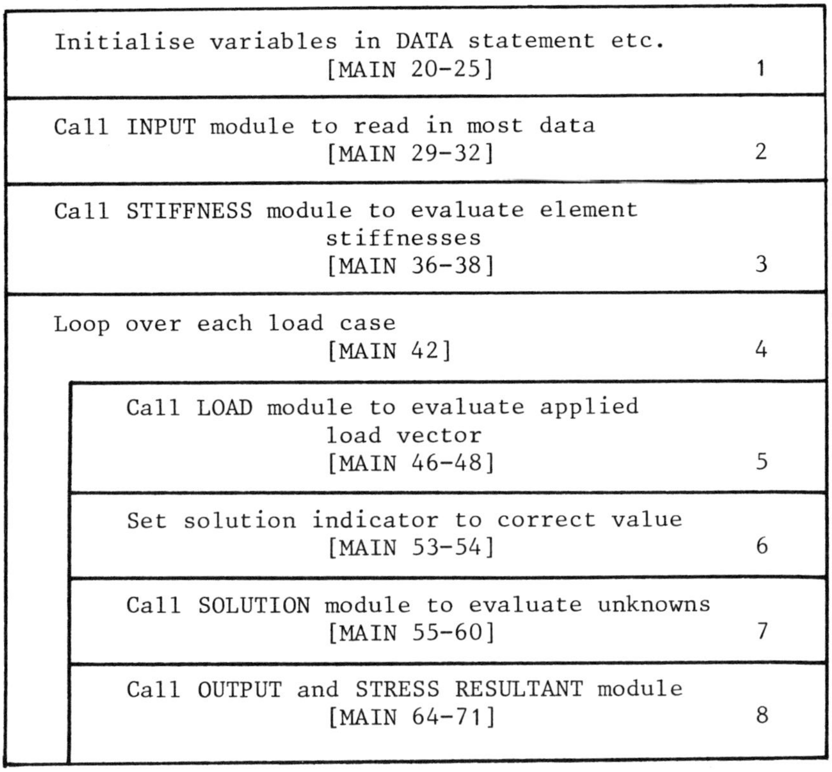

Figure 6.3.1 NS chart for master routine of MINDLIN

```
C***********************************************************************MAIN   1
C                                                                       MAIN   2
C***  ELASTIC ANALYSIS OF MINDLIN PLATES USING THE                      MAIN   3
C***  HETEROSIS ISOPARAMETRIC QUADRILATERAL MINDLIN PLATE ELEMENT       MAIN   4
C                                                                       MAIN   5
C***********************************************************************MAIN   6
      IMPLICIT REAL*8(A-H,O-Z)                                          MAIN   7
      DIMENSION ASDIS(363),COORD(121,2),ELOAD(25,27),ESTIF(27,27),      MAIN   8
     .          EQRHS(10),EQUAT(51,10),FIXED(363),IFFIX(363),           MAIN   9
     .          GLOAD(51),GSTIF(1326),KOUNT(121),LNODS(25,9),           MAIN  10
     .          LOCEL(27),MATNO(25),NACVA(51),NAMEV(10),NDEST(27),      MAIN  11
     .          NOFIX(65),NPIVO(10),PRESC(65,3),PROPS(10,8),            MAIN  12
     .          STRND(121,5),TREAC(65,3),VECRV(51),VIXED(363)           MAIN  13
C                                                                       MAIN  14
```

```
C***********************************************************************
C
C                           DYNAMIC DIMENSIONING
C
C***********************************************************************
      DATA MBUFA/10/,MELEM/25/,MFRON/51/,MMATS/10/,MPOIN/121/,
     .     NDOFN/3/,MVFIX/65/,NDIME/2/,NEVAB/27/,NGAUB/3/,
     .     NGAUS/2/,NNODE/9/
      MSTIF=(MFRON*MFRON-MFRON)/2.0+MFRON
      MTOTV=MPOIN*NDOFN
      MEVAB=NDOFN*9
C
C*** CALL THE SUBROUTINE WHICH READS MOST OF THE PROBLEM DATA
C
      CALL        INPUT      (COORD,FIXED,IFFIX,LNODS,MATNO,MELEM,
     .                        MMATS,MPOIN,MTOTV,MVFIX,NCASE,NDIME,
     .                        NDOFN,NELEM,NMATS,NNODE,NOFIX,NPOIN,
     .                        NTOTV,NVFIX,PRESC,PROPS)
C
C*** COMPUTE THE ELEMENT STIFFNESSES
C
      CALL        STIFPB     (COORD,LNODS,MATNO,MELEM,MEVAB,MMATS,
     .                        MPOIN,NELEM,NEVAB,NGAUB,NGAUS,NNODE,
     .                        PROPS)
C
C*** BEGIN LOAD CASE LOOP
C
      DO 10 ICASE=1,NCASE
C
C*** COMPUTE LOAD AFTER READING RELEVANT EXTRA DATA
C
      CALL        LOADPB     (COORD,ELOAD,ICASE,LNODS,MATNO,MELEM,
     .                        MMATS,MPOIN,NELEM,NEVAB,NGAUB,NNODE,
     .                        NPOIN,PROPS)
C
C
C*** SOLVE EQUATIONS
C
      KRESL=1
      IF(ICASE.GT.1) KRESL=2
      CALL        FRONT      (ASDIS,ELOAD,EQRHS,EQUAT,ESTIF,FIXED,
     .                        IFFIX,GLOAD,GSTIF,KRESL,
     .                        LNODS,LOCEL,MBUFA,MELEM,MEVAB,MFRON,
     .                        MSTIF,MTOTV,MVFIX,NACVA,NAMEV,NDEST,
     .                        NDOFN,NELEM,NEVAB,NNODE,NOFIX,NPIVO,
     .                        NPOIN,NTOTV,TREAC,VECRV,VIXED)
C
C*** OUTPUT DISPLACEMENTS
C
      CALL        OUTDIS     (ASDIS,MTOTV,MVFIX,NOFIX,
     .                        NPOIN,NVFIX,TREAC)
C
C*** OUTPUT STRESS RESULTANTS
C
      CALL        OUTSTR     (ASDIS,COORD,KOUNT,LNODS,MATNO,MELEM,
     .                        MMATS,MPOIN,MTOTV,NDOFN,NELEM,NGAUB,
     .                        NGAUS,NPOIN,PROPS,STRND)
   10 CONTINUE
C
      STOP
      END
```

6.4 Input module

The input module is now presented.

6.4.1 Subroutine INPUT

The main function of INPUT is to read in most of the input data. The NS chart for subroutine INPUT is given in Figure 6.4.1 and then INPUT as listed.

Read and write problem title [INPU 15-18]	1
Read and write control parameters [22-27]	2
Read and write element connections and property numbers [INPU 31-36]	3
Initialise array COORD by calling RAZERO [INPU 40]	4
Read nodal coordinates [INPU 44-48]	5
Interpolate coordinates of midside nodes for straight sides by calling NODEXY [INPU 52-53]	6
Write all nodal coordinates [INPU 54-56]	7
Initialise vector IFFIX by calling IVZERO [INPU 60]	8
Read, store and write fixity conditions [INPU 61-78]	9
Read and write element properties [INPU 82-89]	10

Figure 6.4.1 NS chart for INPUT

```
      SUBROUTINE INPUT     (COORD,FIXED,IFFIX,LNODS,MATNO,MELEM,
     .                      MMATS,MPOIN,MTOTV,MVFIX,NCASE,NDIME,
     .                      NDOFN,NELEM,NMATS,NNODE,NOFIX,NPOIN,
     .                      NTOTV,NVFIX,PRESC,PROPS)
C***********************************************************************
C
C***  INPUT MAIN DATA
C
C***********************************************************************
      IMPLICIT REAL*8(A-H,O-Z)
      DIMENSION COORD(MPOIN,NDIME),FIXED(MTOTV),
     .          IFFIX(MTOTV),LNODS(MELEM,9),
     .          MATNO(MELEM),NOFIX(MVFIX),PRESC(MVFIX,NDOFN),
     .          PROPS(MMATS,8),TITLE(20)
      READ(5,920)  TITLE
      WRITE(6,921) TITLE
  920 FORMAT(20A4)
  921 FORMAT(/////,1X,20A4)
C
C*** READ THE FIRST DATA CARD, AND ECHO IT IMMEDIATELY.
C
      READ(5,900) NPOIN,NELEM,NVFIX,NCASE,NMATS
  900 FORMAT(14I5)
      NTOTV=NPOIN*NDOFN
      WRITE(6,901)NPOIN,NELEM,NVFIX,NCASE,NMATS
  901 FORMAT(//8H NPOIN =,I4,4X,8H NELEM =,I4,4X,8H NVFIX =,I4,4X,
     .8H NCASE =,I4,4X,8H NMATS =,I4)
C
C*** READ THE ELEMENT NODAL CONNECTIONS, AND THE PROPERTY NUMBERS.
C
      WRITE(6,902)
  902 FORMAT(//8H ELEMENT,3X,8HPROPERTY,6X,12HNODE NUMBERS)
      DO 2 IELEM=1,NELEM
      READ(5,900) NUMEL,MATNO(NUMEL),(LNODS(NUMEL,INODE),INODE=1,NNODE)
    2 WRITE(6,903) NUMEL,MATNO(NUMEL),(LNODS(NUMEL,INODE),INODE=1,NNODE)
  903 FORMAT(1X,I5,I9,6X,9I5)
C
C*** ZERO ALL THE NODAL COORDINATES, PRIOR TO READING SOME OF THEM.
C
      CALL RAZERO(COORD,MPOIN,2)
C
C*** READ SOME NODAL COORDINATES, FINISHING WITH THE LAST NODE OF ALL.
C
      WRITE(6,904)
  904 FORMAT(//5H NODE,7X,1HX,9X,1HY)
    4 READ(5,905) IPOIN,(COORD(IPOIN,IDIME),IDIME=1,NDIME)
  905 FORMAT(I5,6F10.5)
      IF(IPOIN.NE.NPOIN) GO TO 4
C
C*** INTERPOLATE COORDINATES OF MID-SIDE NODES
C
      CALL       NODEXY    (COORD,LNODS,MELEM,MPOIN,NDIME,NELEM,
                            NNODE)
      DO 6 IPOIN=1,NPOIN
    6 WRITE(6,906) IPOIN,(COORD(IPOIN,IDIME),IDIME=1,NDIME)
  906 FORMAT(1X,I4,2X,6F10.5)
C
C*** READ THE FIXED VALUES AND INTERPRET FIXITY CODES IN VECTOR FORM
C
      CALL IVZERO(IFFIX,NTOTV)
      WRITE(6,907)
  907 FORMAT(//5H NODE,6X,4HCODE,6X,12HFIXED VALUES)
      DO 10 IVFIX=1,NVFIX
      READ(5,908) NOFIX(IVFIX),IFPRE,
     . (PRESC(IVFIX,IDOFN),IDOFN=1,NDOFN)
      WRITE(6,909) NOFIX(IVFIX),IFPRE,
     . (PRESC(IVFIX,IDOFN),IDOFN=1,NDOFN)
```

```
      NLOCA=(NOFIX(IVFIX)-1)*NDOFN
      IFDOF=10**(NDOFN-1)
      DO 10 IDOFN=1,NDOFN
      NGASH=NLOCA+IDOFN
      IF(IFPRE.LT.IFDOF) GO TO 10
      IFFIX(NGASH)=1
      IFPRE=IFPRE-IFDOF
    8 FIXED(NGASH)=PRESC(IVFIX,IDOFN)
   10 IFDOF=IFDOF/10
  908 FORMAT(1X,I4,I5,5F10.6)
  909 FORMAT(1X,I4,5X,I5,6X,5F10.6)
C
C*** READ THE AVAILABLE SELECTION OF ELEMENT PROPERTIES.
C
      WRITE(6,910)
  910 FORMAT(//7H NUMBER,4X,18HELEMENT PROPERTIES)
      DO 12 IMATS=1,NMATS
      READ(5,900) NUMAT
      READ(5,911) (PROPS(NUMAT,IPROP),IPROP=1,8)
  911 FORMAT(8F10.5)
   12 WRITE(6,912) NUMAT,(PROPS(NUMAT,IPROP),IPROP=1,8)
  912 FORMAT(1X,I4,3X,8E14.6)
      RETURN
      END
```

6.4.2 Subroutines NODEXY, RAZERO and IVZERO

The auxiliary routines NODEXY, RAZERO and IVZERO which are called by INPUT are now listed.

```
      SUBROUTINE NODEXY    (COORD,LNODS,MELEM,MPOIN,NDIME,NELEM,
     .                     NNODE)
C**********************************************************************
C
C***  INTERPOLATES MIDSIDE NODE COORDINATES FOR 8-NODED ELEMENTS
C***  INTERPOLATES CENTRAL AND MIDSIDE NODE COORDINATES FOR
C***  9-NODE ELEMENTS PROVIDED THAT THE SIDES ARE STRAIGHT
C
C**********************************************************************
      IMPLICIT REAL*8(A-H,O-Z)
      DIMENSION COORD(MPOIN,2),LNODS(MELEM,9)
      IF(NNODE.EQ.4) GO TO 50
C
C*** LOOP OVER EACH ELEMENT
C
      DO 40 IELEM=1,NELEM
C
C*** LOOP OVER EACH ELEMENT EDGE
C
      NNOD1=NNODE
      IF(NNODE.EQ.8) NNOD1=9
      DO 20 INODE=1,NNOD1,2
      IF(INODE.EQ.9.AND.NNODE.EQ.8) GO TO 40
      IF(INODE.EQ.9) GO TO 30
C
C*** COMPUTE THE NODE NUMBER OF THE FIRST NODE
C
      NODST=LNODS(IELEM,INODE)
      IGASH=INODE+2
      IF(IGASH.GE.NNODE) IGASH=1
C
C*** COMPUTE THE NODE NUMBER OF THE LAST NODE
C
```

```
      NODFN=LNODS(IELEM,IGASH)
      MIDPT=INODE+1
C
C*** COMPUTE THE NODE NUMBER OF THE INTERMEDIATE NODE
C
      NODMD=LNODS(IELEM,MIDPT)
      TOTAL=ABS(COORD(NODMD,1))+ABS(COORD(NODMD,2))
C
C*** IF THE COORDINATES OF THE INTERMEDIATE NODE ARE BOTH ZERO
C    INTERPOLATE BY A STRAIGHT LINE
C
      IF(TOTAL.GT.0.0) GO TO 20
      KOUNT=1
   10 COORD(NODMD,KOUNT)=(COORD(NODST,KOUNT)+COORD(NODFN,KOUNT))/2.0
      KOUNT=KOUNT+1
      IF(KOUNT.EQ.2) GO TO 10
   20 CONTINUE
   30 LNODE=LNODS(IELEM,INODE)
   40 CONTINUE
   50 CONTINUE
      RETURN
      END
```

```
      SUBROUTINE IVZERO(IVECT,NSIZE)
      IMPLICIT REAL*8(A-H,O-Z)
      DIMENSION IVECT(NSIZE)
C
C*** ZERO INTEGER VECTOR
C
      DO 5 ISIZE=1,NSIZE
      IVECT(ISIZE)=0
 5    CONTINUE
      RETURN
      END
```

```
      SUBROUTINE RAZERO(RVECT,NSIZE,MSIZE)
      IMPLICIT REAL*8(A-H,O-Z)
      DIMENSION RVECT(NSIZE,MSIZE)
C
C*** ZERO REAL ARRAY
C
      DO 5 ISIZE=1,NSIZE
      DO 5 JSIZE=1,MSIZE
      RVECT(ISIZE,JSIZE)=0.0
 5    CONTINUE
      RETURN
      END
```

6.5 Stiffness module

6.5.1 Subroutine STIFPB

A NS chart for STIFPB is listed in Figure 6.5.1. The expanded chart for the section dealing with the numerical integration of bending stiffness matrix $\underline{K}_b$ is given in Figure 6.5.2. The numerical integration of $\underline{K}_s$ is performed in a similar manner.

```
      SUBROUTINE STIFPB    (COORD,LNODS,MATNO,MELEM,
     .                      MEVAB,MMATS,MPOIN,NELEM,
     .                      NEVAB,NGAUB,NGAUS,NNODE,PROPS)
C**********************************************************************
C
C***   EVALUATE STIFFNESS MATRICES FOR ELASTIC
C***   MINDLIN PLATE ELEMENTS
C
C**********************************************************************
      IMPLICIT REAL*8(A-H,O-Z)
      DIMENSION
     .          CARTD(2,9),COORD(MPOIN,2),
     .          DERIV(2,9),ELCOD(2,9),
     .          ESTIF(27,27),GPCOD(2,9),LNODS(MELEM,9),
     .          MATNO(MELEM),POSGP(3),PROPS(MMATS,8),SHAPE(9),STRES(5),
     .          WEIGP(3),
     .          DFLEX(3,3),DSHER(2,2),BFLEI(3,3),BFLEJ(3,3),
     .          BSHEI(2,3),BSHEJ(2,3)
      REWIND 1
C
C*** LOOP OVER EACH ELEMENT
C
      DO 80 IELEM=1,NELEM
      LPROP=MATNO(IELEM)
C
C*** EVALUATE THE COORDINATES OF THE ELEMENT NODAL POINTS
C
      DO 10 INODE=1,NNODE
      LNODE=LNODS(IELEM,INODE)
      LNODE=IABS(LNODE)
      DO 10 IDIME=1,2
   10 ELCOD(IDIME,INODE)=COORD(LNODE,IDIME)
C
C*** INITIALIZE THE ELEMENT STIFFNESS MATRIX
C
      DO 20 IEVAB=1,NEVAB
      DO 20 JEVAB=1,NEVAB
   20 ESTIF(IEVAB,JEVAB)=0.0
C
C*** EVALUATE PART OF STIFFNESS MATRIX
C    ASSOCIATED WITH BENDING DEFORMATION
C
      KGASP=0
C
C*** ENTER LOOPS FOR AREA NUMERICAL INTEGRATION
C
C
C*** SET UP GAUSSIAN INTEGRATION CONSTANTS
C
      CALL      GAUSSQ     (NGAUB,POSGP,WEIGP)
      DO 40 IGAUS=1,NGAUB
      EXISP=POSGP(IGAUS)
      DO 40 JGAUS=1,NGAUB
      KGASP=KGASP+1
      ETASP=POSGP(JGAUS)
C
C*** EVALUATE THE SHAPE FUNCTIONS,ELEMENTAL AREA,ETC
C
      CALL      SFR2       (DERIV,ETASP,EXISP,NNODE,SHAPE)
      CALL      JACOB2     (CARTD,DERIV,DJACB,ELCOD,GPCOD,IELEM,
     .                      KGASP,NNODE,SHAPE)
      DAREA=DJACB*WEIGP(IGAUS)*WEIGP(JGAUS)
C
C*** EVALUATE THE B AND DB MATRICES
C
      CALL      MODPB      (DFLEX,DSHER,LPROP,MMATS,PROPS,
     .                         1,     0)
```

```
C
C
C*** CALCULATE THE ELEMENT STIFFNESSES
C
      DO 30 INODE=1,NNODE
      CALL      BMATPB    (BFLEI,BSHEI,CARTD,INODE,SHAPE,
     .                        1,    0)
      DO 30 JNODE=INODE,NNODE
      CALL      BMATPB    (BFLEJ,BSHEJ,CARTD,JNODE,SHAPE,
     .                        1,    0)
   30 CALL      SUBPB     (BFLEI,BFLEJ,DAREA,DFLEX,ESTIF,INODE,
     .                    JNODE,    3,    3,    3)
   40 CONTINUE
C
C*** EVALUATE PART OF STIFFNESS MATRIX
C    ASSOCIATED WITH SHEAR DEFORMATION
C
      KGASP=0
C
C*** ENTER LOOPS FOR AREA INTEGRATION
C
C
C*** SET UP GAUSSIAN INTEGRATION CONSTANTS
C
      CALL      GAUSSQ      (NGAUS,POSGP,WEIGP)
      DO 60 IGAUS=1,NGAUS
      EXISP=POSGP(IGAUS)
      DO 60 JGAUS=1,NGAUS
      KGASP=KGASP+1
      ETASP=POSGP(JGAUS)
C
C*** EVALUATE THE SHAPE FUNCTIONS,ELEMENTAL AREA,ETC
C
      CALL      SFR2       (DERIV,ETASP,EXISP,NNODE,SHAPE)
      CALL      JACOB2     (CARTD,DERIV,DJACB,ELCOD,GPCOD,IELEM,
     .                      KGASP,NNODE,SHAPE)
      DAREA=DJACB*WEIGP(IGAUS)*WEIGP(JGAUS)
      CALL MODPB(DFLEX,DSHER,LPROP,MMATS,PROPS,0,1)
      DO 50 INODE=1,NNODE
      CALL BMATPB(BFLEI,BSHEI,CARTD,INODE,SHAPE,
     .           0,1)
      DO 50 JNODE=INODE,NNODE
      CALL BMATPB(BFLEJ,BSHEJ,CARTD,JNODE,SHAPE,
     .           0,1)
      CALL SUBPB(BSHEI,BSHEJ,DAREA,DSHER,ESTIF,INODE,
     .          JNODE,3,2,3)
   50 CONTINUE
C
C
   60 CONTINUE
C
C*** CONSTRUCT THE LOWER TRIANGLE OF THE STIFFNESS MATRIX
C
      DO 70 IEVAB=1,NEVAB
      DO 70 JEVAB=IEVAB,NEVAB
   70 ESTIF(JEVAB,IEVAB)=ESTIF(IEVAB,JEVAB)
C
C*** STORE THE STIFFNESS MATRIX,STRESS MATRIX AND SAMPLING POINT
C    COORDINATES FOR EACH ELEMENT ON DISC FILE
C
      WRITE(1) ESTIF
   80 CONTINUE
      RETURN
      END
```

Rewind file 1 ready for writing stiffnesses
[STIF 19] 1

Loop over each element
[STIF 23] 2

Extract element material set number and nodal coordinates
[STIF 24-32] 3

Initialise stiffness matrix array ESTIF and integration point counter
[STIF 36-43] 4

Call GAUSSQ to extract sampling positions and weights for 3-point Gauss-Legendre rule used in the numerical integration of the bending stiffness matrix
[STIF 50] 5

Loop over each Gauss point in ξ direction
[STIF 51] 6

Extract ξ sampling point
[STIF 52] 7

Loop over each Gauss point in η direction
[STIF 53] 8

Extract η sampling point and increment Gauss point counter by 1
[STIF 54-55] 9

Accumulate contributions from $\underline{K}_b$ to the upper triangular part of $\underline{K}$
[STIF 59-80] 10

A

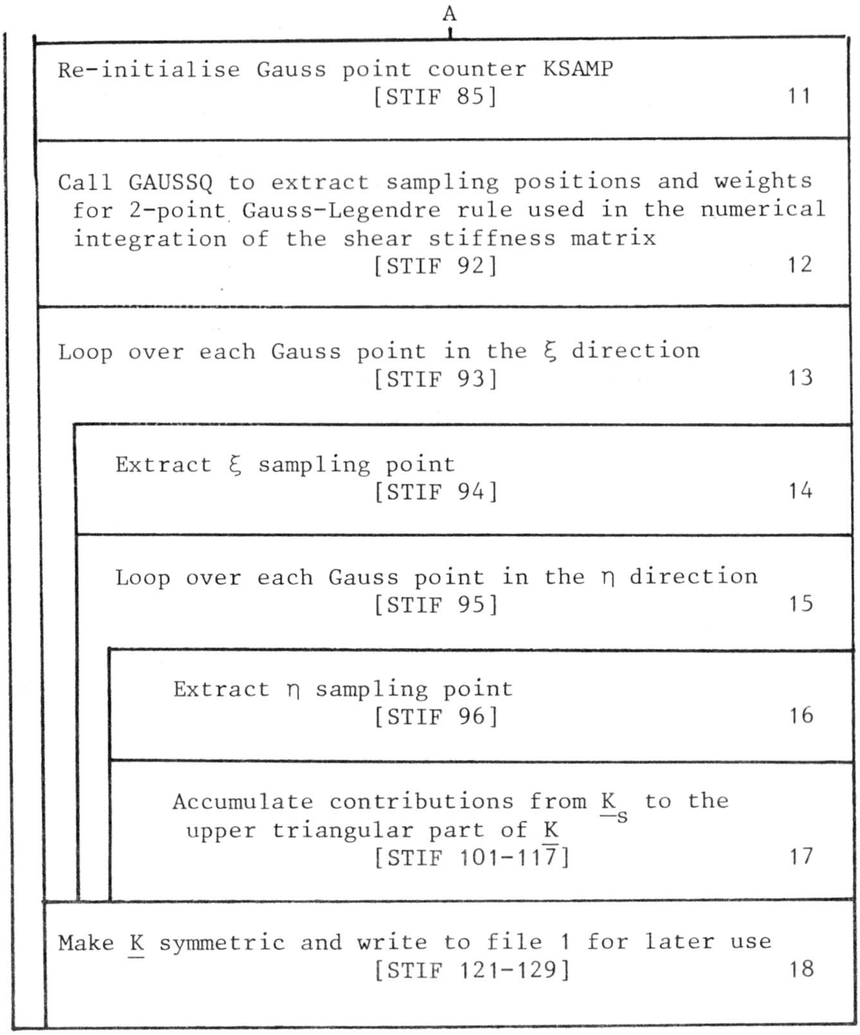

Figure 6.5.1 NS chart for stiffness module

Call SFR2 to evaluate N_i, $\partial N_i/\partial\xi$, $\partial N_i/\partial\eta$ at Gauss points
[STIF 59] 10.1

Call JACOB2 to evaluate $\mathbf{J}$, $\mathbf{J}^{-1}$, det$\mathbf{J}$, $\partial N_i/\partial x$, $\partial N_i/\partial y$ and Gauss point coordinates
[STIF 60-61] 10.2

Evaluate DAREA = det$\mathbf{J}$ * Gauss point weights
[STIF 62] 10.3

Call MODPB to evaluate $\mathbf{D}_b$
[STIF 66-67] 10.4

Loop over inode = 1,9
[STIF 72] 10.5

Call BMATPB to evaluate $\mathbf{B}_b$ (inode)
[STIF 73-74] 10.6

Loop over jnode = inode, 9
[STIF 75] 10.7

Call BMATPB to evaluate $\mathbf{B}_b$ (jnode)
[STIF 76-77] 10.8

Call SUBPB to evaluate contribution to bending stiffness sub-matrix $\mathbf{K}_b$ (inode, jnode)
[STIF 78-79] 10.9

Figure 6.5.2 NS chart for numerical integration of $\mathbf{K}_b$

6.5.2 Auxiliary routines GAUSSQ, SFR2, JACOB2, MODPB, BMATPB and SUBPB

The auxiliary routines used by STIFPB are listed.

```
      SUBROUTINE GAUSSQ    (NGAUS,POSGP,WEIGP)
C***********************************************************************
C
C***  SETS UP GAUSS-LEGENDRE SAMPLING POSITIONS AND WEIGHTS
C
C***********************************************************************
      IMPLICIT REAL*8(A-H,O-Z)
      DIMENSION POSGP(3),WEIGP(3)
      GO TO (1,2,3) NGAUS
    1 POSGP(1)= 0.
      WEIGP(1)= 2.
      RETURN
    2 POSGP(1)=-0.577350269189626
      POSGP(2)= 0.577350269189626
      WEIGP(1)= 1.
      WEIGP(2)= 1.
      RETURN
    3 POSGP(1)=-0.774596669241483
      POSGP(2)= 0.
      POSGP(3)= 0.774596669241483
      WEIGP(1)= 0.555555555555556
      WEIGP(2)= 0.888888888888889
      WEIGP(3)= 0.555555555555556
      RETURN
      END
```

```
      SUBROUTINE SFR2      (DERIV,ETASP,EXISP,NNODE,SHAPE)
C***********************************************************************
C
C***  EVALUATES SHAPE FUNCTIONS AND THEIR DERIVATIVES
C***  FOR 4,8 AND 9 NODED QUADRILATERAL ISOPARAMETRIC
C***  ELEMENTS
C
C***********************************************************************
      IMPLICIT REAL*8(A-H,O-Z)
      DIMENSION DERIV(2,9) ,SHAPE(9)
C
C*** SHAPE FUNCTIONS
C
      S=EXISP
      T=ETASP
      S2=S*2.0
      T2=T*2.0
      SS=S*S
      TT=T*T
      ST=S*T
      SST=S*S*T
      STT=S*T*T
      ST2=S*T*2.0
C
C*** SHAPE FUNCTIONS FOR 8 NODED ELEMENT
C
      SHAPE(1)=(-1.0+ST+SS+TT-SST-STT)/4.0
      SHAPE(2)=(1.0-T-SS+SST)/2.0
      SHAPE(3)=(-1.0-ST+SS+TT-SST+STT)/4.0
      SHAPE(4)=(1.0+S-TT-STT)/2.0
      SHAPE(5)=(-1.0+ST+SS+TT+SST+STT)/4.0
      SHAPE(6)=(1.0+T-SS-SST)/2.0
      SHAPE(7)=(-1.0-ST+SS+TT+SST-STT)/4.0
      SHAPE(8)=(1.0-S-TT+STT)/2.0
C
C*** SHAPE FUNCTION DERIVATIVES
C
      DERIV(1,1)=(T+S2-ST2-TT)/4.0
      DERIV(1,2)=-S+ST
      DERIV(1,3)=(-T+S2-ST2+TT)/4.0
```

```
      DERIV(1,4)=(1.0-TT)/2.0
      DERIV(1,5)=(T+S2+ST2+TT)/4.0
      DERIV(1,6)=-S-ST
      DERIV(1,7)=(-T+S2+ST2-TT)/4.0
      DERIV(1,8)=(-1.0+TT)/2.0
      DERIV(2,1)=(S+T2-SS-ST2)/4.0
      DERIV(2,2)=(-1.0+SS)/2.0
      DERIV(2,3)=(-S+T2-SS+ST2)/4.0
      DERIV(2,4)=-T-ST
      DERIV(2,5)=(S+T2+SS+ST2)/4.0
      DERIV(2,6)=(1.0-SS)/2.0
      DERIV(2,7)=(-S+T2+SS-ST2)/4.0
      DERIV(2,8)=-T+ST
      SHAPE(9)=(1.0-SS)*(1.0-TT)
      DERIV(1,9)=-S2*(1.0-TT)
      DERIV(2,9)=-T2*(1.0-SS)
      RETURN
      END
```

```
      SUBROUTINE JACOB2     (CARTD,DERIV,DJACB,ELCOD,GPCOD,IELEM,
     .                       KGASP,NNODE,SHAPE)
C**********************************************************************
C
C***  EVALUATE JACOBIAN MATRIX AND ITS INVERSE
C***  GAUSS POINT COORDINATES
C***  CARTESIAN SHAPE FUNCTION DERIVATIVES
C***  AT PRESENT SAMPLING POINT
C
C**********************************************************************
      IMPLICIT REAL*8(A-H,O-Z)
      DIMENSION CARTD(2,9),DERIV(2,9),ELCOD(2,9) ,GPCOD(2,9) ,SHAPE(9),
     .          XJACI(2,2),XJACM(2,2)
C
C***  CALCULATE COORDINATES OF SAMPLING POINT
C
      DO 2 IDIME=1,2
      GPCOD(IDIME,KGASP)=0.0
      DO 2 INODE=1,NNODE
      GPCOD(IDIME,KGASP)=GPCOD(IDIME,KGASP)+ELCOD(IDIME,INODE)
     .*SHAPE(INODE)
    2 CONTINUE
C
C***  CREATE JACOBIAN MATRIX XJACM
C
      DO 4 IDIME=1,2
      DO 4 JDIME=1,2
      XJACM(IDIME,JDIME)=0.0
      DO 4 INODE=1,NNODE
      XJACM(IDIME,JDIME)=XJACM(IDIME,JDIME)+DERIV(IDIME,INODE)*
     .ELCOD(JDIME,INODE)
    4 CONTINUE
C
C***  CALCULATE DETERMINANT AND INVERSE OF JACOBIAN MATRIX
C
      DJACB=XJACM(1,1)*XJACM(2,2)-XJACM(1,2)*XJACM(2,1)
      IF(DJACB) 6,6,8
    6 WRITE(6,900) IELEM
      STOP
    8 CONTINUE
      XJACI(1,1)=XJACM(2,2)/DJACB
      XJACI(2,2)=XJACM(1,1)/DJACB
      XJACI(1,2)=-XJACM(1,2)/DJACB
      XJACI(2,1)=-XJACM(2,1)/DJACB
C
C***  CALCULATE CARTESIAN DERIVATIVES
C
```

```
      DO 10 IDIME=1,2
      DO 10 INODE=1,NNODE
      CARTD(IDIME,INODE)=0.0
      DO 10 JDIME=1,2
      CARTD(IDIME,INODE)=CARTD(IDIME,INODE)+XJACI(IDIME,JDIME)*
     .DERIV(JDIME,INODE)
   10 CONTINUE
  900 FORMAT(//,36H PROGRAM HALTED IN SUBROUTINE JACOB2,/,11X,
     .22H ZERO OR NEGATIVE AREA,/,10X,16H ELEMENT NUMBER ,I5)
      RETURN
      END
```

```
      SUBROUTINE MODPB      (DFLEX,DSHER,LPROP,MMATS,PROPS,
     .                       IFFLE,IFSHE)
C***********************************************************************
C
C*** CALCULATES MATRIX OF ELASTIC RIGIDITIES
C*** FOR MINDLIN PLATE
C
C***********************************************************************
      IMPLICIT REAL*8(A-H,O-Z)
      DIMENSION DFLEX(3,3),DSHER(2,2),
     .          PROPS(MMATS,8)
      IF(PROPS(LPROP,4).NE.0.0) GO TO 30
      YOUNG=PROPS(LPROP,1)
      POISS=PROPS(LPROP,2)
      THICK=PROPS(LPROP,3)
C
C*** FORM DFLEX (ISOTROPIC)
C
   10 IF(IFFLE.EQ.0) GO TO 20
      CALL RAZERO(DFLEX,3,3)
      CONST=(YOUNG*THICK**3)/(12.*(1.-POISS*POISS))
      DFLEX(1,1)=CONST
      DFLEX(2,2)=CONST
      DFLEX(1,2)=CONST*POISS
      DFLEX(2,1)=CONST*POISS
      DFLEX(3,3)=CONST*(1.-POISS)/2.
C
C*** FORM DSHER (ISOPTROPIC)
C
   20 IF(IFSHE.EQ.0) RETURN
      CALL RAZERO(DSHER,2,2)
      DSHER(1,1)=(YOUNG*THICK)/(2.4+2.4*POISS)
      DSHER(2,2)=(YOUNG*THICK)/(2.4+2.4*POISS)
      RETURN
C
C*** FORM DFLEX (ORTHOTROPIC)
C
   30 IF(IFFLE.EQ.0) GO TO 40
      CALL RAZERO(DFLEX,3,3)
      DFLEX(1,1)=PROPS(LPROP,1)
      DFLEX(1,2)=PROPS(LPROP,2)
      DFLEX(2,1)=PROPS(LPROP,2)
      DFLEX(2,2)=PROPS(LPROP,3)
      DFLEX(3,3)=PROPS(LPROP,4)
C
C*** FORM DSHER (ORTHOTROPIC)
C
   40 IF(IFSHE.EQ.0) RETURN
      CALL RAZERO(DSHER,2,2)
      DSHER(1,1)=PROPS(LPROP,5)
      DSHER(2,2)=PROPS(LPROP,6)
      RETURN
      END
```

```
      SUBROUTINE BMATPB     (BFLEX,BSHER,CARTD,KNODE,SHAPE,
     .                       IFFLE,IFSHE)
C***********************************************************************
C
C*** EVALUATES STRAIN-DISPLACEMENT MATRIX FOR
C*** FOR MINDLIN PLATE
C
C***********************************************************************
      IMPLICIT REAL*8(A-H,O-Z)
      DIMENSION  BFLEX(3,3),BSHER(2,3),
     .           CARTD(2,9),SHAPE(9)
      DNKDX=CARTD(1,KNODE)
      DNKDY=CARTD(2,KNODE)
C*** FORM BFLEX
   10 IF(IFFLE.EQ.0) GO TO 20
      CALL RAZERO(BFLEX,3,3)
      BFLEX(1,2)=-DNKDX
      BFLEX(2,3)=-DNKDY
      BFLEX(3,2)=-DNKDY
      BFLEX(3,3)=-DNKDX
C*** FORM BSHER
   20 IF(IFSHE.EQ.0) RETURN
      CALL RAZERO(BSHER,2,3)
      BSHER(1,1)=DNKDX
      BSHER(1,2)=-SHAPE(KNODE)
      BSHER(2,1)=DNKDY
      BSHER(2,3)=-SHAPE(KNODE)
      RETURN
      END
```

```
      SUBROUTINE SUBPB      (BIMAT,BJMAT,DAREA,DMATX,ESTIF,INODE,
     .                       JNODE,NCOLI,NROIJ,NCOLJ)
C***********************************************************************
C
C*** CARRY OUT MATRIX MULTIPLICATION
C
C***********************************************************************
      IMPLICIT REAL*8(A-H,O-Z)
      DIMENSION BIMAT(NROIJ,NCOLI),BJMAT(NROIJ,NCOLJ),
     .          DMATX(NROIJ,NROIJ),DBMAT(3,3),
     .          ESTIF(27,27),SBSTF(3,3)
C
C*** EVALUATE  D*BJ
C
      DO 10 J=1,NCOLJ
      DO 10 I=1,NROIJ
      DBMAT(I,J)=0.0
      DO 10 K=1,NROIJ
   10 DBMAT(I,J)=DBMAT(I,J)+DMATX(I,K)*BJMAT(K,J)
C
C*** EVALUATE  BIT*(D*BJ)
C
      DO 20 J=1,NCOLJ
      DO 20 I=1,NCOLI
      SBSTF(I,J)=0.0
      DO 20 K=1,NROIJ
   20 SBSTF(I,J)=SBSTF(I,J)+BIMAT(K,I)*DBMAT(K,J)
C
C*** ASSEMBLE SBSTF INTO ELEMENT STIFFNESS MATRIX
C
      IFROW=0
      JFCOL=0
```

```
      IFROW=(INODE-1)*3+IFROW
      JFCOL=(JNODE-1)*3+JFCOL
      DO 30 I=1,NCOLI
      IRSUB=IFROW+I
      DO 30 J=1,NCOLJ
      JCSUB=JFCOL+J
   30 ESTIF(IRSUB,JCSUB)=ESTIF(IRSUB,JCSUB)+SBSTF(I,J)*DAREA
      RETURN
      END
```

6.6 Load module

6.6.1 Subroutine LOADPB

Subroutine LOADPB evaluates the element nodal forces for

(a) uniformly distributed lateral loading, and

(b) point loads

The NS charts for LOADPB are given in Figures 6.6.1 and 6.6.2. Note that routines RAZERO, GAUSSQ, SFR2 and JACOB2 which are used by LOADPB have already been listed in the previous section.

```
      SUBROUTINE LOADPB   (COORD,ELOAD,ICASE,LNODS,MATNO,MELEM,
     .                      MMATS,MPOIN,NELEM,NEVAB,NGAUB,NNODE,
     .                      NPOIN,PROPS)
C**********************************************************************
C
C***   COMPUTE NODAL FORCES AFTER READING RELEVANT DATA
C***   FOR MINDLIN PLATE ELEMENTS
C
C**********************************************************************
      IMPLICIT REAL*8(A-H,O-Z)
      DIMENSION CARTD(2,9),COORD(MPOIN,2),DERIV(2,9),ELCOD(2,9),
     .          GPCOD(2,9),LNODS(MELEM,9),MATNO(MELEM),
     .          POINT(3),POSGP(3),PROPS(MMATS,8),ELOAD(MELEM,27),
     .          SHAPE(9),TITLE(20),WEIGP(3)
      CALL RAZERO(ELOAD,MELEM,27)
      READ(5,900) TITLE
  900 FORMAT(20A4)
      WRITE(6,905) TITLE
  905 FORMAT(1H0,20A4)
      WRITE(6,910) ICASE
  910 FORMAT(/2X,12HLOAD CASE = ,I5)
C
C*** READ DATA CONTROLLING LOADING TYPES TO BE INPUT
C
        READ(5,915)IPLOD
       WRITE(6,915)IPLOD
  915 FORMAT(4I5)
C
C*** READ NODAL POINT LOADS
C
      IF(IPLOD.EQ.0) GO TO 50
   10 READ(5,920) LODPT,(POINT(IDOFN),IDOFN=1,3)
      WRITE(6,920) LODPT,(POINT(IDOFN),IDOFN=1,3)
  920 FORMAT(I5,3F10.3)
C
C*** ASSOCIATE THE NODAL POINT LOADS WITH AN ELEMENT
```

```
C
      DO 20 IELEM=1,NELEM
      DO 20 INODE=1,NNODE
      NLOCA=IABS(LNODS(IELEM,INODE))
   20 IF(LODPT.EQ.NLOCA) GO TO 30
   30 DO 40 IDOFN=1,3
      NGASH=(INODE-1)*3+IDOFN
   40 ELOAD(IELEM,NGASH)=POINT(IDOFN)
      IF(LODPT.LT.NPOIN) GO TO 10
   50 CONTINUE
C
C***    LOOP OVER EACH ELEMENT
C
      IF(ICASE.GT.1) GO TO 100
       DO 90  IELEM=1,NELEM
       LPROP=MATNO(IELEM)
       UDLOD=PROPS(LPROP,7)
       IF(UDLOD.EQ.0.0)GO TO 90
C
C***  EVALUATE THE COORDINATES OF THE ELEMENT NODAL POINTS
C
      DO 60 INODE=1,NNODE
      LNODE=LNODS(IELEM,INODE)
      LNODE=IABS(LNODE)
      DO 60 IDIME=1,2
      ELCOD(IDIME,INODE)=COORD(LNODE,IDIME)
   60 CONTINUE
      KGASP=0
      CALL GAUSSQ     (NGAUB,POSGP,WEIGP)
C
C***  ENTER  LOOPS  FOR NUMERICAL  INTEGRATION
C
      DO 80 IGAUS=1,NGAUB
      EXISP=POSGP(IGAUS)
      DO 80 JGAUS=1,NGAUB
      ETASP=POSGP(JGAUS)
      KGASP=KGASP+1
C
C***  EVALUATE THE SHAPE FUNCTIONS AT THE SAMPLING
C           POINTS AND ELEMENTAL AREA
C
      CALL        SFR2       (DERIV,ETASP,EXISP,NNODE,SHAPE)
      CALL        JACOB2     (CARTD,DERIV,DJACB,ELCOD,GPCOD,IELEM,
     .                        KGASP,NNODE,SHAPE)
      DAREA=DJACB*WEIGP(IGAUS)*WEIGP(JGAUS)
C
C***   CALCULATE LOADS  AND ASSOCIATE WITH ELEMENT NODALPOINTS
C
      DO 70 INODE=1,NNODE
      NPOSN=(INODE-1)*3+1
      ELOAD(IELEM,NPOSN)=ELOAD(IELEM,NPOSN)+
     . SHAPE(INODE)*UDLOD*DAREA
   70 CONTINUE
   80 CONTINUE
   90 CONTINUE
  100 WRITE(6,925)
  925 FORMAT(1H0,5X,36H TOTAL NODAL FORCES FOR EACH ELEMENT)
      DO 110 IELEM=1,NELEM
  110 WRITE(6,935) IELEM,(ELOAD(IELEM,IEVAB),IEVAB=1,NEVAB)
  935 FORMAT(1X,I4,5X,8E12.4/(10X,8E12.4))
      RETURN
      END
```

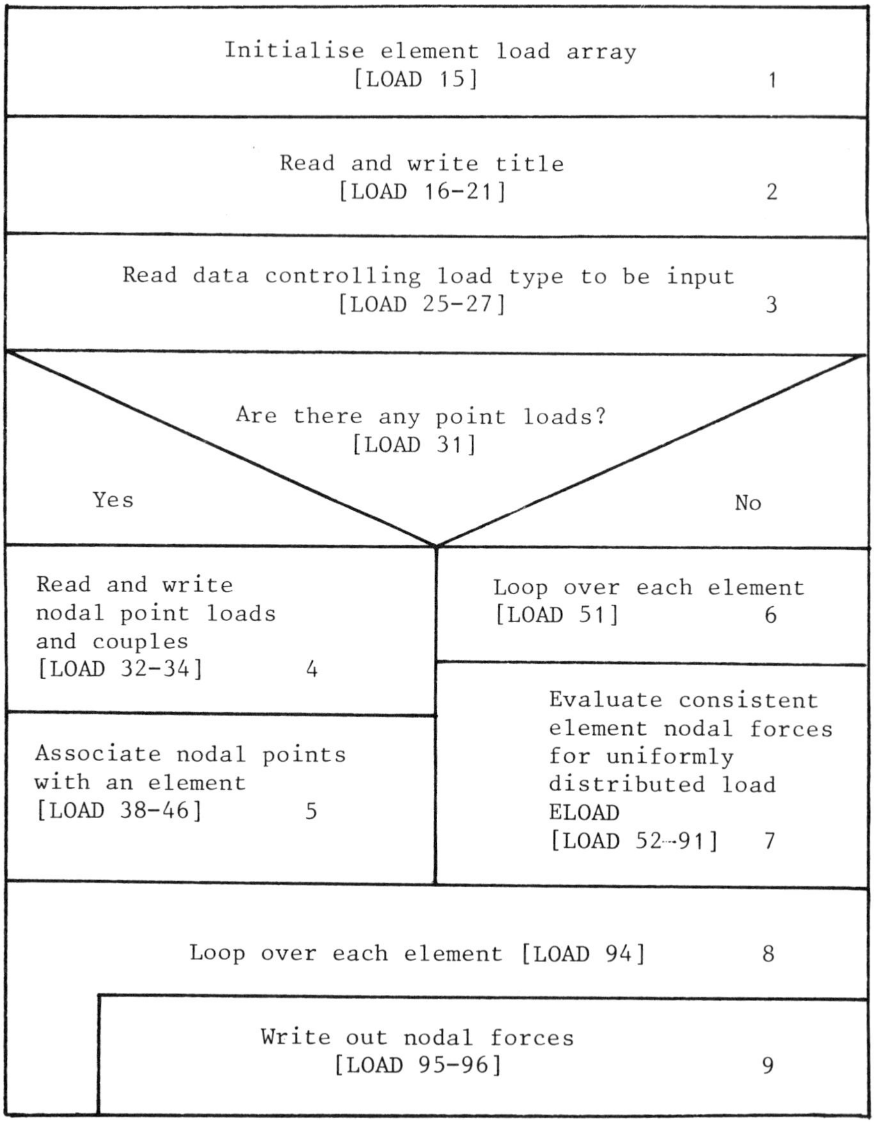

Figure 6.6.1 NS chart for element loads

- Extract element uniform load, material set number and nodal coordinates [LOAD 52-63] 7.1
- Initialise Gauss point counter KGASP [LOAD 64] 7.2
- Call GAUSSQ to extract sampling positions and weights for 3-point Gauss-Legendre rule used to evaluate consistent nodal forces [LOAD 65] 7.3
- Loop over Gauss points in ξ direction [LOAD 69] 7.4
  - Extract ξ sampling position [LOAD 70] 7.5
  - Loop over Gauss point in η direction [LOAD 71] 7.6
    - Extract η sampling position [LOAD 72] 7.7
    - Increment Gauss point counter by 1 [LOAD 73] 7.8
    - Call SFR2 to evaluate N_i, $\partial N_i/\partial\xi$ $\partial N_i/\partial\eta$ [LOAD 78] 7.9
    - Call JACOB2 to evaluate $\det\underline{J}$ [LOAD 79-80] 7.10
    - Evaluate DAREA=$\det\underline{J}$ x Gauss point weights [LOAD 81] 7.11
    - Loop over each element node [LOAD 85] 7.12
      - Accumulate contribution to consistent load vector [LOAD 86-89] 7.13

Figure 6.6.2 NS chart for consistent nodal force evaluation

6.7 Solution module

A frontal solver is used to assemble and solve the stiffness equations. The frontal solver is described in detail elsewhere [21,22] and consequently only a listing is provided in the Appendix of this chapter.

6.8 Output and stress resultant module

6.8.1 Subroutine OUTDIS

Subroutine OUTDIS simply prints out nodal displacements, rotations and reactions and as it is self-explanatory it is simply listed without further comment.

```
      SUBROUTINE OUTDIS(ASDIS,MTOTV,MVFIX,NOFIX,NPOIN,
     .                  NVFIX,TREAC)
      IMPLICIT REAL*8(A-H,O-Z)
      DIMENSION ASDIS(MTOTV),NOFIX(MVFIX),TREAC(MVFIX,3)
C
C*** OUTPUT DISPLACEMENTS
C
      WRITE(6,900)
  900 FORMAT(1H0,5X,13HDISPLACEMENTS)
      WRITE(6,905)
  905 FORMAT(1H0,6X,4HNODE,6X,5HDISP.,8X,7HXZ-ROT.,7X,7HYZ-ROT.)
      DO 5 IPOIN=1,NPOIN
      NGASH=IPOIN*3
      NGISH=NGASH-3+1
    5 WRITE(6,910) IPOIN,(ASDIS(IGASH),IGASH=NGISH,NGASH)
  910 FORMAT(I10,3E14.6)
C
C*** OUTPUT REACTIONS
C
      WRITE(6,915)
  915 FORMAT(1H0,5X,9HREACTIONS)
      WRITE(6,920)
  920 FORMAT(1H0,6X,4HNODE,6X,5HFORCE,6X,9HXZ-MOMENT,5X,
     .9HYZ-MOMENT)
      DO 10 IVFIX=1,NVFIX
   10 WRITE(6,910) NOFIX(IVFIX),(TREAC(IVFIX,IDOFN),IDOFN=1,3)
      RETURN
      END
```

6.8.2 Subroutine OUTSTR

Subroutine OUTSTR evaluates and prints out stress resultants at Gauss-Legendre and nodal points. In particular the following values are provided:-

(a) M_x, M_y, M_{xy} and M_1, M_2, θ at 3x3 Gauss-Legendre points,

(b) M_x, M_y, M_{xy} and M_1, M_2, θ at nodal points,

(c) Q_x, Q_y at 2x2 Gauss-Legendre points,

(d) Extrapolated Q_x, Q_y at nodal points,

and

(e) Averaged nodal values for M_x, M_y, M_{xy} and M_1, M_2, θ and Q_x, Q_y.

Subroutines BENDM, SHEAR and AVERAG are used and they will be described later.

The NS chart for OUTSTR is given in Figure 6.8.1 and then OUTSTR is listed.

```
      SUBROUTINE OUTSTR(ASDIS,COORD,KOUNT,LNODS,MATNO,MELEM,
     .                  MMATS,MPOIN,MTOTV,NDOFN,NELEM,NGAUB,
     .                  NGAUS,NPOIN,PROPS,STRND)
      IMPLICIT REAL*8(A-H,O-Z)
      DIMENSION ASDIS(MTOTV),COORD(MPOIN,2),ELCOD(2,9),
     .ELDIS(3,9),KOUNT(MPOIN),LNODS(MELEM,9),MATNO(MELEM),
     .PROPS(MMATS,8),POSGP(3),POSSP(3),SHEAV(9,2),STRES(5),
     .STRND(MPOIN,5),STRSG(4,2),WEIGP(3),ICODE(9)
      DATA POSSP/-1.0,0.0,1.0/,ICODE/1,8,7,2,9,6,3,4,5/,
     .     NNODE/9/
C
C*** OUTPUT STRESS RESULTANTS AT GAUSS AND NODAL POINTS
C
      WRITE(6,900)
 900  FORMAT(//1H0,5X,17HSTRESS RESULTANTS)
      CALL IVZERO(KOUNT,MPOIN)
      CALL RAZERO(STRND,MPOIN,5)
      DO 35 IELEM=1,NELEM
      LPROP=MATNO(IELEM)
      DO 10 INODE=1,NNODE
      LNODE=LNODS(IELEM,INODE)
      LNODE=IABS(LNODE)
      DO 5 IDIME=1,2
 5    ELCOD(IDIME,INODE)=COORD(LNODE,IDIME)
      NPOSN=(LNODE-1)*3
      DO 10 IDOFN=1,3
      NPOSN=NPOSN+1
      ELDIS(IDOFN,INODE)=ASDIS(NPOSN)
 10   CONTINUE
C
C*** EVALUATE BENDING MOMENTS AT GAUSS POINTS
C
      KSAMP=0
      CALL GAUSSQ(NGAUB,POSGP,WEIGP)
C     WRITE(6,905)IELEM
C905  FORMAT(//1H0,5X,13HELEMENT NO. =,I5)
C     WRITE(6,910)
C910  FORMAT(/,1X,31HBENDING MOMENTS AT GAUSS POINTS,/)
C     WRITE(6,915)
C915  FORMAT(1H0,2X,4HS.P.,2X,8HX-COORD.,1X,8HY-COORD.,3X,
C    .8HX-MOMENT,4X,8HY-MOMENT,3X,9HXY-MOMENT,4X,
C    .8HMAX P.M.,4X,8HMIN P.M.,6X,5HANGLE)
      DO 15 IGAUB=1,NGAUB
      EXISP=POSGP(IGAUB)
```

```
      DO 15 JGAUB=1,NGAUB
      ETASP=POSGP(JGAUB)
      CALL BENDM(ELCOD,ELDIS,ETASP,EXISP,IELEM,KSAMP,LPROP,
     .MMATS,NDOFN,PROPS,STRES)
 15   CONTINUE
C
C*** EVALUATE BENDING MOMENTS AT NODAL POINTS
C
      KSAMP=0
C     WRITE(6,920)
C920  FORMAT(/,1X,31HBENDING MOMENTS AT NODAL POINTS,/)
C     WRITE(6,915)
      DO 20 ISAMP=1,3
      EXISP=POSSP(ISAMP)
      DO 20 JSAMP=1,3
      ETASP=POSSP(JSAMP)
      CALL BENDM(ELCOD,ELDIS,ETASP,EXISP,IELEM,KSAMP,LPROP,
     .MMATS,NDOFN,PROPS,STRES)
      INODE=ICODE(KSAMP)
      LNODE=LNODS(IELEM,INODE)
      LNODE=IABS(LNODE)
      KOUNT(LNODE)=KOUNT(LNODE)+1
      DO 20 ISTRE=1,3
      STRND(LNODE,ISTRE)=STRND(LNODE,ISTRE)
     .+STRES(ISTRE)
 20   CONTINUE
C
C*** EVALUATE THE SHEAR FORCES AT GAUSS POINTS
C
      KSAMP=0
      CALL GAUSSQ(NGAUS,POSGP,WEIGP)
C     WRITE(6,925)
C925  FORMAT(/,1X,28HSHEAR FORCES AT GAUSS POINTS)
C     WRITE(6,930)
C930  FORMAT(/,2X,4HS.P.,2X,8HX-COORD.,2X,8HY-COORD.,
C    .4X,8HXZ-SHEAR,4X,8HYZ-SHEAR)
      DO 25 IGAUS=1,NGAUS
      EXISP=POSGP(IGAUS)
      DO 25 JGAUS=1,NGAUS
      ETASP=POSGP(JGAUS)
      CALL SHEAR(ELCOD,ELDIS,ETASP,EXISP,IELEM,KSAMP,
     .LPROP,MMATS,NDOFN,PROPS,STRES,STRSG)
 25   CONTINUE
C
C*** EXTRAPOLATE SHEAR FORCES TO NODAL POINTS
C
      CALL EXTRA(SHEAV,STRSG)
      DO 30 ISAMP=1,9
      INODE=ICODE(ISAMP)
      LNODE=LNODS(IELEM,INODE)
      LNODE=IABS(LNODE)
      DO 30 ISTRE=1,2
      KSTRE=ISTRE+3
 30   STRND(LNODE,KSTRE)=STRND(LNODE,KSTRE)+
     .SHEAV(ISAMP,ISTRE)
 35   CONTINUE
C
C*** EVALUATE AND PRINT AVERAGE NODAL VALUES
C
      CALL AVERAG(KOUNT,MPOIN,NPOIN,STRES,STRND)
      RETURN
      END
```

Write general title for stress results and initialise KOUNT and STRND
[OUTS 14-17] 1

Loop over each element
[OUTS 18] 2

> Extract element material set, nodal coordinates and displacements
> [OUTS 19-29] 3
>
> Evaluate and print out M_x, M_y, M_{xy} and M_1, M_2, θ at 3x3 Gauss-Legendre points by calling BENDM
> [OUTS 33-49] 4*
>
> Evaluate and print out M_x, M_y, M_{xy} and M_1, M_2, θ at nodal points by calling BENDM. Accumulate bending moments at nodal points in STRND for future averaging
> [OUTS 53-70] 5*
>
> Evaluate and print out shear forces Q_x and Q_y at 2x2 Gauss-Legendre points by calling SHEAR
> [OUTS 74-87] 6*
>
> Evaluate (by extrapolation using EXTRA) and print out shear forces at nodal points. Accumulate shear forces at nodal points in STRND for future averaging
> [OUTS 91-100] 7*
>
> Call AVERAG to calculate and print out the average nodal values for M_x, M_y, M_{xy} and M_1, M_2, θ and Q_x, Q_y.
> [OUTS 104] 8

Figure 6.8.1 NS chart for evaluation and printing of stress resultants at Gauss-Legendre and nodal points

---

*printing suppressed by appropriate use of c in column 1

6.8.3 Subroutine BENDM

This routine evaluates and prints out bending and principal moments at specified values of (ξ,η). The routine uses subroutines SFR2, JACOB2, MODPB, STRPB and PRINC. Subroutines STRPB and PRINC have not been described before and will be discussed and listed after BENDM has been dealt with.

The NS chart for BENDM is given in Figure 6.8.2. after which BENDM is listed.

Call SFR to evaluate shape function N_i and derivatives $N_{i,\xi}$ and $N_{i,\eta}$	[BEND 13]	1
Call JACOB2 to evaluate cartesian shape function derivatives $N_{i,x}$ and $N_{i,y}$	[BEND 14-15]	2
Call MODPB to evaluate $\underline{D}_b$	[BEND 16-17]	3
Call STRPB to evaluate M_x, M_y, M_{xy}	[BEND 18-19]	4
Call PRINC to evaluate M_1, M_2, θ	[BEND 20]	5
Print out sampling point coordinates, M_x, M_y, M_{xy}, M_1, M_2, θ	[BEND 21-23]	6

Figure 6.8.2 NS chart for BENDM

```
      SUBROUTINE BENDM(ELCOD,ELDIS,ETASP,EXISP,IELEM,                   BEND   1
     .KSAMP,LPROP,MMATS,NDOFN,PROPS,STRES)                              BEND   2
      IMPLICIT REAL*8(A-H,O-Z)                                          BEND   3
      DIMENSION CARTD(2,9),DERIV(2,9),DFLEX(3,3),DGRAD(6),              BEND   4
     .DSHER(2,2),ELCOD(2,9),ELDIS(3,9),                                 BEND   5
     .PROPS(MMATS,8),SHAPE(9),SPCOD(2,9),STRES(5)                       BEND   6
      DATA NNODE/9/                                                     BEND   7
C                                                                       BEND   8
C*** EVALUATES AND PRINTS BENDING MOMENTS MX MY MXY AND                 BEND   9
C*** PRINCIPAL MOMENTS M1 M2 AND ANGLE.                                 BEND  10
C                                                                       BEND  11
```

```
      KSAMP=KSAMP+1
      CALL SFR2(DERIV,ETASP,EXISP,NNODE,SHAPE)
      CALL JACOB2(CARTD,DERIV,DJACB,ELCOD,SPCOD,
     .IELEM,KSAMP,NNODE,SHAPE)
      CALL MODPB(DFLEX,DSHER,LPROP,MMATS,
     .           PROPS,1,0)
      CALL STRPB(CARTD,DFLEX,DGRAD,DSHER,ELDIS,NDOFN,
     .           NNODE,SHAPE,STRES,1,0)
      CALL PRINC(ANGLE,PMMAX,PMMIN,STRES)
C     WRITE(6,900)KSAMP,(SPCOD(IDIME,KSAMP),IDIME=1,2),
C    .(STRES(ISTRE),ISTRE=1,3),PMMAX,PMMIN,ANGLE
C900  FORMAT(1X,I5,2F10.4,6E12.4)
      RETURN
      END
```

6.8.4 Subroutine STRPB

This routine evaluates bending moments M_x, M_y, M_{xy} or shear forces. It uses subroutine RAZERO and GRADPB. After the program is listed below, subroutine GRADPB will be considered.

```
      SUBROUTINE STRPB      (CARTD,DFLEX,DGRAD,DSHER,ELDIS,NDOFN,
     .                       NNODE,SHAPE,STRES,IFFLE,IFSHE)
C***********************************************************************
C
C*** EVALUATES STRESS RESULTANTS FOR MINDLIN PLATE
C
C***********************************************************************
      IMPLICIT REAL*8(A-H,O-Z)
      DIMENSION CARTD(2,9),DFLEX(3,3),DGRAD(6),DSHER(2,2),
     .          ELDIS(3,9),SHAPE(9),STRES(5)
C*** ZERO STRESS VECTOR
      CALL RAZERO(STRES,5,1)
      CALL GRADPB(CARTD,DGRAD,ELDIS,NDOFN,NNODE)
C
C*** EVALUATE ROTATIONS AT GAUSS POINT , IF NEEDED
C
      IF(IFSHE.EQ.0) GO TO 20
      XZROT=0.0
      YZROT=0.0
      DO 10 INODE=1,NNODE
      XZROT=XZROT+SHAPE(INODE)*ELDIS(2,INODE)
   10 YZROT=YZROT+SHAPE(INODE)*ELDIS(3,INODE)
C
C*** EVALUATE BENDING STRESS RESULTANTS
C
   20 IF(IFFLE.EQ.0) GOTO 30
      EFLXX=-DGRAD(2)
      EFLYY=-DGRAD(6)
      EFLXY=-(DGRAD(3)+DGRAD(5))
      STRES(1)=DFLEX(1,1)*EFLXX+DFLEX(1,2)*EFLYY
      STRES(2)=DFLEX(2,1)*EFLXX+DFLEX(2,2)*EFLYY
      STRES(3)=DFLEX(3,3)*EFLXY
C
C*** EVALUATE SHEAR STRESS RESULTANTS
C
   30 IF(IFSHE.EQ.0) RETURN
      ESHXX=DGRAD(1)-XZROT
      ESHYY=DGRAD(4)-YZROT
      STRES(4)=DSHER(1,1)*ESHXX
      STRES(5)=DSHER(2,2)*ESHYY
      RETURN
      END
```

6.8.5 Subroutine GRADPB

This routine, which simply calculates the displacement gradients $w_{,x}$, $w_{,y}$, $\theta_{x,x}$, $\theta_{x,y}$, $\theta_{y,x}$ and $\theta_{y,y}$ is now listed.

```
      SUBROUTINE GRADPB    (CARTD,DGRAD,ELDIS,NDOFN,NNODE)
C**********************************************************************
C
C*** FORM TOTAL DISPLACEMENTS GRADIENTS
C
C**********************************************************************
      IMPLICIT REAL*8(A-H,O-Z)
      DIMENSION CARTD(2,9),DGRAD(6),ELDIS(3,9)
C
C*** ZERO DGRAD
C
      CALL RAZERO(DGRAD,6,1)
C
C*** FORM TOTAL DISPLACEMENTS GRADIENTS
C
      DO 10 INODE=1,NNODE
      DNIDX=CARTD(1,INODE)
      DNIDY=CARTD(2,INODE)
      DO 10 IDOFN=1,NDOFN
      IPOSN=NDOFN+IDOFN
      CONST=ELDIS(IDOFN,INODE)
      DGRAD(IDOFN)=DGRAD(IDOFN)+DNIDX*CONST
   10 DGRAD(IPOSN)=DGRAD(IPOSN)+DNIDY*CONST
      RETURN
      END
```

6.8.6 Subroutine PRINC

This routine, which simply evaluates the principal moments and their direction, is now listed.

```
      SUBROUTINE PRINC(ANGLE,PMMAX,PMMIN,STRES)
      IMPLICIT REAL*8(A-H,O-Z)
      DIMENSION STRES(5)
C
C*** EVALUATE PRINCIPAL MOMENTS
C
      XXMOM=STRES(1)
      YYMOM=STRES(2)
      XYMOM=STRES(3)
      XXAYY=0.5*(XXMOM+YYMOM)
      XXMYY=0.5*(XXMOM-YYMOM)
      IF(XXMYY.EQ.0.0)XXMYY=0.000000001
      SUMMM=SQRT(XXMYY*XXMYY+XYMOM*XYMOM)
      PMMAX=XXAYY+SUMMM
      PMMIN=XXAYY-SUMMM
      ANGLE=ATAN(XYMOM/XXMYY)*0.5
      ANGLE=ANGLE*(180.0/3.141593)
      RETURN
      END
```

6.8.7 Subroutine SHEAR

Subroutine SHEAR evaluates and prints out shear forces Q_x and Q_y at specified values of (ξ,η). This routine uses subroutines SFR2, JACOB2, MODPB and STRPB and is very similar in structure to BENDM. The routine is now listed.

```
      SUBROUTINE SHEAR(ELCOD,ELDIS,ETASP,EXISP,IELEM,KSAMP,
     .LPROP,MMATS,NDOFN,PROPS,STRES,STRSG)
      IMPLICIT REAL*8(A-H,O-Z)
      DIMENSION CARTD(2,9),DERIV(2,9),DFLEX(3,3),DGRAD(6),
     .DSHER(2,2),ELCOD(2,9),ELDIS(3,9),
     .PROPS(MMATS,8),SHAPE(9),SPCOD(2,9),STRES(5),STRSG(4,2)
      DATA NNODE/9/
C
C*** EVALUATES AND PRINTS SHEAR FORCES AT GAUSS POINTS
C
      KSAMP=KSAMP+1
      CALL SFR2(DERIV,ETASP,EXISP,NNODE,SHAPE)
      CALL JACOB2(CARTD,DERIV,DJACB,ELCOD,SPCOD,IELEM,
     .KSAMP,NNODE,SHAPE)
      CALL MODPB(DFLEX,DSHER,LPROP,MMATS,PROPS,0,1)
      CALL STRPB(CARTD,DFLEX,DGRAD,DSHER,ELDIS,NDOFN,NNODE,
     .SHAPE,STRES,0,1)
C     WRITE(6,900)KSAMP,(SPCOD(IDIME,KSAMP),IDIME=1,2),
C    .(STRES(ISTRE),ISTRE=4,5)
C900  FORMAT(1X,I5,2F10.4,2E12.5)
      STRSG(KSAMP,1)=STRES(4)
      STRSG(KSAMP,2)=STRES(5)
      RETURN
      END
```

6.8.8 Subroutine EXTRA

Using a bilinear function this routine extrapolates the shear forces at the Gauss points to the nodal points.

```
      SUBROUTINE EXTRA(SHEAV,STRSG)
      IMPLICIT REAL*8(A-H,O-Z)
      DIMENSION SHEAV(9,2),STRSG(4,2)
C
C*** EXTRAPOLATES SHEAR FORCES FROM GAUSS POINTS TO
C*** NODAL POINTS
C
      FACT1=1.0+0.5*SQRT(3.0)
      FACT2=-0.5
      FACT3=-0.5
      FACT4=1.0-0.5*SQRT(3.0)
      DO 5 ISTRE=1,2
      SHEAV(1,ISTRE)=STRSG(1,ISTRE)*FACT1+STRSG(2,ISTRE)*
     .FACT2+STRSG(3,ISTRE)*FACT3+STRSG(4,ISTRE)*FACT4
      SHEAV(3,ISTRE)=STRSG(2,ISTRE)*FACT1+STRSG(3,ISTRE)*
     .FACT4+STRSG(1,ISTRE)*FACT2+STRSG(4,ISTRE)*FACT3
      SHEAV(7,ISTRE)=STRSG(3,ISTRE)*FACT1+STRSG(1,ISTRE)*
     .FACT2+STRSG(4,ISTRE)*FACT3+STRSG(2,ISTRE)*FACT4
      SHEAV(9,ISTRE)=STRSG(4,ISTRE)*FACT1+STRSG(1,ISTRE)*
     .FACT4+STRSG(2,ISTRE)*FACT2+STRSG(3,ISTRE)*FACT3
      SHEAV(2,ISTRE)=0.5*(SHEAV(1,ISTRE)+SHEAV(3,ISTRE))
      SHEAV(4,ISTRE)=0.5*(SHEAV(1,ISTRE)+SHEAV(7,ISTRE))
      SHEAV(6,ISTRE)=0.5*(SHEAV(3,ISTRE)+SHEAV(9,ISTRE))
      SHEAV(8,ISTRE)=0.5*(SHEAV(7,ISTRE)+SHEAV(9,ISTRE))
```

```
      SHEAV(5,ISTRE)=0.25*(STRSG(1,ISTRE)+STRSG(2,ISTRE)
     .+STRSG(3,ISTRE)+STRSG(4,ISTRE))
    5 CONTINUE
      RETURN
      END
```

6.8.9 Subroutine AVERAG

Subroutine AVERAG evaluates averaged nodal bending moments and shear forces and prints them out.

```
      SUBROUTINE AVERAG(KOUNT,MPOIN,NPOIN,STRES,STRND)
      IMPLICIT REAL*8(A-H,O-Z)
      DIMENSION KOUNT(MPOIN),STRES(5),STRND(MPOIN,5)
C
C*** EVALUATES AND PRINTS AVERAGED NODAL STRESS RESULTANTS
C
      WRITE(6,900)
  900 FORMAT(/1X,24HBENDING MOMENTS AT NODES)
      WRITE(6,905)
  905 FORMAT(1X,5H NODE,2X,9HMX-MOMENT,4X,9HMY-MOMENT,
     .3X,10HMXY-MOMENT,2X,8HMAX-P.M.,4X,8HMIN-P.M.,
     .6X,5HANGLE)
      DO 10 IPOIN=1,NPOIN
      EJOIN=FLOAT(KOUNT(IPOIN))
      DO 5 ISTRE=1,3
    5 STRES(ISTRE)=STRND(IPOIN,ISTRE)/EJOIN
      CALL PRINC(ANGLE,PMMAX,PMMIN,STRES)
      WRITE(6,910)IPOIN,(STRES(ISTRE),ISTRE=1,3),
     .PMMAX,PMMIN,ANGLE
  910 FORMAT(1X,I5,6E12.4)
   10 CONTINUE
      WRITE(6,915)
  915 FORMAT(/1X,21HSHEAR FORCES AT NODES)
      WRITE(6,920)
  920 FORMAT(1X,5H NODE,4X,8HXZ-SHEAR,4X,8HYZ-SHEAR)
      DO 20 IPOIN=1,NPOIN
      EJOIN=FLOAT(KOUNT(IPOIN))
      DO 15 ISTRE=4,5
   15 STRES(ISTRE)=STRND(IPOIN,ISTRE)/EJOIN
   20 WRITE(6,925)IPOIN,(STRES(ISTRE),ISTRE=4,5)
  925 FORMAT(1X,I5,2E12.4)
      RETURN
      END
```

7. USER INSTRUCTIONS FOR PROGRAM MINDLIN

These now follows a set of user instructions for the preparation of input data for program MINDLIN.

DATA SET 1 PROBLEM TITLE (20A4) 1 line

Cols. 1-80	TITLE	problem title

DATA SET 2 CONTROL DATA (5I5) 1 line

Cols. 1- 5	NPOIN	number of nodal points
6-10	NELEM	number of elements
11-15	NVFIX	number of restrained nodes
16-20	NCASE	number of load cases
21-25	NMATS	number of material sets

DATA SET 3 ELEMENT CONNECTIONS (11I5) 1 line

Cols. 1- 5	NUMEL	element number
6-10	MATNO(NUMEL)	material set number
11-15	LNODS(NUMEL,1)	1st nodal connection number
16-20	LNODS(NUMEL,2)	2nd nodal connection number
.	.	. .
.	.	. .
.	.	. .
51-55	LNODS(NUMEL,9)	9th nodal connection number

DATA SET 4 NODAL COORDINATES (I5,2F10.5)
(1 line for each node unless it is a central node in which case the coordinates should be omitted as a central node is a hierarchical node. Note that coordinates for midside nodes on straight edges may be omitted)

Cols. 1- 5	IPOIN	nodal point number
6-15	COORD(IPOIN,1)	x-coordinate of node
16-25	COORD(IPOIN,2)	y-coordinate of node

DATA SET 5 RESTRAINED NODES (2I5,3F10.6) NVFIX lines

Cols. 1- 5	NOFIX(IVFIX)	restrained node number
8-10	IFPRE	condition of restraint ABC for each degree of freedom 0 - free 1 - fixed
11-20	PRESC(IVFIX,1)	prescribed value for w
21-30	PRESC(IVFIX,2)	prescribed value for ψ_x
31-40	PRESC(IVFIX,3)	prescribed value for ψ_y

DATA SET 6	MATERIAL SET DATA 1	(I5)	1 line

Cols. 1- 5	NUMAT	material set number

DATA SET 7	MATERIAL SET DATA 2	(8F10.5)	1 line

Cols. 1-10	PROPS(NUMAT,1)	1st material property E or D_{11}
11-20	PROPS(NUMAT,2)	2nd material property ν or D_{12}
21-30	PROPS(NUMAT,3)	3rd material property t or D_{22}
31-40	PROPS(NUMAT,4)	4th material property 0 or D_{66}
41-50	PROPS(NUMAT,5)	5th material property 0 or D_{55}
51-60	PROPS(NUMAT,6)	6th material property 0 or D_{44}
61-70	PROPS(NUMAT,7)	uniform load intensity q

Repeat data sets 6 and 7 NMATS.

DATA SET 8	LOAD CASE TITLE	(20A4)	1 line

Cols. 1-80	TITLE	load case title

DATA SET 9	LOAD TYPE	(I5)	1 line

Cols. 1- 5	IPLOD	load type indicator 0 - uniform load 1 - point loads

DATA SET 10	NODAL LOADS	(I5,3F10.3)	1 line for each loaded node

Cols. 1- 5	LODPT	node number
6-15	POINT(1)	lateral force
16-25	POINT(2)	couple in xz-plane
26-35	POINT(3)	couple in yz-plane

(N.B. This data set should be terminated by the highest node number even if it is not loaded).

Repeat data sets 8-10 NCASE times.

8. SAMPLE SOLUTIONS

8.1 Introduction

This section contains the following examples analysed using program MINDLIN,

(a) a constant curvature patch test on a 2x2 mesh,

(b) a simply supported, uniformly loaded square plate,

(c) a clamped, uniformly loaded circular plate,

(d) fibre reinforced plates, and

(e) three-ply laminates.

8.2 Test Example No.1 - Patch Test

The patch test is taken as a test example for MINDLIN as it proves convergence of the heterosis element used in MINDLIN. The patch test appropriate for Mindlin plates is one which provides constant curvature and zero shear strain. Here displacements and rotations corresponding to a state of constant curvature and zero shear strain are applied at the nodes of the boundary of the mesh of four heterosis plate elements shown in Figure 8.2.1. Note that this example has the somewhat unusual feature of no applied loading.

The expressions for the displacement and normal rotation fields used in the patch test are given as

$$w(x,y) = \frac{\bar{a}}{2} x^2 + \bar{b}_{xy} + \bar{c}x + \bar{e}y + \frac{\bar{d}}{2} y^2 \qquad (8.2.1)$$

$$\psi_x(x,y) = - \theta_x(x,y) = \bar{a}x + \bar{b}y + \bar{c} \qquad 0 < x < a$$

$$\psi_y(x,y) = - \theta_y(x,y) = \bar{b}x + \bar{d}y + \bar{e} \qquad 0 < y < a$$

Note that these particular fields are chosen so that constant curvatures $-\psi_{x,x}, -\psi_{y,y}, (\psi_{x,y} + \psi_{y,x})$ and zero shear strains $(w_{,x} - \psi_x, w_{,y} - \psi_y)$ are obtained.

With only the prescribed nodal values corresponding to (8.2.1) applied on the nodes of the boundary of the mesh shown in Figure 8.2.1 and no applied forces or couples, MINDLIN should give nodal values at the internal nodes

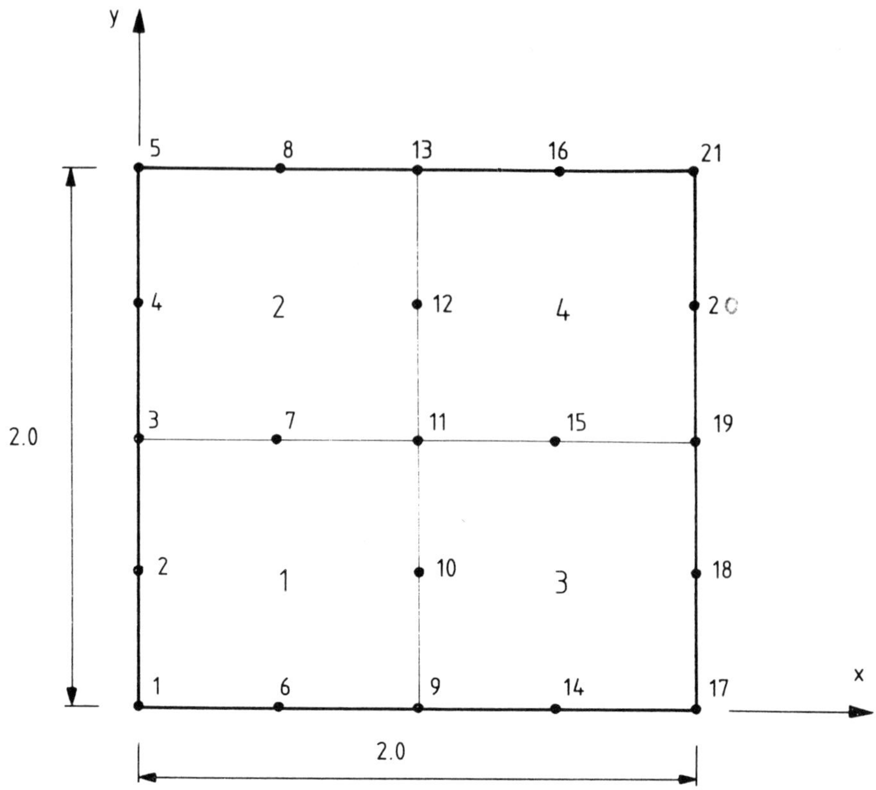

Figure 8.2.1 Constant curvature patch test
- geometry and mesh

corresponding to (8.2.1) and also curvatures and shear strains and hence bending moments and shear forces which have the following values,

$$M_x = -D(\psi_{x,x} + \nu\psi_{y,y}) = -D(\bar{a} + \nu\bar{d})$$

$$M_y = -D(\psi_{y,y} + \nu\psi_{x,x}) = -D(\bar{d} + \nu\bar{a})$$

$$M_{xy} = -\frac{(1-\nu)}{2} D(\psi_{x,y} + \psi_{y,x}) = -(1-\nu)\ D\bar{b}$$

$$Q_x = Q_y = 0 \qquad (8.2.2)$$

```
 PATCH TEST - CONSTANT CURVATURE
25    4   20    1    1
 1    1    1    6    9   10   11    7    3    2   22
 2    1    3    7   11   12   13    8    5    4   23
 3    1    9   14   17   18   19   15   11   10   24
 4    1   11   15   19   20   21   16   13   12   25
 1       0.0       0.0
 3       0.0       1.0
 5       0.0       2.0
 9       1.0       0.0
11       1.0       1.0
13       1.0       2.0
17       2.0       0.0
19       2.0       1.0
21       2.0       2.0
25       0.0       0.0
  1  111       0.0       4.0       7.0
  2  111      4.25       5.5      10.0
  3  111      10.0       7.0      13.0
  4  111     17.25       8.5      16.0
  5  111      26.0      10.0      19.0
  6  111      2.25       5.0       8.5
  8  111     31.25      11.0      20.5
  9  111       5.0       6.0      10.0
 13  111      37.0      12.0      22.0
 14  111      8.25       7.0      11.5
 16  111     43.25      13.0      23.5
 17  111      12.0       8.0      13.0
 18  111     19.25       9.5      16.0
 19  111      28.0      11.0      19.0
 20  111     38.25      12.5      22.0
 21  111      50.0      14.0      25.0
 22  100
 23  100
 24  100
 25  100
  1
 10920.0       0.3      0.01       0.0       0.0       0.0       0.0
 NO LOADS
  0
```

Figure 8.2.2 Input data for example 1

In the present case the following numerical values are chosen to define the patch test:

Side length $a = 2.0$

Flexural rigidity $D = 0.001$ (i.e. Young's modulus $E = 10.92 \times 10^3$, plate thickness $t = 0.01$)

Poisson's ratio $\nu = 0.3$

Patch test parameters $\bar{a} = -2.0$, $\bar{b} = -3.0$, $\bar{c} = -4.0$, $\bar{d} = -6.0$, $\bar{e} = -7.0$

Thus the displacements and normal rotation fields have the following form,

$$w = -(x^2 + 3xy + 4x + 7y + 3y^2)$$

$$\psi_x = -\theta_x = -(2x + 3y + 4)$$

$$\psi_y = -\theta_y = -(3x + 6y + 7) \qquad (8.2.3)$$

Typical input data for the constant curvature patch test is shown in Figure 8.2.2.

Using equations (8.2.2) the following stress resultants are expected at all points within the finite element mesh.

$$M_x = -.0038,\ M_y = -.0066,\ M_{xy} = -.0021,\ Q_x = 0,\ Q_y = 0$$

Typical output from MINDLIN, shown in Figure 8.2.3, confirms that the element passes the patch test and hence convergence is proved.

8.3 Test Example No.2 - Simply Supported, Uniformly Loaded, Square Slab

The purpose of this example of a simply supported, uniformly loaded square slab is to illustrate the notation adopted in MINDLIN for displacements, rotations and the stress resultants and also the accuracy of the heterosis element.

The slab and the 2x2 finite element mesh used in a symmetric quadrant are illustrated in Figure 8.3.1. The following properties are chosen for the slab:

Side length	$a = 1.0$	
Uniform load intensity	$q = 1.0$	
Flexural rigidity	$D = 1.0$	(i.e. isotropic material with Young's modulus $e = 10.92 \times 10^3$ and plate thickness $t = 0.1$)
Poisson's ratio	$\nu = 0.3$	

Typical input data for this problem is given in Figure 8.3.2.

The following values are expected from the Mindlin plate closed form solution:

DISPLACEMENTS

NODE	DISP.	XZ-ROT.	YZ-ROT.
1	0.000000E+00	0.400000E+01	0.700000E+01
2	0.425000E+01	0.550000E+01	0.100000E+02
3	0.100000E+02	0.700000E+01	0.130000E+02
4	0.172500E+02	0.850000E+01	0.160000E+02
5	0.260000E+02	0.100000E+02	0.190000E+02
6	0.225000E+01	0.500000E+01	0.850000E+01
7	0.137500E+02	0.800000E+01	0.145000E+02
8	0.312500E+02	0.110000E+02	0.205000E+02
9	0.500000E+01	0.600000E+01	0.100000E+02
10	0.107500E+02	0.750000E+01	0.130000E+02
11	0.180000E+02	0.900000E+01	0.160000E+02
12	0.267500E+02	0.105000E+02	0.190000E+02
13	0.370000E+02	0.120000E+02	0.220000E+02
14	0.825000E+01	0.700000E+01	0.115000E+02
15	0.227500E+02	0.100000E+02	0.175000E+02
16	0.432500E+02	0.130000E+02	0.235000E+02
17	0.120000E+02	0.800000E+01	0.130000E+02
18	0.192500E+02	0.950000E+01	0.160000E+02
19	0.280000E+02	0.110000E+02	0.190000E+02
20	0.382500E+02	0.125000E+02	0.220000E+02
21	0.500000E+02	0.140000E+02	0.250000E+02
22	0.000000E+00	0.305868E-12	-0.476226E-12
23	0.000000E+00	-0.320409E-12	0.467983E-12
24	0.000000E+00	-0.323712E-12	0.470296E-12
25	0.000000E+00	0.310449E-12	-0.473560E-12

REACTIONS

NODE	FORCE	XZ-MOMENT	YZ-MOMENT
1	-0.473173E-13	-0.983333E-03	-0.145000E-02
2	0.300325E-13	-0.253333E-02	-0.140000E-02
3	0.726247E-13	-0.126667E-02	-0.700000E-03
4	-0.522956E-14	-0.253333E-02	-0.140000E-02
5	-0.177161E-13	-0.283333E-03	0.750000E-03
6	0.209507E-13	-0.140000E-02	-0.440000E-02
8	-0.752257E-13	0.140000E-02	0.440000E-02
9	-0.445160E-13	-0.700000E-03	-0.220000E-02
13	0.935142E-13	0.700000E-03	0.220000E-02
14	-0.291546E-13	-0.140000E-02	-0.440000E-02
16	-0.777188E-13	0.140000E-02	0.440000E-02
17	0.862237E-14	0.283333E-03	-0.750000E-03
18	-0.180683E-13	0.253333E-02	0.140000E-02
19	0.532907E-14	0.126667E-02	0.700000E-03
20	-0.126478E-12	0.253333E-02	0.140000E-02
21	-0.375716E-13	0.983333E-03	0.145000E-02
22	0.595080E-13	0.000000E+00	0.000000E+00
23	-0.698766E-13	0.000000E+00	0.000000E+00
24	0.232513E-13	0.000000E+00	0.000000E+00
25	0.145928E-12	0.000000E+00	0.000000E+00

Figure 8.2.3(a) Output for examples
- displacements and reactions

STRESS RESULTANTS

BENDING MOMENTS AT NODES

NODE	MX-MOMENT	MY-MOMENT	MXY-MOMENT	MAX-P.M.	MIN-P.M.	ANGLE
1	-0.3800E-02	-0.6600E-02	-0.2100E-02	-0.2676E-02	-0.7724E-02	-0.2815E+02
2	-0.3800E-02	-0.6600E-02	-0.2100E-02	-0.2676E-02	-0.7724E-02	-0.2815E+02
3	-0.3800E-02	-0.6600E-02	-0.2100E-02	-0.2676E-02	-0.7724E-02	-0.2815E+02
4	-0.3800E-02	-0.6600E-02	-0.2100E-02	-0.2676E-02	-0.7724E-02	-0.2815E+02
5	-0.3800E-02	-0.6600E-02	-0.2100E-02	-0.2676E-02	-0.7724E-02	-0.2815E+02
6	-0.3800E-02	-0.6600E-02	-0.2100E-02	-0.2676E-02	-0.7724E-02	-0.2815E+02
7	-0.3800E-02	-0.6600E-02	-0.2100E-02	-0.2676E-02	-0.7724E-02	-0.2815E+02
8	-0.3800E-02	-0.6600E-02	-0.2100E-02	-0.2676E-02	-0.7724E-02	-0.2815E+02
9	-0.3800E-02	-0.6600E-02	-0.2100E-02	-0.2676E-02	-0.7724E-02	-0.2815E+02
10	-0.3800E-02	-0.6600E-02	-0.2100E-02	-0.2676E-02	-0.7724E-02	-0.2815E+02
11	-0.3800E-02	-0.6600E-02	-0.2100E-02	-0.2676E-02	-0.7724E-02	-0.2815E+02
12	-0.3800E-02	-0.6600E-02	-0.2100E-02	-0.2676E-02	-0.7724E-02	-0.2815E+02
13	-0.3800E-02	-0.6600E-02	-0.2100E-02	-0.2676E-02	-0.7724E-02	-0.2815E+02
14	-0.3800E-02	-0.6600E-02	-0.2100E-02	-0.2676E-02	-0.7724E-02	-0.2815E+02
15	-0.3800E-02	-0.6600E-02	-0.2100E-02	-0.2676E-02	-0.7724E-02	-0.2815E+02
16	-0.3800E-02	-0.6600E-02	-0.2100E-02	-0.2676E-02	-0.7724E-02	-0.2815E+02
17	-0.3800E-02	-0.6600E-02	-0.2100E-02	-0.2676E-02	-0.7724E-02	-0.2815E+02
18	-0.3800E-02	-0.6600E-02	-0.2100E-02	-0.2676E-02	-0.7724E-02	-0.2815E+02
19	-0.3800E-02	-0.6600E-02	-0.2100E-02	-0.2676E-02	-0.7724E-02	-0.2815E+02
20	-0.3800E-02	-0.6600E-02	-0.2100E-02	-0.2676E-02	-0.7724E-02	-0.2815E+02
21	-0.3800E-02	-0.6600E-02	-0.2100E-02	-0.2676E-02	-0.7724E-02	-0.2815E+02
22	-0.3800E-02	-0.6600E-02	-0.2100E-02	-0.2676E-02	-0.7724E-02	-0.2815E+02
23	-0.3800E-02	-0.6600E-02	-0.2100E-02	-0.2676E-02	-0.7724E-02	-0.2815E+02
24	-0.3800E-02	-0.6600E-02	-0.2100E-02	-0.2676E-02	-0.7724E-02	-0.2815E+02
25	-0.3800E-02	-0.6600E-02	-0.2100E-02	-0.2676E-02	-0.7724E-02	-0.2815E+02

SHEAR FORCES AT NODES

NODE	XZ-SHEAR	YZ-SHEAR
1	0.1310E-12	0.3081E-13
2	0.5294E-13	0.1308E-13
3	-0.4853E-13	-0.8389E-13
4	0.6349E-14	-0.5346E-13
5	0.8465E-13	0.5620E-13
6	0.6640E-13	0.3327E-13
7	-0.6458E-13	-0.4038E-13
8	0.9384E-13	0.4749E-13
9	0.1086E-13	0.4365E-13
10	-0.2349E-13	0.1393E-13
11	0.1196E-13	-0.3100E-13
12	0.4692E-13	-0.1872E-13
13	0.1210E-13	0.8775E-14
14	0.6834E-13	0.4170E-13
15	-0.3986E-13	-0.1488E-13
16	0.2692E-13	-0.4165E-14
17	0.1168E-12	0.3185E-13
18	0.2123E-13	0.4843E-13
19	-0.1843E-12	0.3535E-13
20	-0.8076E-13	0.9296E-14
21	0.1327E-12	0.1290E-13
22	0.5829E-14	0.9714E-14
23	0.9714E-14	-0.9714E-14
24	0.7772E-14	0.3497E-13
25	0.1972E-30	-0.3109E-13

Figure 8.2.3(b) Output for example 1
- bending moments and shear forces

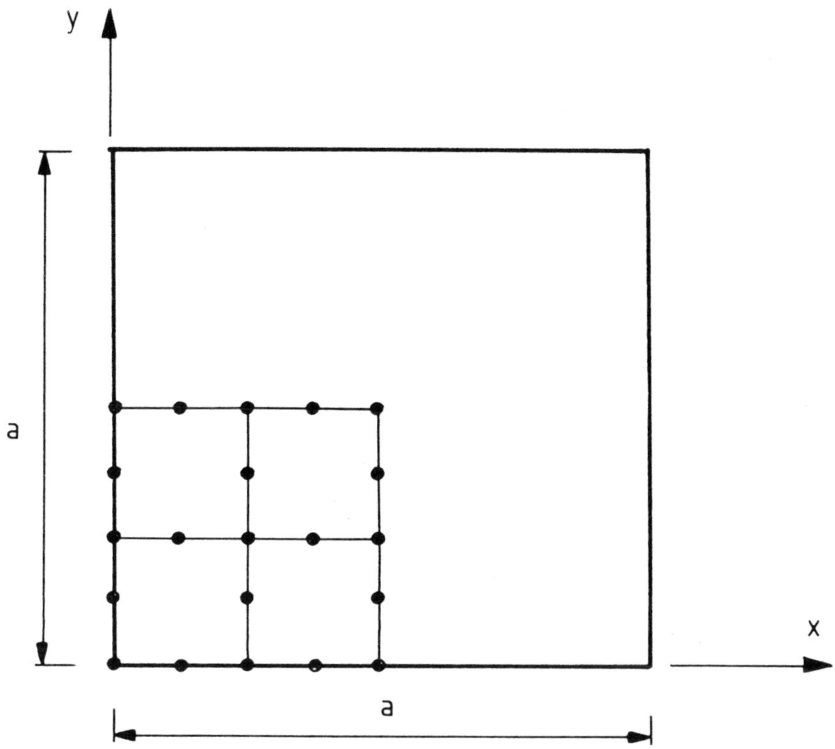

$E = 10.92 \times 10^3$
$\nu = 0.3$
$t = 0.01$
$a = 1.00$

Figure 8.3.1 Simply supported uniformly loaded, square slab - geometry and mesh

```
  SIMPLY SUPPORTED SQUARE PLATE, 2X2 MESH, D/H=0.1
 25    4   20    1    1
  1    1    1    6    9   10   11    7    3    2   22
  2    1    3    7   11   12   13    8    5    4   23
  3    1    9   14   17   18   19   15   11   10   24
  4    1   11   15   19   20   21   16   13   12   25
  1        0.0       0.0
  3        0.0       0.25
  5        0.0       0.5
  9        0.25      0.0
 11        0.25      0.25
 13        0.25      0.5
 17        0.5       0.0
 19        0.5       0.25
 21        0.5       0.5
 25        0.0       0.0
  1  111
  2  101
  3  101
  4  101
  5  101
  6  110
  8  001
  9  110
 13  001
 14  110
 16  001
 17  110
 18  010
 19  010
 20  010
 21  011
 22  100
 23  100
 24  100
 25  100
  1
 10920.0       0.3       0.1       0.0       0.0       0.0       1.0
UDL
  0    1
```

Figure 8.3.2 Input data for example 2

Lateral displacement $w = 0.00427$ at centre of plate

Bending moments $M_x = M_y = 0.0479$ at centre of plate

Shear force $Q_x = 0.338$ at middle edge of plate

Twisting moment $M_{xy} = 0.0325$ at corner of plate

Typical output from MINDLIN, shown in Figure 8.3.3, shows the excellent accuracy achieved by this coarse mesh of heterosis elements.

DISPLACEMENTS

NODE	DISP.	XZ-ROT.	YZ-ROT.
1	0.000000E+00	0.000000E+00	0.000000E+00
2	0.000000E+00	0.543055E-02	0.000000E+00
3	0.000000E+00	0.989111E-02	0.000000E+00
4	0.000000E+00	0.125661E-01	0.000000E+00
5	0.000000E+00	0.135156E-01	0.000000E+00
6	0.000000E+00	0.000000E+00	0.543055E-02
7	0.126444E-02	0.878433E-02	0.345732E-02
8	0.172200E-02	0.121085E-01	0.000000E+00
9	0.000000E+00	0.000000E+00	0.989111E-02
10	0.126444E-02	0.345732E-02	0.878433E-02
11	0.225748E-02	0.630609E-02	0.630609E-02
12	0.289606E-02	0.811601E-02	0.326844E-02
13	0.309963E-02	0.877420E-02	0.000000E+00
14	0.000000E+00	0.000000E+00	0.125661E-01
15	0.289606E-02	0.326844E-02	0.811601E-02
16	0.398362E-02	0.455949E-02	0.000000E+00
17	0.000000E+00	0.000000E+00	0.135156E-01
18	0.172200E-02	0.000000E+00	0.121085E-01
19	0.309963E-02	0.000000E+00	0.877420E-02
20	0.398362E-02	0.000000E+00	0.455949E-02
21	0.426902E-02	0.000000E+00	0.000000E+00
22	0.000000E+00	0.636367E-04	0.636367E-04
23	0.000000E+00	0.740310E-04	-0.115907E-04
24	0.000000E+00	-0.115907E-04	0.740310E-04
25	0.000000E+00	0.243240E-04	0.243240E-04

REACTIONS

NODE	FORCE	XZ-MOMENT	YZ-MOMENT
1	-0.695702E-03	-0.135815E-02	-0.135815E-02
2	-0.349264E-01	0.000000E+00	-0.469070E-02
3	-0.204734E-01	0.000000E+00	-0.172762E-02
4	-0.559913E-01	0.000000E+00	-0.169526E-02
5	-0.132611E-01	0.000000E+00	-0.135465E-03
6	-0.349264E-01	-0.469070E-02	0.000000E+00
8	0.000000E+00	0.000000E+00	-0.331962E-02
9	-0.204734E-01	-0.172762E-02	0.000000E+00
13	0.000000E+00	0.000000E+00	-0.305359E-02
14	-0.559913E-01	-0.169526E-02	0.000000E+00
16	0.000000E+00	0.000000E+00	-0.736452E-02
17	-0.132611E-01	-0.135465E-03	0.000000E+00
18	0.000000E+00	-0.331962E-02	0.000000E+00
19	0.000000E+00	-0.305359E-02	0.000000E+00
20	0.000000E+00	-0.736452E-02	0.000000E+00
21	0.000000E+00	-0.204364E-02	-0.204364E-02
22	-0.769965E-02	0.000000E+00	0.000000E+00
23	-0.718324E-03	0.000000E+00	0.000000E+00
24	-0.718324E-03	0.000000E+00	0.000000E+00
25	-0.976152E-03	0.000000E+00	0.000000E+00

Figure 8.3.3(a) Output for example 2
- displacement and reactions

STRESS RESULTANTS

BENDING MOMENTS AT NODES

NODE	MX-MOMENT	MY-MOMENT	MXY-MOMENT	MAX-P.M.	MIN-P.M.	ANGLE
1	0.0000E+00	0.0000E+00	-0.3313E-01	0.3313E-01	-0.3313E-01	-0.4500E+02
2	0.1389E-02	0.4167E-03	-0.2871E-01	0.2962E-01	-0.2781E-01	-0.4452E+02
3	0.3368E-02	0.1011E-02	-0.2105E-01	0.2327E-01	-0.1889E-01	-0.4340E+02
4	0.3421E-02	0.1026E-02	-0.1044E-01	0.1273E-01	-0.8282E-02	-0.4173E+02
5	0.3549E-02	0.1065E-02	-0.2433E-03	0.3572E-02	0.1041E-02	-0.5542E+01
6	0.4167E-03	0.1389E-02	-0.2871E-01	0.2962E-01	-0.2781E-01	0.4452E+02
7	0.1846E-01	0.1805E-01	-0.1824E-01	0.3650E-01	0.1106E-04	-0.4467E+02
8	0.2334E-01	0.2026E-01	-0.2113E-03	0.2335E-01	0.2024E-01	-0.3908E+01
9	0.1011E-02	0.3368E-02	-0.2105E-01	0.2327E-01	-0.1889E-01	0.4340E+02
10	0.1805E-01	0.1846E-01	-0.1824E-01	0.3650E-01	0.1106E-04	0.4467E+02
11	0.3165E-01	0.3165E-01	-0.1380E-01	0.4545E-01	0.1784E-01	0.4500E+02
12	0.3854E-01	0.3452E-01	-0.7128E-02	0.4394E-01	0.2912E-01	-0.3711E+02
13	0.4148E-01	0.3708E-01	-0.2305E-03	0.4149E-01	0.3707E-01	-0.2989E+01
14	0.1026E-02	0.3421E-02	-0.1044E-01	0.1273E-01	-0.8282E-02	0.4173E+02
15	0.3452E-01	0.3854E-01	-0.7128E-02	0.4394E-01	0.2912E-01	0.3711E+02
16	0.4564E-01	0.4568E-01	-0.5884E-04	0.4573E-01	0.4560E-01	0.3539E+02
17	0.1065E-02	0.3549E-02	-0.2433E-03	0.3572E-02	0.1041E-02	0.5542E+01
18	0.2026E-01	0.2334E-01	-0.2113E-03	0.2335E-01	0.2024E-01	0.3908E+01
19	0.3708E-01	0.4148E-01	-0.2305E-03	0.4149E-01	0.3707E-01	0.2989E+01
20	0.4568E-01	0.4564E-01	-0.5884E-04	0.4573E-01	0.4560E-01	-0.3539E+02
21	0.4921E-01	0.4921E-01	0.0000E+00	0.4921E-01	0.4921E-01	0.0000E+00
22	0.1026E-01	0.1026E-01	-0.2460E-01	0.3486E-01	-0.1434E-01	0.4500E+02
23	0.2195E-01	0.1917E-01	-0.9230E-02	0.2989E-01	0.1123E-01	-0.4072E+02
24	0.1917E-01	0.2195E-01	-0.9230E-02	0.2989E-01	0.1123E-01	0.4072E+02
25	0.4220E-01	0.4220E-01	-0.3615E-02	0.4582E-01	0.3859E-01	-0.4500E+02

SHEAR FORCES AT NODES

NODE	XZ-SHEAR	YZ-SHEAR
1	0.3586E-02	0.3586E-02
2	0.1528E+00	0.2467E-01
3	0.2954E+00	0.5309E-02
4	0.3136E+00	0.3630E-03
5	0.3386E+00	0.3586E-01
6	0.2467E-01	0.1528E+00
7	0.1901E+00	0.5441E-01
8	0.2376E+00	0.2903E-02
9	0.5309E-02	0.2954E+00
10	0.5441E-01	0.1901E+00
11	0.9968E-01	0.9968E-01
12	0.1171E+00	0.5139E-01
13	0.1384E+00	-0.1171E-01
14	0.3630E-03	0.3136E+00
15	0.5139E-01	0.1171E+00
16	0.6680E-01	0.4417E-04
17	0.3586E-01	0.3386E+00
18	0.2903E-02	0.2376E+00
19	-0.1171E-01	0.1384E+00
20	0.4417E-04	0.6680E-01
21	-0.6545E-02	-0.6545E-02
22	0.1076E+00	0.1076E+00
23	0.2136E+00	0.2463E-01
24	0.2463E-01	0.2136E+00
25	0.6035E-01	0.6035E-01

Figure 8.3.3(b) Output for example 2
– bending moments and shear forces

8.4 Test Example No.3 - Clamped, Uniformly Loaded, Circular Slab

In this example, a clamped, uniformly loaded, circular plate is analysed using the 12 element mesh shown in Figure 8.4.1. Table 8.4.1 compares displacements, bending moments and shear forces with thin plate solutions.

Typical input and output data for a plate of thickness-to-radius t/a = 0.1 is given in Figures 8.4.2 and 8.4.3 respectively.

TABLE 8.4.1 Results for circular clamped plate under uniformly distributed load

t/a	$w\ qa^4/D$	$M_x\ qa^2$		$Q_x\ qa$
	r=0	r=0	r=1	r=1
0.01	.01562	.0912	-.1213	-.5644
0.02	.01564	.0874	-.1222	-.5512
0.10	.01633	.0832	-.1238	-.5194
thin plate	.01565	.0813	-.1254	-.5000

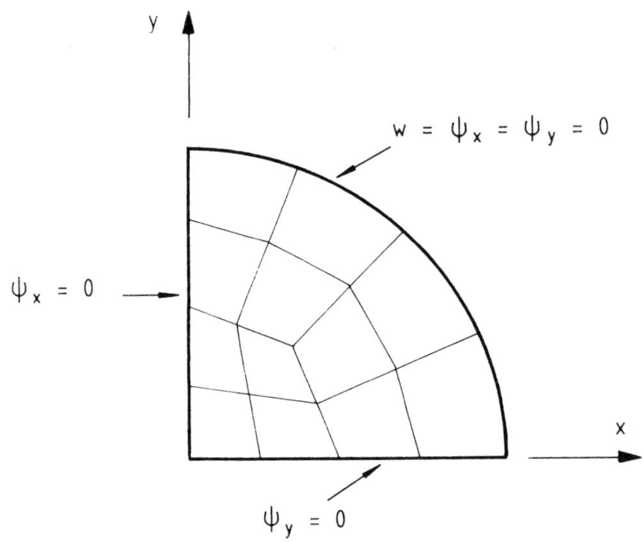

Figure 8.4.1 Mesh for symmetric quadrant of clamped circular plate

```
CLAMPED CIRCULAR PLATE - UDL - 12 ELEMENTS
 61   12   24    1    1
  1    1    1    2    3   12   21   20   19   10   11
  2    1    3    4    5   14   23   22   21   12   13
  3    1   19   20   21   30   39   38   37   28   29
  4    1   21   22   23   32   41   40   39   30   31
  5    1    5    6    7   16   25   24   23   14   15
  6    1   23   24   25   34   43   42   41   32   33
  7    1   39   40   41   42   43   50   55   48   49
  8    1   37   38   39   48   55   54   53   46   47
  9    1    7    8    9   18   27   26   25   16   17
 10    1   25   26   27   36   45   44   43   34   35
 11    1   55   50   43   44   45   52   57   56   51
 12    1   53   54   55   56   57   60   61   58   59
  1      0.0         0.0
  3     0.25         0.0
  5      0.5         0.0
  7     0.75         0.0
  9      1.0         0.0
 18   0.9808      0.1951
 19      0.0        0.25
 21   0.1768      0.1768
 23   0.4268      0.1768
 25   0.6402      0.2652
 27   0.9239      0.3827
 36   0.8315      0.5556
 37      0.0         0.5
 39   0.1768      0.4268
 41   0.3535      0.3535
 43   0.5303      0.5303
 45   0.7071      0.7071
 52   0.5556      0.8315
 53      0.0        0.75
 55   0.2652      0.6402
 57   0.3827      0.9239
 60   0.1951      0.9808
 61      0.0         1.0
  1  011
  2  001
  3  001
  4  001
  5  001
  6  001
  7  001
  8  001
  9  111
 10  010
 18  111
 19  010
 27  111
 28  010
 36  111
 37  010
 45  111
 46  010
 52  111
 53  010
 57  111
 58  010
 60  111
 61  111
  1
 10920.0       0.3       0.1       0.0       0.0       0.0       1.0
UDL
  0
```

Figure 8.4.2 Input for example 3

DISPLACEMENTS

NODE	DISP.	XZ-ROT.	YZ-ROT.
1	0.163331E-01	0.000000E+00	0.000000E+00
2	0.158449E-01	-0.766091E-02	0.000000E+00
3	0.143907E-01	-0.146436E-01	0.000000E+00
4	0.121552E-01	-0.201110E-01	0.000000E+00
5	0.931651E-02	-0.234016E-01	0.000000E+00
6	0.624192E-02	-0.238727E-01	0.000000E+00
7	0.327366E-02	-0.205919E-01	0.000000E+00
8	0.102590E-02	-0.128853E-01	0.000000E+00
9	0.000000E+00	0.000000E+00	0.000000E+00
10	0.158449E-01	0.000000E+00	-0.766091E-02
11	-0.372569E-04	-0.358313E-05	-0.358313E-05
12	0.146824E-01	-0.126303E-01	-0.520617E-02
13	-0.190760E-04	-0.351540E-04	-0.238898E-05
14	0.100050E-01	-0.224855E-01	-0.431918E-02
15	-0.739055E-05	0.307003E-04	0.788088E-04
16	0.425633E-02	-0.216229E-01	-0.421793E-02
17	-0.741032E-04	-0.316500E-03	0.109017E-03
18	0.000000E+00	0.000000E+00	0.000000E+00
19	0.143907E-01	0.000000E+00	-0.146436E-01
20	0.146824E-01	-0.520617E-02	-0.126303E-01
21	0.143948E-01	-0.103308E-01	-0.103308E-01
22	0.126645E-01	-0.165206E-01	-0.969623E-02
23	0.102163E-01	-0.209971E-01	-0.871337E-02
24	0.742494E-02	-0.222302E-01	-0.920119E-02
25	0.459547E-02	-0.208062E-01	-0.861940E-02
26	0.146274E-02	-0.139359E-01	-0.577132E-02
27	0.000000E+00	0.000000E+00	0.000000E+00
28	0.121552E-01	0.000000E+00	-0.201110E-01
29	-0.190760E-04	-0.238898E-05	-0.351540E-04
30	0.126645E-01	-0.969623E-02	-0.165206E-01
31	-0.358743E-04	0.190124E-04	0.190124E-04
32	0.100073E-01	-0.189526E-01	-0.128579E-01
33	-0.689338E-05	0.767748E-04	-0.326698E-04
34	0.425727E-02	-0.182741E-01	-0.123101E-01
35	-0.742454E-04	-0.145999E-03	-0.299207E-03
36	0.000000E+00	0.000000E+00	0.000000E+00
37	0.931651E-02	0.000000E+00	-0.234016E-01
38	0.100050E-01	-0.431918E-02	-0.224855E-01
39	0.102163E-01	-0.871337E-02	-0.209971E-01
40	0.100073E-01	-0.128579E-01	-0.189526E-01
41	0.932144E-02	-0.165477E-01	-0.165477E-01
42	0.624455E-02	-0.168857E-01	-0.168857E-01
43	0.327542E-02	-0.145628E-01	-0.145628E-01
44	0.102658E-02	-0.911476E-02	-0.911476E-02
45	0.000000E+00	0.000000E+00	0.000000E+00
46	0.624192E-02	0.000000E+00	-0.238727E-01
47	-0.739055E-05	0.788088E-04	0.307003E-04
48	0.742494E-02	-0.920119E-02	-0.222302E-01
49	-0.689338E-05	-0.326698E-04	0.767748E-04
50	0.425727E-02	-0.123101E-01	-0.182741E-01
51	-0.742454E-04	-0.299207E-03	-0.145999E-03
52	0.000000E+00	0.000000E+00	0.000000E+00
53	0.327366E-02	0.000000E+00	-0.205919E-01
54	0.425633E-02	-0.421793E-02	-0.216229E-01
55	0.459547E-02	-0.861940E-02	-0.208062E-01
56	0.146274E-02	-0.577132E-02	-0.139359E-01
57	0.000000E+00	0.000000E+00	0.000000E+00
58	0.102590E-02	0.000000E+00	-0.128853E-01
59	-0.741032E-04	0.109017E-03	-0.316500E-03
60	0.000000E+00	0.000000E+00	0.000000E+00
61	0.000000E+00	0.000000E+00	0.000000E+00

Figure 8.4.3(a) Output from example 3 - displacements

REACTIONS

NODE	FORCE	XZ-MOMENT	YZ-MOMENT
1	0.000000E+00	0.341395E-02	0.341395E-02
2	0.000000E+00	0.000000E+00	0.131685E-01
3	0.000000E+00	0.000000E+00	0.622694E-02
4	0.000000E+00	0.000000E+00	0.106782E-01
5	0.000000E+00	0.000000E+00	0.437875E-02
6	0.000000E+00	0.000000E+00	0.568645E-02
7	0.000000E+00	0.000000E+00	0.135413E-02
8	0.000000E+00	0.000000E+00	-0.180994E-02
9	-0.348519E-01	0.835757E-02	-0.160143E-02
10	0.000000E+00	0.131685E-01	0.000000E+00
18	-0.130906E+00	0.319313E-01	0.594752E-02
19	0.000000E+00	0.622694E-02	0.000000E+00
27	-0.611628E-01	0.152403E-01	0.631327E-02
28	0.000000E+00	0.106782E-01	0.000000E+00
36	-0.130893E+00	0.267790E-01	0.183761E-01
37	0.000000E+00	0.437875E-02	0.000000E+00
45	-0.697691E-01	0.118212E-01	0.118212E-01
46	0.000000E+00	0.568645E-02	0.000000E+00
52	-0.130893E+00	0.183761E-01	0.267790E-01
53	0.000000E+00	0.135413E-02	0.000000E+00
57	-0.611628E-01	0.631327E-02	0.152403E-01
58	0.000000E+00	-0.180994E-02	0.000000E+00
60	-0.130906E+00	0.594752E-02	0.319313E-01
61	-0.348519E-01	-0.160143E-02	0.835757E-02

STRESS RESULTANTS

BENDING MOMENTS AT NODES

NODE	MX-MOMENT	MY-MOMENT	MXY-MOMENT	MAX-P.M.	MIN-P.M.	ANGLE
1	0.8320E-01	0.8320E-01	0.0000E+00	0.8320E-01	0.8320E-01	0.0000E+00
2	0.7742E-01	0.8039E-01	-0.1711E-04	0.8039E-01	0.7742E-01	0.3304E+00
3	0.7060E-01	0.7519E-01	0.2460E-03	0.7521E-01	0.7059E-01	-0.3060E+01
4	0.5151E-01	0.6544E-01	-0.5350E-04	0.6544E-01	0.5151E-01	0.2200E+00
5	0.3273E-01	0.5389E-01	0.1430E-03	0.5390E-01	0.3273E-01	-0.3870E+00
6	0.4727E-03	0.3567E-01	-0.5916E-03	0.3568E-01	0.4628E-03	0.9628E+00
7	-0.3176E-01	0.1879E-01	-0.7953E-03	0.1880E-01	-0.3177E-01	0.9011E+00
8	-0.7770E-01	-0.9167E-02	-0.4082E-03	-0.9165E-02	-0.7771E-01	0.3412E+00
9	-0.1238E+00	-0.3714E-01	-0.7626E-04	-0.3714E-01	-0.1238E+00	0.5042E-01
10	0.8039E-01	0.7742E-01	-0.1711E-04	0.8039E-01	0.7742E-01	-0.3304E+00
11	0.7663E-01	0.7663E-01	-0.9751E-03	0.7761E-01	0.7566E-01	0.4500E+02
12	0.7263E-01	0.7398E-01	-0.1421E-02	0.7488E-01	0.7173E-01	0.3230E+02
13	0.5543E-01	0.6520E-01	-0.2637E-02	0.6587E-01	0.5477E-01	0.1418E+02
14	0.3780E-01	0.5410E-01	-0.3265E-02	0.5473E-01	0.3717E-01	0.1091E+02
15	0.9389E-02	0.3778E-01	-0.5629E-02	0.3885E-01	0.8314E-02	0.1082E+02
16	-0.1904E-01	0.1986E-01	-0.7415E-02	0.2123E-01	-0.2040E-01	0.1043E+02
17	-0.6805E-01	-0.8787E-02	-0.1223E-01	-0.6363E-02	-0.7047E-01	0.1121E+02
18	-0.1192E+00	-0.3983E-01	-0.1607E-01	-0.3669E-01	-0.1223E+00	0.1103E+02
19	0.7519E-01	0.7060E-01	0.2460E-03	0.7521E-01	0.7059E-01	0.3060E+01
20	0.7398E-01	0.7263E-01	-0.1421E-02	0.7488E-01	0.7173E-01	-0.3230E+02
21	0.7291E-01	0.7291E-01	-0.2847E-02	0.7575E-01	0.7006E-01	0.4500E+02
22	0.5829E-01	0.6488E-01	-0.4919E-02	0.6750E-01	0.5566E-01	0.2810E+02
23	0.4215E-01	0.5445E-01	-0.6362E-02	0.5714E-01	0.3945E-01	0.2298E+02
24	0.1692E-01	0.3918E-01	-0.1114E-01	0.4380E-01	0.1230E-01	0.2252E+02
25	-0.8616E-02	0.2024E-01	-0.1438E-01	0.2618E-01	-0.1456E-01	0.2245E+02
26	-0.5795E-01	-0.1197E-01	-0.2299E-01	-0.2451E-02	-0.6748E-01	0.2250E+02
27	-0.1105E+00	-0.4956E-01	-0.3047E-01	-0.3694E-01	-0.1231E+00	0.2249E+02
28	0.6544E-01	0.5151E-01	-0.5350E-04	0.6544E-01	0.5151E-01	-0.2200E+00
29	0.6520E-01	0.5543E-01	-0.2637E-02	0.6587E-01	0.5477E-01	-0.1418E+02
30	0.6488E-01	0.5829E-01	-0.4919E-02	0.6750E-01	0.5566E-01	-0.2810E+02

Figure 8.4.3(b) Output from example 3
- reactions and bending moments

31	0.5417E-01	0.5417E-01	-0.7007E-02	0.6117E-01	0.4716E-01	0.4500E+02
32	0.4291E-01	0.4874E-01	-0.8453E-02	0.5477E-01	0.3688E-01	0.3549E+02
33	0.1798E-01	0.2927E-01	-0.1422E-01	0.3892E-01	0.8325E-02	0.3418E+02
34	-0.7048E-02	0.7840E-02	-0.1943E-01	0.2121E-01	-0.2041E-01	0.3452E+02
35	-0.5064E-01	-0.2618E-01	-0.2963E-01	-0.6352E-02	-0.7047E-01	0.3379E+02
36	-0.9558E-01	-0.6342E-01	-0.3967E-01	-0.3669E-01	-0.1223E+00	0.3397E+02
37	0.5389E-01	0.3273E-01	0.1430E-03	0.5390E-01	0.3273E-01	0.3870E+00
38	0.5410E-01	0.3780E-01	-0.3265E-02	0.5473E-01	0.3717E-01	-0.1091E+02
39	0.5445E-01	0.4215E-01	-0.6362E-02	0.5714E-01	0.3945E-01	-0.2298E+02
40	0.4874E-01	0.4291E-01	-0.8453E-02	0.5477E-01	0.3688E-01	-0.3549E+02
41	0.4330E-01	0.4330E-01	-0.1055E-01	0.5385E-01	0.3275E-01	-0.4500E+02
42	0.1816E-01	0.1816E-01	-0.1764E-01	0.3579E-01	0.5218E-03	-0.4500E+02
43	-0.6494E-02	-0.6494E-02	-0.2528E-01	0.1879E-01	-0.3177E-01	-0.4500E+02
44	-0.4343E-01	-0.4343E-01	-0.3428E-01	-0.9151E-02	-0.7770E-01	-0.4500E+02
45	-0.8050E-01	-0.8050E-01	-0.4335E-01	-0.3715E-01	-0.1238E+00	0.4500E+02
46	0.3567E-01	0.4727E-03	-0.5916E-03	0.3568E-01	0.4628E-03	-0.9628E+00
47	0.3778E-01	0.9389E-02	-0.5629E-02	0.3885E-01	0.8314E-02	-0.1082E+02
48	0.3918E-01	0.1692E-01	-0.1114E-01	0.4380E-01	0.1230E-01	-0.2252E+02
49	0.2927E-01	0.1798E-01	-0.1422E-01	0.3892E-01	0.8325E-02	-0.3418E+02
50	0.7840E-02	-0.7048E-02	-0.1943E-01	0.2121E-01	-0.2041E-01	-0.3452E+02
51	-0.2618E-01	-0.5064E-01	-0.2963E-01	-0.6352E-02	-0.7047E-01	-0.3379E+02
52	-0.6342E-01	-0.9558E-01	-0.3967E-01	-0.3669E-01	-0.1223E+00	-0.3397E+02
53	0.1879E-01	-0.3176E-01	-0.7953E-03	0.1880E-01	-0.3177E-01	-0.9011E+00
54	0.1986E-01	-0.1904E-01	-0.7415E-02	0.2123E-01	-0.2040E-01	-0.1043E+02
55	0.2024E-01	-0.8616E-02	-0.1438E-01	0.2618E-01	-0.1456E-01	-0.2245E+02
56	-0.1197E-01	-0.5795E-01	-0.2299E-01	-0.2451E-02	-0.6748E-01	-0.2250E+02
57	-0.4956E-01	-0.1105E+00	-0.3047E-01	-0.3694E-01	-0.1231E+00	-0.2249E+02
58	-0.9167E-02	-0.7770E-01	-0.4082E-03	-0.9165E-02	-0.7771E-01	-0.3412E+00
59	-0.8787E-02	-0.6805E-01	-0.1223E-01	-0.6363E-02	-0.7047E-01	-0.1121E+02
60	-0.3983E-01	-0.1192E+00	-0.1607E-01	-0.3669E-01	-0.1223E+00	-0.1103E+02
61	-0.3714E-01	-0.1238E+00	-0.7626E-04	-0.3714E-01	-0.1238E+00	-0.5042E-01

SHEAR FORCES AT NODES

NODE	XZ-SHEAR	YZ-SHEAR
1	-0.3655E-03	-0.3655E-03
2	-0.6177E-01	-0.3655E-03
3	-0.1194E+00	-0.2550E-02
4	-0.1841E+00	0.9506E-03
5	-0.2432E+00	-0.1048E-02
6	-0.3156E+00	0.3945E-02
7	-0.3785E+00	-0.8405E-02
8	-0.4395E+00	0.7798E-03
9	-0.5194E+00	0.3499E-01
10	-0.3655E-03	-0.6177E-01
11	-0.5355E-01	-0.5355E-01
12	-0.1083E+00	-0.4431E-01
13	-0.1718E+00	-0.4355E-01
14	-0.2305E+00	-0.4896E-01
15	-0.2913E+00	-0.5073E-01
16	-0.3469E+00	-0.7133E-01
17	-0.4173E+00	-0.7834E-01
18	-0.4962E+00	-0.6137E-01
19	-0.2550E-02	-0.1194E+00
20	-0.4431E-01	-0.1083E+00
21	-0.8867E-01	-0.8867E-01
22	-0.1483E+00	-0.8917E-01
23	-0.2144E+00	-0.9316E-01
24	-0.2653E+00	-0.1087E+00
25	-0.3167E+00	-0.1320E+00
26	-0.3928E+00	-0.1626E+00
27	-0.4593E+00	-0.1905E+00
28	0.9506E-03	-0.1841E+00
29	-0.4355E-01	-0.1718E+00
30	-0.8917E-01	-0.1483E+00
31	-0.1413E+00	-0.1413E+00
32	-0.1971E+00	-0.1316E+00
33	-0.2427E+00	-0.1690E+00
34	-0.2953E+00	-0.1958E+00
35	-0.3505E+00	-0.2395E+00
36	-0.3939E+00	-0.3079E+00
37	-0.1048E-02	-0.2432E+00
38	-0.4896E-01	-0.2305E+00
39	-0.9316E-01	-0.2144E+00
40	-0.1316E+00	-0.1971E+00
41	-0.1748E+00	-0.1748E+00
42	-0.2239E+00	-0.2239E+00
43	-0.2672E+00	-0.2672E+00
44	-0.3109E+00	-0.3109E+00
45	-0.3674E+00	-0.3674E+00
46	0.3945E-02	-0.3156E+00
47	-0.5073E-01	-0.2913E+00
48	-0.1087E+00	-0.2653E+00
49	-0.1690E+00	-0.2427E+00
50	-0.1958E+00	-0.2953E+00
51	-0.2395E+00	-0.3505E+00
52	-0.3079E+00	-0.3939E+00
53	-0.8405E-02	-0.3785E+00
54	-0.7133E-01	-0.3469E+00
55	-0.1320E+00	-0.3167E+00
56	-0.1626E+00	-0.3928E+00
57	-0.1905E+00	-0.4593E+00
58	0.7798E-03	-0.4395E+00
59	-0.7834E-01	-0.4173E+00
60	-0.6137E-01	-0.4962E+00
61	0.3499E-01	-0.5194E+00

Figure 8.4.3(c) Output from example 3
Bending moments and shear forces

8.5 TEST EXAMPLE No.4 - FIBRE REINFORCED PLATES

The flexure of uniformly loaded, square, boron/epoxy and glass/epoxy composite simply supported plates is considered. General details for these plates have already been presented in Chapter 1. Results obtained by Hussainy and Srinivas [23] are compared with those obtained by MINDLIN in Table 8.5.1.

TABLE 8.5.1 Results for uniformly-loaded, square, boron-epoxy and glass-epoxy composite, simply supported plates

Orientation	RESULTS AT CENTRE OF PLOT					
	Ef_w/hq		M_x/h^2q		M_y/h^2q	
	Hussainy and Srinivas	FE Element	Hussainy and Srinivas	FE Element	Hussainy and Srinivas	FE Element
			Boron epoxy			
30°	2124.0	2123.7	5.821	5.830	2.228	2.228
45°	1960.5	1960.6	3.533	3.533	3.533	3.533
60°	2124.0	2123.7	2.228	2.228	5.821	5.830
			Glass epoxy			
30°	1283.7	1283.5	5.501	5.519	3.072	3.080
45°	1205.0	1205.0	3.953	3.962	3.953	3.962

8.6 TEST EXAMPLE No.5 - THREE PLY LAMINATES

Values for the central deflection of uniformly loaded, square, simply supported, three ply laminated plates (described earlier in Chapter 1) are presented in Table 8.6.1. Results are compared with those given by Srinivas and Rao [24] which are based on 'exact' analysis. If the transverse shear rigidities are multiplied by 10^4 then the thin plate theory results are obtained.

TABLE 8.6.1 Central deflection for three-ply square laminated plates under uniform load q

C	Analytical Method	$wQ_{66}(2)/hq$
1	Exact theory	181.05
	FE analysis ($k_1^2=k_2^2=0.83$)	181.19
	Thin plate theory	168.38
	FE analysis ($G_{13}, G_{23} \times 10^4$)	166.95
10	Exact theory	41.906
	FE analysis ($k_1^2=k_2^2=0.3521$)	41.963
	Thin plate theory	31.241
	FE analysis ($G_{13}, G_{23} \times 10^4$)	30.980
50	Exact theory	16.753
	FE analysis ($k_1^2=k_2^2=0.0938$)	16.817
	Thin plate theory	6.762
	FE analysis ($G_{13}, G_{23} \times 10^4$)	6.713

$$(Q_{11}: Q_{12}: Q_{22}: Q_{66}: C_{44}: C_{55}) = (3.802: 0.879: 1.996: 1: 0.608: 1.015)$$

$$a/b = 1,\ h/a = 0.1;\ C = \frac{Q_{66}(1)}{Q_{66}(2)} = \frac{Q_{66}(3)}{Q_{66}(2)}$$

REFERENCES

1. MINDLIN, R.D.
Influence of rotary inertia and shear on flexural motions of isotropic elastic plates
J. Appl. Mech., 18, 31-38 (1951).

2. ZIENKIEWICZ, O.C., TAYLOR, R.L. and TOO, J.M.
Reduced integration in general analysis of plates and shells
Int. J. Num. Meth. Engng., 3, 275-290 (1971).

3. HINTON, E., RAZZAQUE, A., ZIENKIEWICZ, O.C. and DAVIES, J.D.
Simple finite element solution for plates of homogeneous, sandwich and cellular construction
Proc. ICE, Part II, 59, 43-65, (1975).

4. ZIENKIEWICZ, O.C. and HINTON, E.
Reduced integration, function smoothing and non-conformity in finite element analysis (with special reference to thick plates)
J. Franklin Institute, 302,

5. PUGH, E.D.L., HINTON, E. and ZIENKIEWICZ, O.C.
A study of quadrilateral plate bending elements with reduced integration
Int. J. Num. Meth. Engng., 12, 1059-1079, (1978).

6. HUGHES, T.J.R., COHEN, M. and HAROUN, M.
Reduced and selective integration techniques in the finite element analysis of plates
Nucl. Engng. and Design, 46, 203-222 (1978).

7. HINTON, E., SALONEN, E.M. and BICANIC, N.
A study of locking phenomena in isoparametric
3rd MAFELAP Conf. Brunel Univ., Uxbridge 1978, Proc. (ed. J.R. Whiteman), Academic Press, London (1979).

8. MALKUS, D.S. and HUGHES, T.J.R.
Mixed finite elements - reduced and selective integration techniques: a unification of concepts
Comp. Meth. in Appl. Mech. Engng., 15, 63-81 (1978).

9. HUGHES, T.J.R. and TEZDUYAR, T.E.
Finite elements based upon Mindlin plate theory with particular reference to the four-node bilinear isoparametric element
ASME. J. Appl. Mech., 48, 587-596 (1981).

10. TESSLER, A.
On a conforming Mindlin-type plate element
Fourth MAFELAP Conf., Brunel Univ., Uxbridge, 1981
Proc. (ed. J.R. Whiteman), Academic Press, London (1982).

11. MacNEAL, R.H.
A simple quadrilateral shell element
Comput. and Structures, 8, 175-183 (1978).

12. TESSLER, A. and HUGHES, T.J.R.
An improved treatment of transverse shear in the Mindlin-type four-node quadrilateral element
Comp. Meth. in Appl. Mech. Engng., 39, 311-335 (1983).

13. DVORKIN, E.N. and BATHE, K.J.
A continuum mechanics based four-node shell element for general non-linear analysis
Eng. Comput., 1, 77-88, (1984).

14. BATHE, K.J. and DVORKIN, E.N.
A four-node plate bending element based on Mindlin/Reissner plate theory and a mixed interpolation
Int. J. Num. Meth. Engng. (in press).

15. CRISFIELD, M.A.
A quadratic Mindlin element using shear constraints
Computers and Structures (in press).

16. PARK, K.C. and FLAGGS, D.L.
A rank-sufficient four-noded plate element with one-point integration, part I: element design
Comp. Meth. in Appl. Mech. and Engng. (in press).

17. BELYTSCHKO, T., TSAY, C.S. and LIU,W.K.
A stabilisation matrix for the bilinear Mindlin plate element
Comp. Meth. in Appl. Mech. and Engng., 29, 313-327, (1981).

18. BATHE, K.J.
Finite element procedures in engineering analysis
Prentice-Hall, Englewood Cliffs, New Jersey (1982).

19. HUGHES, T.J.R. and COHEN, M.
The 'heterosis' finite element for plate bending
Comput. and Structures, 9, 445-450 (1978).

20. HINTON, E., SCOTT, F.C. and RICKETTS, R.E.
Local least squares smoothing for parabolic isoparametric elements
Int. J. Num. Meth. Engng., 9, 235-238 (1975).

21. HINTON, E. and OWEN,D.R.J.
Finite element programming
Academic Press, London (1977).

22. OWEN,D.R.J. and HINTON, E.
Finite elements in plasticity: theory and practice
Pineridge Press, Swansea (1981).

23. HUSSAINY, S.A. and SRINIVAS, S.
Flexure of rectangular composite plates
Fibre Science and Technology, 8, 59-76, 1975.

24. SRINIVAS, S. and RAO, A.K.
Bending, vibration and buckling of simply supported thick orthotropic rectangular plates and laminates
Int. J. Solids Structures, 6, 1463-1481 (1970).

APPENDIX - SUBROUTINE FRONT

Subroutine FRONT is now listed. For a description of a related solver, see references [21] and [22]

```
      SUBROUTINE FRONT     (ASDIS,ELOAD,EQRHS,EQUAT,ESTIF,FIXED,
     .                      IFFIX,GLOAD,GSTIF,KRESL,
     .                      LNODS,LOCEL,MBUFA,MELEM,MEVAB,MFRON,
     .                      MSTIF,MTOTV,MVFIX,NACVA,NAMEV,NDEST,
     .                      NDOFN,NELEM,NEVAB,NNODE,NOFIX,NPIVO,
     .                      NPOIN,NTOTV,TREAC,VECRV,VIXED)
C***********************************************************************
C
C***  FRONTAL SOLVER FOR SYMMETRIC POSITIVE DEFINITE MATRICES
C
C***********************************************************************
      IMPLICIT REAL*8(A-H,O-Z)
      DIMENSION ASDIS(MTOTV),ELOAD(MELEM,MEVAB),EQRHS(MBUFA),
     .          EQUAT(MFRON,MBUFA),ESTIF(MEVAB,MEVAB),FIXED(MTOTV),
     .          IFFIX(MTOTV)                ,GLOAD(MFRON),
     .          GSTIF(MSTIF),LNODS(MELEM,9) ,LOCEL(MEVAB),NACVA(MFRON),
     .          NAMEV(MBUFA),NDEST(MEVAB),NOFIX(MVFIX),NPIVO(MBUFA),
     .          TREAC(MVFIX,NDOFN),
     .          VECRV(MFRON),VIXED(MTOTV)
      NFUNC(I,J)=(J*J-J)/2+I
      CALL RAZERO(TREAC,MVFIX,NDOFN)
C
C*** CHANGE THE SIGN OF THE LAST APPEARANCE OF EACH NODE
C
      IF(KRESL.GT.1) GO TO 455
      DO 140 IPOIN=1,NPOIN
      KLAST=0
      DO 130 IELEM=1,NELEM
      DO 120 INODE=1,NNODE
      IF(LNODS(IELEM,INODE).NE.IPOIN) GO TO 120
      KLAST=IELEM
      NLAST=INODE
  120 CONTINUE
  130 CONTINUE
      IF(KLAST.NE.0) LNODS(KLAST,NLAST)=-IPOIN
  140 CONTINUE
  455 CONTINUE
C
C*** START BY INITIALIZING EVERYTHING THAT MATTERS TO ZERO
C
      DO 450 IBUFA=1,MBUFA
  450 EQRHS(IBUFA)=0.0
      DO 150 ISTIF=1,MSTIF
  150 GSTIF(ISTIF)=0.0
      DO 160 IFRON=1,MFRON
      GLOAD(IFRON)=0.0
      VECRV(IFRON)=0.0
      NACVA(IFRON)=0
      DO 160 IBUFA=1,MBUFA
  160 EQUAT(IFRON,IBUFA)=0.0
C
C*** AND PREPARE FOR DISC READING AND WRITING OPERATIONS
C
      NBUFA=0
      IF(KRESL.GT.1) NBUFA=MBUFA
      REWIND 1
      REWIND 2
      REWIND 4
      REWIND 8
C
C*** ENTER MAIN ELEMENT ASSEMBLY-REDUCTION LOOP
C
```

```
      NFRON=0
      KELVA=0
      DO 320 IELEM=1,NELEM
      IF(KRESL.GT.1) GO TO 400
      KEVAB=0
      READ(1) ESTIF
      DO 170 INODE=1,NNODE
      DO 170 IDOFN=1,NDOFN
      NPOSI=(INODE-1)*NDOFN+IDOFN
      LOCNO=LNODS(IELEM,INODE)
      IF(LOCNO.GT.0) LOCEL(NPOSI)=(LOCNO-1)*NDOFN+IDOFN
      IF(LOCNO.LT.0) LOCEL(NPOSI)=(LOCNO+1)*NDOFN-IDOFN
  170 CONTINUE
C
C*** START BY LOOKING FOR EXISTING DESTINATIONS
C
      DO 210 IEVAB=1,NEVAB
      NIKNO=IABS(LOCEL(IEVAB))
      KEXIS=0
      DO 180 IFRON=1,NFRON
      IF(NIKNO.NE.NACVA(IFRON)) GO TO 180
      KEVAB=KEVAB+1
      KEXIS=1
      NDEST(KEVAB)=IFRON
  180 CONTINUE
      IF(KEXIS.NE.0) GO TO 210
C
C*** WE NOW SEEK NEW EMPTY PLACES FOR DESTINATION VECTOR
C
      DO 190 IFRON=1,MFRON
      IF(NACVA(IFRON).NE.0) GO TO 190
      NACVA(IFRON)=NIKNO
      KEVAB=KEVAB+1
      NDEST(KEVAB)=IFRON
      GO TO 200
  190 CONTINUE
C
C*** THE NEW PLACES MAY DEMAND AN INCREASE IN CURRENT FRONTWIDTH
C
  200 IF(NDEST(KEVAB).GT.NFRON) NFRON=NDEST(KEVAB)
  210 CONTINUE
      WRITE(8) LOCEL,NDEST,NACVA,NFRON
  400 IF(KRESL.GT.1) READ(8) LOCEL,NDEST,NACVA,NFRON
C
C*** ASSEMBLE ELEMENT LOADS
C
      DO 220 IEVAB=1,NEVAB
      IDEST=NDEST(IEVAB)
      GLOAD(IDEST)=GLOAD(IDEST)+ELOAD(IELEM,IEVAB)
C
C*** ASSEMBLE THE ELEMENT STIFFNESSES-BUT NOT IN RESOLUTION
C
      IF(KRESL.GT.1) GO TO 402
      DO 222 JEVAB=1,IEVAB
      JDEST=NDEST(JEVAB)
      NGASH=NFUNC(IDEST,JDEST)
      NGISH=NFUNC(JDEST,IDEST)
      IF(JDEST.GE.IDEST) GSTIF(NGASH)=GSTIF(NGASH)+ESTIF(IEVAB,JEVAB)
      IF(JDEST.LT.IDEST) GSTIF(NGISH)=GSTIF(NGISH)+ESTIF(IEVAB,JEVAB)
  222 CONTINUE
  402 CONTINUE
  220 CONTINUE
C
C*** RE-EXAMINE EACH ELEMENT NODE, TO ENQUIRE WHICH CAN BE ELIMINATED
C
      DO 310 IEVAB=1,NEVAB
      NIKNO=-LOCEL(IEVAB)
      IF(NIKNO.LE.0) GO TO 310
```

```
C
C*** FIND POSITIONS OF VARIABLES READY FOR ELIMINATION
C
      DO 300 IFRON=1,NFRON
      IF(NACVA(IFRON).NE.NIKNO) GO TO 300
      NBUFA=NBUFA+1
C
C*** WRITE EQUATIONS TO DISC OR TO TAPE
C
      IF(NBUFA.LE.MBUFA) GO TO 406
      NBUFA=1
      IF(KRESL.GT.1) GO TO 408
      WRITE(2) EQUAT,EQRHS,NPIVO,NAMEV
      GO TO 406
  408 WRITE(4) EQRHS
      READ(2) EQUAT,EQRHS,NPIVO,NAMEV
  406 CONTINUE
C
C*** EXTRACT THE COEFFICIENTS OF THE NEW EQUATION FOR ELIMINATION
C
      IF(KRESL.GT.1) GO TO 404
      DO 230 JFRON=1,MFRON
      IF(IFRON.LT.JFRON) NLOCA=NFUNC(IFRON,JFRON)
      IF(IFRON.GE.JFRON) NLOCA=NFUNC(JFRON,IFRON)
      EQUAT(JFRON,NBUFA)=GSTIF(NLOCA)
  230 GSTIF(NLOCA)=0.0
  404 CONTINUE
C
C*** AND EXTRACT THE CORRESPONDING RIGHT HAND SIDES
C
      EQRHS(NBUFA)=GLOAD(IFRON)
      GLOAD(IFRON)=0.0
      KELVA=KELVA+1
      NAMEV(NBUFA)=NIKNO
      NPIVO(NBUFA)=IFRON
C
C*** DEAL WITH PIVOT
C
      PIVOT=EQUAT(IFRON,NBUFA)
      IF(PIVOT.GT.0.0) GO TO 235
      WRITE(6,900) NIKNO,PIVOT
  900 FORMAT(1H0,3X,52HNEGATIVE OR ZERO PIVOT ENCOUNTERED FOR VARIABLE N
     .O. ,I4,10H OF VALUE ,E17.6)
      STOP
  235 CONTINUE
      EQUAT(IFRON,NBUFA)=0.0
C
C*** ENQUIRE WHETHER PRESENT VARIABLE IS FREE OR PRESCRIBED
C
      IF(IFFIX(NIKNO).EQ.0) GO TO 250
C
C*** DEAL WITH A PRESCRIBED DEFLECTION
C
      DO 240 JFRON=1,NFRON
  240 GLOAD(JFRON)=GLOAD(JFRON)-FIXED(NIKNO)*EQUAT(JFRON,NBUFA)
      GO TO 280
C
C*** ELIMINATE A FREE VARIABLE - DEAL WITH THE RIGHT HAND SIDE FIRST
C
  250 DO 270 JFRON=1,NFRON
      GLOAD(JFRON)=GLOAD(JFRON)-EQUAT(JFRON,NBUFA)*EQRHS(NBUFA)/PIVOT
C
C*** NOW DEAL WITH THE COEFFICIENTS IN CORE
C
      IF(KRESL.GT.1) GO TO 418
      IF(EQUAT(JFRON,NBUFA).EQ.0.0) GO TO 270
      NLOCA=NFUNC(0,JFRON)
      CUREQ=EQUAT(JFRON,NBUFA)
```

```
      DO 260 LFRON=1,JFRON
      NGASH=LFRON+NLOCA
  260 GSTIF(NGASH)=GSTIF(NGASH)-CUREQ*EQUAT(LFRON,NBUFA)
     . /PIVOT
  418 CONTINUE
  270 CONTINUE
  280 EQUAT(IFRON,NBUFA)=PIVOT
C
C*** RECORD THE NEW VACANT SPACE, AND REDUCE FRONTWIDTH IF POSSIBLE
C
      NACVA(IFRON)=0
      GO TO 290
C
C*** COMPLETE THE ELEMENT LOOP IN THE FORWARD ELIMINATION
C
  300 CONTINUE
  290 IF(NACVA(NFRON).NE.0) GO TO 310
      NFRON=NFRON-1
      IF(NFRON.GT.0) GO TO 290
  310 CONTINUE
  320 CONTINUE
      IF(KRESL.EQ.1) WRITE(2) EQUAT,EQRHS,NPIVO,NAMEV
      BACKSPACE 2
C
C*** ENTER BACK-SUBSTITUTION PHASE. LOOP BACKWARDS THROUGH VARIABLES
C
      DO 340 IELVA=1,KELVA
C
C***READ A NEW BLOCK OF EQUATIONS - IF NEEDED
C
      IF(NBUFA.NE.0) GO TO 412
      BACKSPACE 2
      READ(2) EQUAT,EQRHS,NPIVO,NAMEV
      BACKSPACE 2
      NBUFA=MBUFA
      IF(KRESL.EQ.1) GO TO 412
      BACKSPACE 4
      READ(4) EQRHS
      BACKSPACE 4
  412 CONTINUE
C
C*** PREPARE TO BACK-SUBSTITUTE FROM THE CURRENT EQUATION
C
      IFRON=NPIVO(NBUFA)
      NIKNO=NAMEV(NBUFA)
      PIVOT=EQUAT(IFRON,NBUFA)
      IF(IFFIX(NIKNO).NE.0) VECRV(IFRON)=FIXED(NIKNO)
      IF(IFFIX(NIKNO).EQ.0) EQUAT(IFRON,NBUFA)=0.0
C
C*** BACK-SUBSTITUTE IN THE CURRENT EQUATION
C
      DO 330 JFRON=1,MFRON
  330 EQRHS(NBUFA)=EQRHS(NBUFA)-VECRV(JFRON)*EQUAT(JFRON,NBUFA)
C
C*** PUT THE FINAL VALUES WHERE THEY BELONG
C
      IF(IFFIX(NIKNO).EQ.0) VECRV(IFRON)=EQRHS(NBUFA)/PIVOT
      IF(IFFIX(NIKNO).NE.0) VIXED(NIKNO)=-EQRHS(NBUFA)
      NBUFA=NBUFA-1
      ASDIS(NIKNO)=VECRV(IFRON)
  340 CONTINUE
C
C*** STORE REACTIONS FOR PRINTING LATER
C
      KBOUN=1
      DO 370 IPOIN=1,NPOIN
      NLOCA=(IPOIN-1)*NDOFN
```

```
      DO 350 IDOFN=1,NDOFN                                              FRON 266
      NGUSH=NLOCA+IDOFN                                                 FRON 267
      IF(IFFIX(NGUSH).GT.0) GO TO 360                                   FRON 268
  350 CONTINUE                                                          FRON 269
      GO TO 370                                                         FRON 270
  360 DO 510 IDOFN=1,NDOFN                                              FRON 271
      NGASH=NLOCA+IDOFN                                                 FRON 272
  510 TREAC(KBOUN,IDOFN)=TREAC(KBOUN,IDOFN)+VIXED(NGASH)                FRON 273
      KBOUN=KBOUN+1                                                     FRON 274
  370 CONTINUE                                                          FRON 275
C                                                                       FRON 276
C*** ADD REACTIONS INTO THE TOTAL LOAD ARRAY                            FRON 277
C                                                                       FRON 278
      RETURN                                                            FRON 279
      END                                                               FRON 280
```

CHAPTER 4

ANALYSIS OF ELASTO-PLASTIC AND GEOMETRICALLY NONLINEAR ANISOTROPIC PLATES AND SHELLS

J. A. FIGUEIRAS* and D. R. J. OWEN**

*Department of Civil Engineering, University of Oporto, Portugal
**Department of Civil Engineering, University of Wales, Swansea U.K.

1. INTRODUCTION

The development of appropriate methods for the analysis of shell structures is increasingly demanded to ensure the integrity of structural designs. Analytical solutions to shell structures are limited in scope and in general are not applicable to arbitrary shapes, load conditions, irregular stiffening and support conditions, cutouts and many other aspects of practical design. The finite element method has consequently become prominent in the analysis of such shells in view of the ease with which such complexities can be dealt with.

Although finite element procedures have been applied to shell analysis for over twenty years, the search for suitable elements which are readily applicable to general shell structures is still in progress [1-4]. Three distinct approaches to the finite element representation of shell structures have been employed based on (i) flat triangular or quadrilateral elements, (ii) curved elements formulated on the basis of various shell theories and (iii) elements derived from three-dimensional elements by the use of degeneration methods.

The first numerical studies of shell problems involved facetting the shell by plane triangular plate elements, onto which a membrane stiffness was superposed. The results obtained were found to be satisfactory, but depending on the problem, very fine meshes had to be employed. A number of difficulties and shortcomings arise from the application of flat elements to curved shells [5], such as the presence of 'discontinuity' bending moments, which do not appear in the continuously curved actual structure. Despite these problems, which can be surmounted through various artifices and additional computational effort, simple triangular elements have been recently developed and applied to the nonlinear analysis of shells of arbitrary shape [6-8]. The problem that remains is to determine if the loss in solution accuracy (as compared with curved elements) is properly balanced by the lower computational cost of these simple elements.

In the second approach a classical concept is employed, whereby a shell theory is used as the starting point of the finite element formulation. Different shell theories have been developed from the three-dimensional field equations by incorporating various assumptions appropriate to the structural behaviour. A variety of finite elements with different degrees of complexity have been formulated for deep and shallow shells [1,9,10]. Curved shell elements based on the Kirchhoff-Love theory guarantee a high solution accuracy but are complicated by convergence and compatibility requirements. In the case of a fully nonlinear analysis, additional difficulties arise in the formulation of efficient finite elements due to the unavailability of a general nonlinear shell theory.

The third approach employed in the formulation of shell elements avoids the complexities of fully general shell theories by discretizing directly the three-dimensional equations of continuum mechanics. Isoparametric elements with independent rotational and displacement degrees of freedom are employed, in which the three dimensional stress and strain conditions are degenerated to shell behaviour. This procedure was originally introduced by Ahmad et al. [11] for the linear analysis of moderately thick shells. Although this type of element seemed very promising when it was introduced, difficulties later appeared, due to the degeneration process, as the thickness of the element was reduced. A great improvement of the model was achieved by means of the so-called reduced integration technique [12,13].

Since then, the degenerate concept of formulating general shell elements (for thick and thin structures) has been adopted by many investigators to derive different types of elements [14-16]. This formulation has been recently applied to the geometric [17,18] and material [19,20] nonlinear analysis of shells, as well as to combined geometric and material nonlinearities [15,16,21-23]. The beneficial effects of reduced integration techniques on isoparametric plate and shell elements, and the problems arising from the use of such techniques have been discussed in several works [24-27].

Recent interest and increased usage of advanced fibrous composite materials has made the assessment of the effects of anisotropy on the behaviour of structural elements mandatory. The excellent strength to weight properties of modern fibre-reinforced materials, and the improved characteristics of composite laminate structures, justify the increased application of these heterogeneous anisotropic materials. The designer has to be able to account not only for given loadings but also for the directional stiffness requirements by changing the orientation of the various lamina. A numerical model is indispensable for this task.

Various theories, based on the classical Kirchhoff hypotheses, have been established to analyse composite plates [29-30]. For anisotropic composite laminates, where the ratio of Young's modulus to shear modulus can be very large, a theory analogous to those presented by Reissner [31] and Mindlin [32] for isotropic plates assumes greater importance. Experimental results as well as exact elasticity solutions have indicated that the shear correction may be significant even for span-to-depth ratios as high as 30 [33-35]. Heterogeneous and anisotropic plate theories accounting for shear deformation have been developed by several investigators. Constant transverse shear strain through the thickness is assumed in some models [36,37] and the assumption of constant transverse shear strain is made in others [38]. The former violates transverse shear stress continuity at layer interface, and the latter, although satisfying interface continuity, violates point equilibrium conditions as well as boundary conditions on external surfaces, which are usually assumed to be free from tangential tractions. However, a general theory for laminated anisotropic plates and shells, taking into account shear deformations, is not yet well established.

Although isoparametric elements of the Serendipity family give satisfactory solutions for thick and moderately thin shells [12] the results exhibit 'locking' for an increasing ratio of span to thickness [24-26]. To develop a numerical model which can analyse thick or thin plates and shells, the nine-node Lagrangian element and the Heterosis element [16,27,44,45] are also introduced into the present model. The accommodation of variable material properties, not only along the surface of the structure but also through the thickness, is made possible by a discrete layered approach. The assumption of a constant transverse shear strain is made, and a correction shear factor derived in cylindrical bending is used to approximate the real shear strain energy component [46,47].

The nonlinear stress-strain response of anisotropic materials is also of interest to designers. The Huber-Mises yield criterion has been generalized by Hill [39,40] for anisotropic metals. Hill's theory was extended by Hu [41] and Whang [42] to consider strain-hardening behaviour [43]. A full three-dimensional Huber-Mises yield criterion, in which the stress components are modified by introducing anisotropic parameters, is employed. The tangential/initial stiffness approach to numerical solution is adopted and an associated flow rule, in which the plastic potential is assumed to be the anisotropic yield function, is used to define the elasto-plastic incremental constitutive relation [48,49].

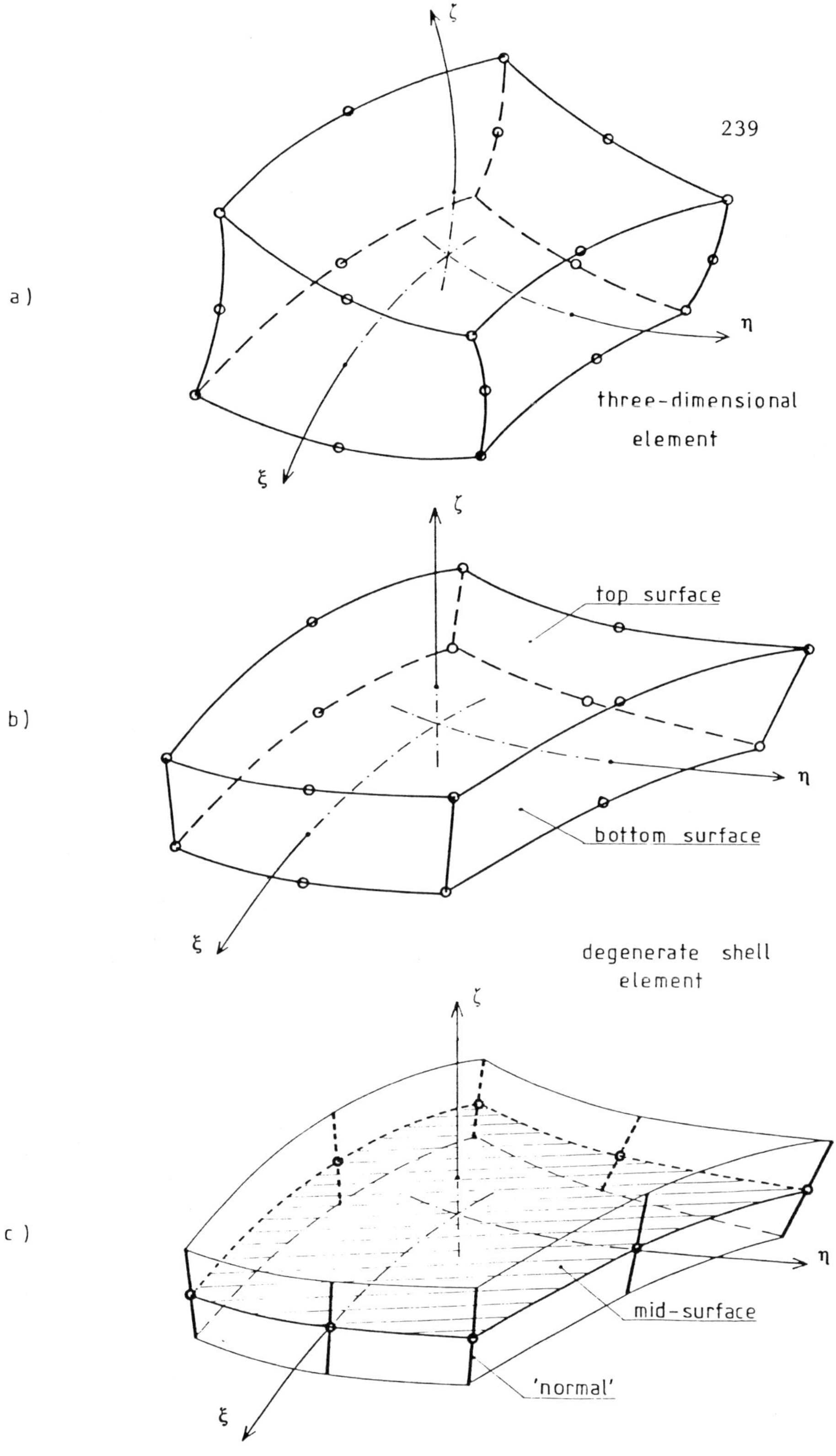

Fig. 1 **(a) Quadratic solid three-dimensional element, (b) and (c) the corresponding degenerate shell element.**

In the remainder of this chapter a general description of the degenerate shell concept is given.

2. DEGENERATE ISOPARAMETRIC ELEMENTS

2.1 General

Figure 1(a) shows a solid three-dimensional element based on a quadratic displacement field and Figures 2(b) and (c) illustrate the corresponding quadratic degenerate shell element. Two basic assumptions are adopted in this process: Firstly it is assumed that, even for thick shells, 'normals' to the middle surface remain practically straight after deformation. Secondly, the strain energy corresponding to stresses perpendicular to the middle surface is disregarded, i.e. the stress component normal to the shell mid-surface is constrained to be zero in the constitutive equations.

Five degrees of freedom are specified at each nodal point, corresponding to its three displacements and the two rotations of the 'normal' at the node. The definition of independent rotational and displacement degrees of freedom permits transverse shear deformation to be taken into account, since rotations are not tied to the slope of the mid-surface. This approach is equivalent to using a general shell theory and reduces to the hypotheses of Reissner [31] and Mindlin [32] when applied to plates.

2.2 Coordinate Systems

In the formulation of degenerate curved shell elements different coordinate systems have to be used. The four coordinate sets employed in the present formulation (see Fig. 2(a) and (b)) are now described.

1. Global Coordinate Set - x_i

This is a Cartesian coordinate system, freely chosen, in relation to which the geometry of the structure is defined in space. Nodal coordinates and displacements, as well as the global stiffness matrix and applied force vector are referred

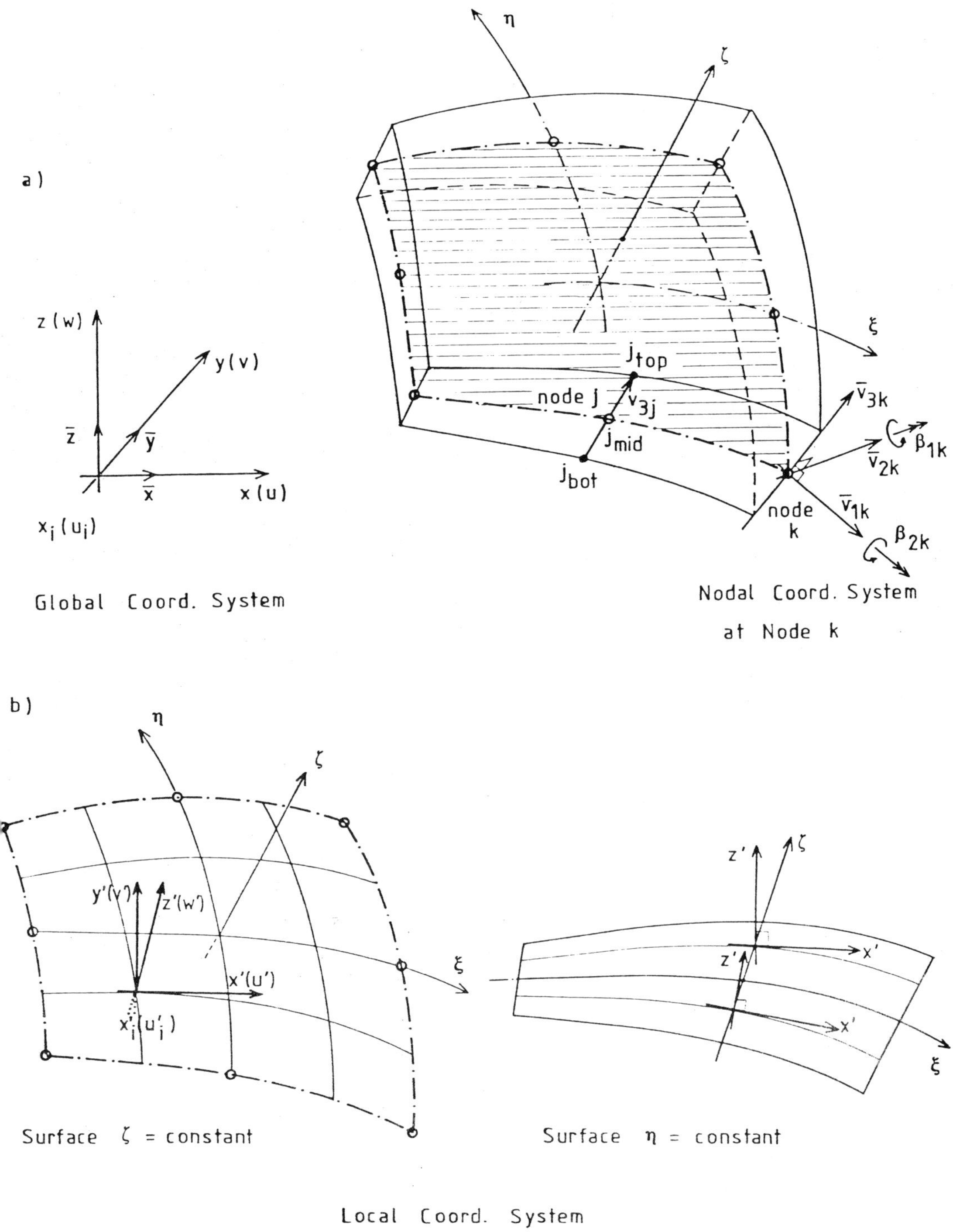

Fig. 2 Coordinate systems: (a) nodal and curvilinear systems, (b) local system of axes.

to this system.

The following notation is used (Fig. 2(a)):

x_i (i=1,3) and $x_1 = x$, $x_2 = y$, $x_3 = z$

u_i (i=1,3) and $u_1 = u$, $u_2 = v$, $u_3 = w$

$\bar{x}_i$ (i=1,3) is a unit vector in the x_i direction

2. Nodal Coordinate Set - $\underline{v}_{ik}$

A nodal coordinate system is defined at each nodal point (see Fig. 2(a)) with origin at the reference surface (mid-surface). The vector $\underline{v}_{3k}$ is constructed from the nodal coordinates of the top and bottom surfaces at node k,

$$\underline{v}_{3k} = \underline{x}_k^{top} - \underline{x}_k^{bot} \qquad (1)$$

where $\underline{x}_k = [x_k \; y_k \; z_k]^T$

The vector $\underline{v}_{1k}$ is perpendicular to $\underline{v}_{3k}$ and parallel to the global xz-plane,

$$v_{1k}^x = v_{3k}^z, \; v_{1k}^y = 0.0, \; v_{1k}^z = - v_{3k}^x$$

or, if $\underline{v}_{3k}$ is in the y-direction ($v_{3k}^x = v_{3k}^z = 0.0$) (2)

$$v_{1k}^x = - v_{3k}^y, \; v_{1k}^y = v_{1k}^z = 0.0 \text{ (x-direction)}$$

where the superscripts refer to the vector components in the global coordinate system. The vector $\underline{v}_{2k}$ is perpendicular to the plane defined by $\underline{v}_{1k}$ and $\underline{v}_{3k}$,

$$\underline{v}_{2k} = \underline{v}_{3k} \times \underline{v}_{1k} \qquad (3)$$

The unit vectors in the directions of $\underline{v}_{1k}$, $\underline{v}_{2k}$ and $\underline{v}_{3k}$ are represented by $\bar{\underline{v}}_{1k}$, $\bar{\underline{v}}_{2k}$ and $\bar{\underline{v}}_{3k}$ respectively (Fig. 2(a)).

The vector $\bar{\underline{v}}_{3k}$ defines the direction of the 'normal' at node k, which is not necessarily perpendicular to the mid-surface at k. Vectors $\bar{\underline{v}}_{1k}$ and $\bar{\underline{v}}_{2k}$ define the rotations (β_{2k} and β_{1k} respectively) of the corresponding 'normal'. The advantage of this definition for the vector $\bar{\underline{v}}_{3k}$ (not necessarily perpendicular to the shell mid-surface) is that, as a consequence, there are no gaps or overlaps along element boundaries. The definition used for $\bar{\underline{v}}_{1k}$ provides an easy identification of its direction in general curved structures.

3. Curvilinear Coordinate Set - ξ, η, ζ

In this system ξ, η are two curvilinear coordinates in the middle plane of the shell element and ζ is a linear coordinate in the thickness direction (Fig. 2.2). It is assumed that ξ, η and ζ vary between -1 and +1 on the respective faces of the element. The relations between the curvilinear coordinates (ξ, η, ζ) and the global coordinates (x, y, z) are given by the same expressions (9), which later define the element geometry. It should be noted that the ζ-direction is only approximately perpendicular to the shell mid-surface, since ζ is defined as a function of $\bar{\underline{v}}_{3k}$.

4. Local coordinate Set - x'_i

This is a Cartesian coordinate system defined at the sampling points wherein stresses and strains are to be calculated. The direction x'_3 (z') is taken perpendicular to the surface ζ = constant (Fig. 2(b)), being obtained by the cross product of the ξ and η directions,

$$\underline{x}'_3 = \begin{bmatrix} \frac{\partial x}{\partial \xi} \\ \frac{\partial y}{\partial \xi} \\ \frac{\partial z}{\partial \xi} \end{bmatrix} \times \begin{bmatrix} \frac{\partial x}{\partial \eta} \\ \frac{\partial y}{\partial \eta} \\ \frac{\partial z}{\partial \eta} \end{bmatrix} \tag{4}$$

The direction x'_1 (x') can be taken tangent to the ξ-direction at the sampling point,

$$\underline{x}'_1 = \begin{bmatrix} \frac{\partial x}{\partial \xi} \\ \frac{\partial y}{\partial \eta} \\ \frac{\partial z}{\partial \xi} \end{bmatrix} \tag{5}$$

However a different definition of the x'_1-direction is required when distorted meshes are employed for anisotropic structures (for instance, fibre composite and reinforced concrete shells), in order to define the material directions in relation to the local system of axes in a consistent manner. A definition similar to that of the v_{1k} direction is then adopted.

The direction x_2' (y') is defined by the cross product of the x_3' and x_1' directions,

$$\underline{x}_2' = \underline{x}_3' \times \underline{x}_1' \tag{6}$$

This local coordinate system varies along the thickness for any 'normal' with this variation depending on the shell curvature and variable thickness (see Fig. 2(b)).

The direction cosine matrix [θ] which relates the transformations between the local and the global coordinate system is defined by,

$$[\theta] = [\bar{\underline{x}}', \bar{\underline{y}}', \bar{\underline{z}}'] \tag{7}$$

where $\bar{\underline{x}}'$, $\bar{\underline{y}}'$, and $\bar{\underline{z}}'$ are unit vectors in the direction of the x', y' and z' axes respectively.

2.3 Element Geometry

The global coordinates of pairs of points on the top and bottom surface at each node (Fig. 2(a)) are usually input to define the element geometry. Alternatively, the mid-surface nodal coordinates and the corresponding directional thickness can be furnished. In the isoparametric formulation the coordinates of a point within the element are obtained by applying the element shape functions to the nodal coordinates,

$$x_i = \sum_{k=1}^{n} N_k \frac{1+\zeta}{2} x_{ik}^{top} + \sum_{k=1}^{n} N_k \frac{1-\zeta}{2} x_{ik}^{bot} \tag{8}$$

or

$$x_i = \sum_{k=1}^{n} N_k x_{ik}^{mid} + \sum_{k=1}^{n} N_k \frac{h_k}{2} \zeta \bar{v}_{3k}^{i} \tag{9}$$

where i = 1,3 refers to the three global directions,

n - is the number of nodes per element,

$N_k = N_k(\xi,\eta)$ (k = 1,n) are the element shape functions corresponding to the surface ζ = constant,

h_k - is the shell thickness at node k, i.e. the respective 'normal' length,

ξ, η, ζ - are the curvilinear coordinates of the point under consideration.

The first term on the right hand side of (9) represents the intercept of the 'normal' with the mid-surface whereas the second term defines the position of the point along this 'normal'. More explicitly, expression (9) can be written,

$$\begin{bmatrix} x \\ y \\ z \end{bmatrix} = \sum_{k=1}^{n} N_k \begin{bmatrix} x_k \\ y_k \\ z_k \end{bmatrix}_{mid} + \sum_{j=1}^{n} N_k \frac{h_k}{2} \zeta \begin{bmatrix} \bar{v}^x_{3k} \\ \bar{v}^y_{3k} \\ \bar{v}^z_{3k} \end{bmatrix} \tag{10}$$

2.4 Displacement Field

Taking into consideration the two shell assumptions of the degeneration process, the displacement field is described by the five degrees of freedom of a 'normal'; the three displacements of its mid-point (u^{mid}_{ik}) and two rotations (β_{1k}, β_{2k}) (see Fig. 3). The displacements of a point on the 'normal' resulting from the two rotations are calculated as illustrated in Figure 3,

$$\begin{aligned} \delta_{1k} &= h\ \beta_{1k} \\ \delta_{2k} &= h\ \beta_{2k} \end{aligned} \tag{11}$$

where δ_{1k} is a displacement in the $\underline{\bar{v}}_{1k}$ direction and

δ_{2k} is a displacement in the negative $\underline{\bar{v}}_{2k}$ direction.

The corresponding displacement components u_i are obtained by,

$$\begin{aligned} (u_i)_{\beta_{1k}} &= \delta_{1k}\ \bar{v}^i_{1k} \\ (u_i)_{\beta_{2k}} &= \delta_{2k}\ (-\ \bar{v}^i_{2k}) \end{aligned} \tag{12}$$

In expression (11) a linear relation was employed, which assumes that the rotations (the incremental angles) are small; to admit large increments of rotations, expression (11) should be substituted by a nonlinear rotation-displacement relationship [16, 18,22].

The element displacement field can then be expressed by,

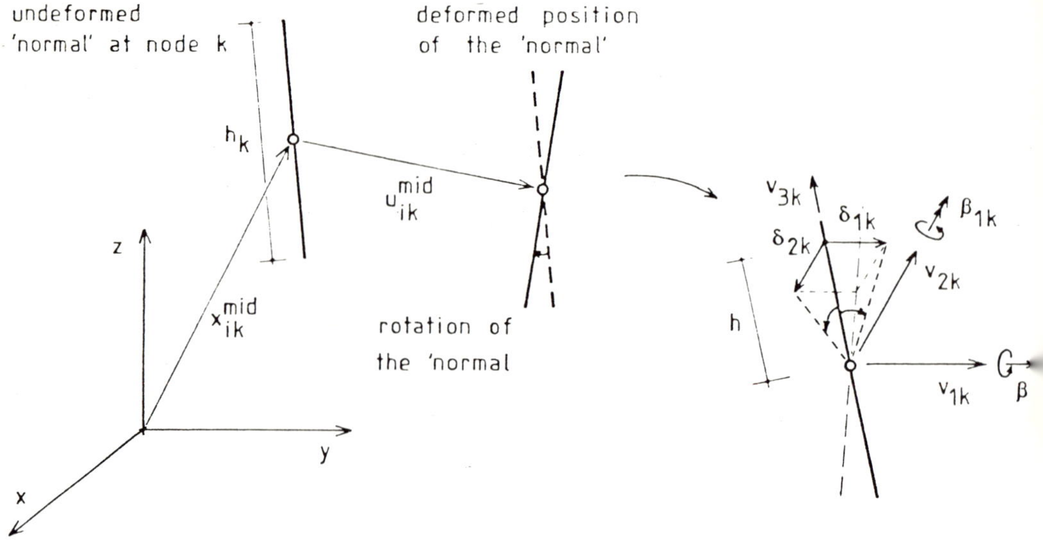

Fig. 3   Displacements of a point on the "normal" at node k.

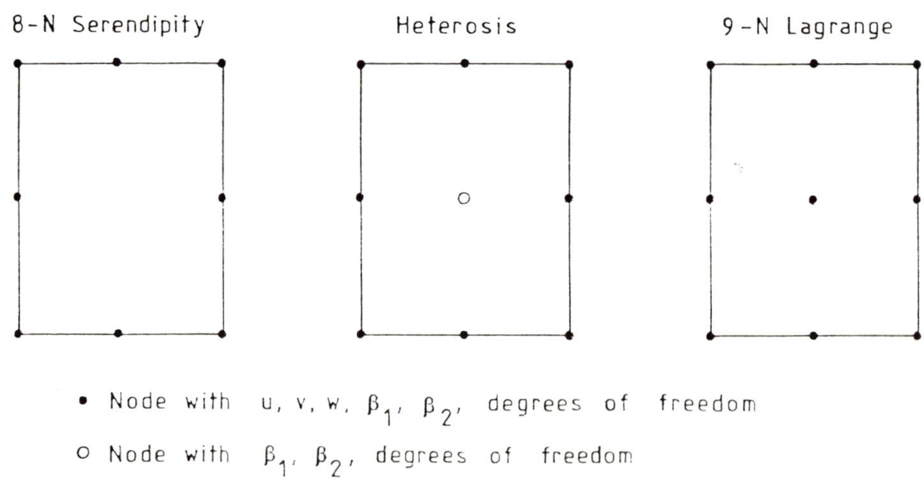

Fig. 4   Nodal configuration of the three quadratic shell elements.

$$u_i = \sum_{k=1}^{n} N_k \, u_{ik}^{mid} + \sum_{k=1}^{n} N_k \, \zeta \, \frac{h_k}{2} [\bar{v}_{1k}, -\bar{v}_{2k}] \begin{bmatrix} \beta_{1k} \\ \beta_{2k} \end{bmatrix} \qquad (13)$$

or

$$\begin{bmatrix} u \\ v \\ w \end{bmatrix} = \sum_{k=1}^{n} N_k \begin{bmatrix} u_k \\ v_k \\ w_k \end{bmatrix}_{mid} + \sum_{k=1}^{n} N_k \, \zeta \, \frac{h_k}{2} \begin{bmatrix} v_{1k}^x, & -v_{2k}^x \\ v_{1k}^y, & -v_{2k}^y \\ v_{1k}^z, & -v_{2k}^z \end{bmatrix} \begin{bmatrix} \beta_{1k} \\ \beta_{2k} \end{bmatrix} \qquad (14)$$

The contribution to the global displacements from a given node k is:

$$\begin{bmatrix} u \\ v \\ w \end{bmatrix} = \begin{bmatrix} N_k & 0 & 0 & N_k \zeta \frac{h_k}{2} v_{1k}^x & - N_k \zeta \frac{h_k}{2} v_{2k}^x \\ 0 & N_k & 0 & N_k \zeta \frac{h_k}{2} v_{1k}^y & - N_k \zeta \frac{h_k}{2} v_{2k}^y \\ 0 & 0 & N_k & N_k \zeta \frac{h_k}{2} v_{1k}^z & - N_k \zeta \frac{h_k}{2} v_{2k}^z \end{bmatrix} \begin{bmatrix} u_k \\ v_k \\ w_k \\ \beta_{1k} \\ \beta_{2k} \end{bmatrix} \qquad (15)$$

or

$$\underline{u}_k = \underline{N}_k \, \underline{\delta}_k$$

For the complete element we have

$$\underline{u} = \underline{N} \, \underline{\delta} \qquad (16)$$

where $\underline{N} = [\underline{N}_1, \ldots, \underline{N}_k, \ldots, \underline{N}_n]$ is the shape function matrix of the degenerate element, and

$\delta^T = [\underline{\delta}_1^T, \ldots, \underline{\delta}_k^T, \ldots, \underline{\delta}_n^T]$ is the vector of the element nodal variables.

2.5 Definition of Strains

In order to easier deal with the shell assumption of zero normal stress in the z'-direction ($\sigma_{z'} = 0$), the strain components should be defined in terms of the local system of axes x_i' (where $x_3' = z'$ is perpendicular to the $\xi\eta$-plane). The local system of axes is also the most convenient system for expressing the stress components (and their resultants) for shell analysis and design. The five significant strain components are,

$$\underline{\varepsilon} = \begin{bmatrix} \varepsilon_{x'} \\ \varepsilon_{y'} \\ \gamma_{x'y'} \\ \gamma_{x'z'} \\ \gamma_{y'z'} \end{bmatrix} = \begin{bmatrix} \frac{\partial u'}{\partial x'} \\ \frac{\partial v'}{\partial y'} \\ \frac{\partial u'}{\partial y'} + \frac{\partial v'}{\partial x'} \\ \frac{\partial u'}{\partial z'} + \frac{\partial w'}{\partial x'} \\ \frac{\partial v'}{\partial z'} + \frac{\partial w'}{\partial y'} \end{bmatrix} \tag{17}$$

where u', v' and w' are the displacement components in the local system x_i'. These local derivatives are obtained from the global derivatives of the displacements u, v, and w by the following operation,

$$\begin{bmatrix} \frac{\partial u'}{\partial x'} & \frac{\partial v'}{\partial x'} & \frac{\partial w'}{\partial x'} \\ \frac{\partial u'}{\partial y'} & \frac{\partial v'}{\partial y'} & \frac{\partial w'}{\partial y'} \\ \frac{\partial u'}{\partial z'} & \frac{\partial v'}{\partial z'} & \frac{\partial w'}{\partial z'} \end{bmatrix} = [\theta]^T \begin{bmatrix} \frac{\partial u}{\partial x} & \frac{\partial v}{\partial x} & \frac{\partial w}{\partial x} \\ \frac{\partial u}{\partial y} & \frac{\partial v}{\partial y} & \frac{\partial w}{\partial y} \\ \frac{\partial u}{\partial z} & \frac{\partial v}{\partial z} & \frac{\partial w}{\partial z} \end{bmatrix} [\theta] \tag{18}$$

where $[\theta]$ is a transformation matrix defined by (7).

The derivatives of the displacements with respect to the global coordinates are given by,

$$\begin{bmatrix} \frac{\partial u}{\partial x} & \frac{\partial v}{\partial x} & \frac{\partial w}{\partial x} \\ \frac{\partial u}{\partial y} & \frac{\partial v}{\partial y} & \frac{\partial w}{\partial y} \\ \frac{\partial u}{\partial z} & \frac{\partial v}{\partial z} & \frac{\partial w}{\partial z} \end{bmatrix} = \underline{J}^{-1} \begin{bmatrix} \frac{\partial u}{\partial \xi} & \frac{\partial v}{\partial \xi} & \frac{\partial w}{\partial \xi} \\ \frac{\partial u}{\partial \eta} & \frac{\partial v}{\partial \eta} & \frac{\partial w}{\partial \eta} \\ \frac{\partial u}{\partial \zeta} & \frac{\partial v}{\partial \zeta} & \frac{\partial w}{\partial \zeta} \end{bmatrix} \tag{19}$$

where $\underline{J}$ is the Jacobian matrix.

$$\underline{J} = \begin{bmatrix} \frac{\partial x}{\partial \xi} & \frac{\partial y}{\partial \xi} & \frac{\partial z}{\partial \xi} \\ \frac{\partial x}{\partial \eta} & \frac{\partial y}{\partial \eta} & \frac{\partial z}{\partial \eta} \\ \frac{\partial x}{\partial \zeta} & \frac{\partial y}{\partial \zeta} & \frac{\partial z}{\partial \zeta} \end{bmatrix} \tag{20}$$

In expression (19) the displacement derivatives referred to the curvilinear coordinates are obtained from equation (13), whereas the Jacobian matrix results from (8).

The strain matrix $\underline{B}$, relating the strain components in the local system to the element nodal variables, can then be constructed as,

$$\underline{\varepsilon} = \underline{B}\ \underline{\delta} \tag{21}$$

where $\underline{\varepsilon}$ and $\underline{\delta}$ are defined in (17) and (16) respectively and $\underline{B}$ is a matrix with five rows and a number of columns equal to the number of element nodal variables.

2.6 Definition of Stresses

Taking into consideration the assumption of zero stress in the direction perpendicular to the shell mid-surface ($\sigma_{z'} = 0$), the five stress components in the local system are:

$$\underline{\sigma} = \begin{bmatrix} \sigma_{x'} \\ \sigma_{y'} \\ \tau_{x'y'} \\ \tau_{x'z'} \\ \tau_{y'z'} \end{bmatrix} = \underline{D}\ (\underline{\varepsilon} - \underline{\varepsilon}_o) \tag{22}$$

where $\underline{\varepsilon}$ is defined in (17) and $\underline{\varepsilon}_o$ is an initial strain vector which may represent, for instance, the expansion due to thermal load. The elasticity matrix $\underline{D}$ is a 5x5 matrix, whose specific form is considered in the following section.

2.7 The elastic constitutive relation

2.7.1 The generalized Hooke's law

Although for the most general cases of anisotropy the number of independent elastic constants is 21, this number is considerably reduced if the material internal composition possesses symmetry of any kind [50]. We shall assume a state of anisotropy which possesses three mutually orthogonal planes of symmetry at each layer. Fibre-reinforced composite structures almost invariably possess this kind of symmetry, with the mid-layer surface being a symmetric plane at each point. If the reference system of orthogonal axes (x_i') is parallel to the principal material axes (1,2,3) we obtain the following strain-stress relationships:

$$\begin{aligned}
\varepsilon_1 &= C_{11}\sigma_1 + C_{12}\sigma_2 + C_{13}\sigma_3 \\
\varepsilon_2 &= C_{12}\sigma_1 + C_{22}\sigma_2 + C_{23}\sigma_3 \\
\varepsilon_3 &= C_{13}\sigma_1 + C_{23}\sigma_2 + C_{33}\sigma_3 \\
\gamma_{12} &= C_{44}\tau_{12} \\
\gamma_{13} &= C_{55}\tau_{13} \\
\gamma_{23} &= C_{66}\tau_{23}
\end{aligned} \tag{23}$$

where the nine elastic constants C_{ij} can be expressed as functions of Young's moduli, Poisson's ratios and the shear moduli. Noting the zero normal stress assumption $\sigma_{z'} = \sigma_3 = 0$, we obtain the elasticity matrix $\underline{\bar{D}}$ relating stresses and strains, to be

$$\underline{\sigma}_{1,2,3} = \underline{\bar{D}}\ \underline{\varepsilon}_{1,2,3} \tag{24}$$

where

$$\begin{aligned}
\underline{\sigma}_{1,2,3} &= \{\sigma_1, \sigma_2, \tau_{12}, \tau_{13}, \tau_{23}\}^T \\
\underline{\varepsilon}_{1,2,3} &= \{\varepsilon_1, \varepsilon_2, \gamma_{12}, \gamma_{13}, \gamma_{23}\}^T
\end{aligned} \tag{25}$$

$$\underline{\bar{D}} = \begin{bmatrix} \bar{D}_1 & \bar{D}_{12} & 0 & 0 & 0 \\ \bar{D}_{12} & \bar{D}_2 & 0 & 0 & 0 \\ 0 & 0 & \bar{D}_3 & 0 & 0 \\ 0 & 0 & 0 & \bar{D}_4 & 0 \\ 0 & 0 & 0 & 0 & \bar{D}_5 \end{bmatrix}$$

and

$$\begin{aligned}
\bar{D}_1 &= E_1/(1-\nu_{12}\nu_{21}) & \bar{D}_3 &= G_{12} \\
\bar{D}_2 &= E_2/(1-\nu_{12}\nu_{21}) & \bar{D}_4 &= K_1 * G_{13} \\
\bar{D}_{12} &= E_2\nu_{12}/(1-\nu_{12}\nu_{21}) & \bar{D}_5 &= K_2 * G_{23}
\end{aligned} \tag{26}$$

The terms K_1 and K_2 are shear correction factors in the $\overline{13}$ and $\overline{23}$ planes, respectively, which are defined later in the next section.

If the principal axes of anisotropy 1, 2 do not coincide with the reference axes x', y', but are rotated by a certain angle θ, the new elasticity matrix $\underline{D}$ is determined from the following transformation:

$$\underline{\sigma}_{1,2,3} = \underline{T}\,\underline{\sigma}$$
$$\underline{\varepsilon}_{1,2,3} = \underline{T}'\,\underline{\varepsilon} \tag{27}$$

where

$$\underline{T} = \begin{bmatrix} \underline{T}_1 & 0 \\ 0 & \underline{T}_2 \end{bmatrix} \tag{28}$$

in which

$$\underline{T}_1 = \begin{bmatrix} \cos^2\theta & \sin^2\theta & 2\sin\theta\cos\theta \\ \sin^2\theta & \cos^2\theta & -2\sin\theta\cos\theta \\ -\sin\theta\cos\theta & \cos\theta\sin\theta & (\cos^2\theta - \sin^2\theta) \end{bmatrix} \tag{29a}$$

$$\underline{T}_2 = \begin{bmatrix} \cos\theta & \sin\theta \\ -\sin\theta & \cos\theta \end{bmatrix} \tag{29b}$$

and

$$\underline{T}' = \begin{bmatrix} \underline{T}_1' & 0 \\ 0 & \underline{T}_2 \end{bmatrix} \tag{30a}$$

in which

$$\underline{T}_1' = \begin{bmatrix} \cos^2\theta & \sin^2\theta & \sin\theta\cos\theta \\ \sin^2\theta & \cos^2\theta & -\sin\theta\cos\theta \\ -2\sin\theta\cos\theta & 2\sin\theta\cos\theta & (\cos^2\theta - \sin^2\theta) \end{bmatrix} \tag{30b}$$

From (24) and (27) we can write

$$\underline{\sigma} = \underline{D}\,\underline{\varepsilon} \tag{31}$$

where $\underline{D}$ is the general elasticity matrix and $\underline{T}^{-1} \equiv \underline{T}'^{T}$,

$$\underline{D} = \underline{T}'^T \underline{\bar{D}T}' = \begin{bmatrix} D_1 & D_{12} & D_{13} & 0 & 0 \\ D_{12} & D_2 & D_{23} & 0 & 0 \\ D_{13} & D_{23} & D_3 & 0 & 0 \\ 0 & 0 & 0 & D_4 & D_{45} \\ 0 & 0 & 0 & D_{45} & D_5 \end{bmatrix}$$

2.7.2 Shear correction

At layer interfaces, continuity of both the transverse shear stresses and displacements is required. Assuming that 'normals' to the middle surface remain straight but do not necessarily remain normal to the middle surface of the shell, the latter continuity condition is satisfied but the former is not. This assumption makes the transverse shear strain constant through the thickness, which is a coarse approximation to the actual variation, even for a homogeneous cross-section. For homogeneous cross-sections, the shear strain distribution is commonly accepted to be a parabolic function of z'. Therefore, a correction shear factor K must be introduced in order to approximate, on an average basis, the transverse shear strain energy. We will assume, in order to calculate K, a heterogeneous plate in which the external surfaces are free from tangential tractions. The equilibrium equation in the x direction ($x \equiv 1$, $y \equiv 2$ and $z \equiv 3$) can be expressed as:

$$\frac{\partial \sigma_x}{\partial x} + \frac{\partial \tau_{xy}}{\partial y} + \frac{\partial \tau_{xz}}{\partial z} = 0 \tag{32}$$

Assuming cylindrical bending, with $z = h/2$ and $z = - h/2$ at the top and bottom surfaces, respectively, then

$$\begin{aligned} \tau_{xz} &= - \int_{-h/2}^{z} \frac{\partial \sigma_x}{\partial x} \, d\bar{z} = - \int_{-h/2}^{z} \frac{\partial M_x}{\partial x} \frac{D_1(\bar{z})}{R_1} \bar{z} \, d\bar{z} \\ &= - \frac{Q_x}{R_1} \int_{-h/2}^{z} D_1(\bar{z})\bar{z} \, d\bar{z} = \frac{Q_x}{R_1} . \, g(z) \end{aligned} \tag{33}$$

where

Q_x is the shear force on the $\overline{xz}$ plane

$R_1 = \int_{-h/2}^{h/2} D_1(\bar{z})z^2 \, dz$ is the flexural plate stiffness in the x direction

$\bar{z}$ is a variable co-ordinate through the thickness

$g(z) = -\int_{-h/2}^{z} D_1(\bar{z})\bar{z} \, d\bar{z}$ is the shear stress shape function.

The function g(z) which determines the shape of the shear stress diagram is independent of the loadings for the conditions specified. It becomes the well-known parabolic function $g(z) = [(D_1 h^2)/8][1-4(z/h)^2]$ for the case of a homogeneous cross-section. The strain energy component (per unit mid-plane surface) will be given by

$$w_s = \int_{-h/2}^{h/2} \tau_{xz}^2/G_{13}(z)dz \tag{34}$$

and substituting (33) into (34) results in

$$w_s = \frac{Q_x^2}{R_1^2} \int_{-h/2}^{h/2} g^2(z)/G_{13}(z)dz \tag{35}$$

where $G_{13}(z)$ is the (variable) shear modulus, in the $\overline{xz}$ plane.

On the other hand, the strain energy component, under the assumption of constant shear strain, is

$$\begin{aligned} \bar{w}_s &= \int_{-h/2}^{h/2} \bar{\gamma}_{xz} \cdot G_{13}(z) \, \bar{\gamma}_{xz} \, dz \\ &= \frac{Q_x^2}{h^2 \bar{G}_1^2} \cdot h\bar{G}_1 = \frac{Q_x^2}{h\bar{G}_1} \end{aligned} \tag{36}$$

where

$$h\bar{G}_1 = \int_{-h/2}^{h/2} G_{13}(z)dz$$

and $\bar{\gamma}_{xz}$ is the mean value of the shear strain. Equating expressions (35) and (36) we obtain the correction shear factor K_1 in the $\overline{xz}$ plane to be

$$K_1 = \bar{w}_s/w_s = R_1^2[h.\bar{G}_1 \int_{-h/2}^{h/2} g^2(z)/G_{13}(z)dz]^{-1} \tag{37}$$

For a homogeneous cross-section the shear correction factor is equal to 5/6. To obtain K_2, the shear correction factor in the $\overline{yz}$ plane, we proceed in the same way. This calculation is valid for general bending situations provided that the function g(z) approximately retains its shape, and can be applied to symmetric or non-symmetric sections.

A realistic value of the shear stress should be used in the yield criterion which can be evaluated according to

$$\tau_{xz}^{COR.} = \bar{G}_{13} \cdot \gamma_{xz} \frac{g(z)}{\bar{g}} \qquad (38)$$

where

$$\bar{g} = 1/h \int_{-h/2}^{h/2} g(z)\, dz$$

For numerical implementation all integrals will be replaced by summation over the layer thicknesses in the case of composite laminated structures with different material layers.

2.8 Quadratic shell elements

A brief presentation of the quadratic elements used in this work is given in this section. The elements considered are the 8-node Serendipity, the 9-node Lagrangian and the Heterosis element.

1. 8-node Serendipity

The 8-node Serendipity element (Fig. 4) was the simplest element considered in the original work of Ahmad [11] on the degenerate shell concept, and probably has been the most popular element of the isoparametric family [52]. The results initially obtained with this element [11] by using a normal integration rule (3x3) showed that the solution became increasingly stiff as the shell thickness was reduced. This so-called locking phenomenon is exhibited by this element even in moderately thin situations. A great improvement of the results was achieved by using a reduced integration rule (2x2) [12,13]. However, for very thin situations the results given by the 8-node

Serendipity element with reduced integration exhibit increasing divergence, i.e. locking occurs again.

2. 9-node Lagrangian

Experiments carried out by Pugh et al. [24] with linear, quadratic and cubic elements of the Serendipity and Lagrangian families showed that the 9-node Lagrangian element (Fig. 4) is nearly optimal as a general plate element. Similar opinion was expressed in Ref. [25] when reporting experiments on clamped square Mindlin plates; the performance exhibited by Lagrangian elements appeared to be superior to Serendipity elements when reduced or selective integration was adopted. However, a serious problem may arise with this element when reduced integration is employed; the stiffness matrix exhibits rank deficiency which originates the appearance of spurious mechanisms, i.e. zero-energy modes [24-27]. These communicable mechanisms may result in very erratic solutions, specially when the restraints imposed by the boundary conditions are minimal. This drawback is not exhibited by the 8-node Serendipity with reduced integration.

3. Heterosis Element

The shortcomings pointed out above (locking and spurious mechanisms) led to the development of the Heterosis element (Fig. 4). This 9-node quadrilateral, initially developed for plates [44] employs Serendipity shape functions for the transverse displacement, w, and Lagrange shape functions for the rotations, β_1, β_2. The Heterosis element exhibits improved characteristics when compared with each of the previous quadratic elements. When selectively integrated, its stiffness matrix possesses correct rank and the element does not lock in very thin situations. Adopting uniform reduced integration, the element behaves better than the 9-node Lagrangian, with the appearance of spurious mechanisms being improbable for practical applications.

The Heterosis concept has been extended to shell analysis in Ref. [44] (using the Serendipity shape functions for the lateral displacement, w, only) and in Ref. [16] (using the Serendipity shape functions for all translational degrees of freedom u, v, w). This last approach corresponds to employing the Serendipity shape functions in the first term (right hand side) of (13) and the Lagrangian shape functions in the second term of the same expression.

4. Hierarchical Formulation

A hierarchical formulation is adopted in the present work to implement the 9-node Lagrangian and Heterosis elements. The shape functions of the eight boundary nodes are the Serendipity shape functions:

(a) for corner nodes

$$N_k = \frac{1}{4}(1 + \xi\xi_k)(1 + \eta\eta_k)(\xi\xi_k + \eta\eta_k - 1) \qquad (39)$$

(b) for midside nodes

$$N_k = \frac{\xi_k^2}{2}(1 + \xi\xi_k)(1 - \eta^2) + \frac{\eta_k^2}{2}(1 + \eta\eta_k)(1 - \xi^2) \qquad (40)$$

and the hierarchic shape function for the central 9th node is the bubble function,

$$N_9 = (1 - \xi^2)(1 - \eta^2) \qquad (41)$$

Note that in this formulation, the variables associated with the central node represent 'relative displacements', i.e. they are departures from the interpolated Serendipity values. The actual generalized displacement vector $\underline{a}$ (displacements and rotations) at the central node would be given by

$$a_{i9} = \sum_{k=1}^{8} N_k(0,0)\, a_{ik} + \Delta a_{i9} \qquad (i=1,5) \qquad (42)$$

where N_k (k=1,8) are the Serendipity shape functions

a_{ik} are the corresponding 'displacements' of the boundary nodes

Δa_{i9} are the 'displacements' associated with the degrees of freedom of the central node.

The geometry of all three quadratic elements is interpolated by using the eight Serendipity shape functions only. The displacement field of the 9-node Lagrangian element is obtained by interpolating the variables associated with the eight boundary nodes using the corresponding Serendipity shape functions and the 'relative displacements' Δa_{i9} associated with the 9th node using the bubble function. The displacement field of the Heterosis element is recovered from that of the 9-node Lagrangian element simply by constraining either the transverse displacement of the central node ($\Delta w_9 = 0$) or all three translational displacements ($\Delta u_9 = \Delta v_9 = \Delta w_9 = 0$).

2.9 Numerical Integration

Numerical integration of isoparametric shell elements has been widely considered in the literature since it is an important feature of the element efficiency. In the through thickness direction, wherein a linear variation of strain is assumed, 2 Gauss points are sufficient to capture the bending behaviour in linear material problems. High order Gaussian quadrature rules (for instance 5-7 Gauss points) have been advocated for non-linear material problems [19], but a simple and general procedure to discretize and integrate through the thickness is offered by the so-called layered model (Section 2.10). In the shell 'plane' (surfaces ζ = constant) the normal (full) integration rule consists of mxm Gauss points where m is the number of nodes along each element side. Nevertheless, when degenerate shell elements are fully integrated, they exhibit shear and membrane locking [4, 24-27], which affects the majority of applications. Therefore, the beneficial effects of reduced integration techniques ((m-1)x(m-1) Gauss points) on isoparametric shell elements have been established in a number of investigations. For the lower-order elements, reduced integration appears to be absolutely essential for good behaviour in thin shell applications; for higher order elements, significant improvements in accuracy are attained with reduced integration. However, uniform application of reduced integration may occasionally lead to spurious zero-energy modes appearing in the global equations. An alternative procedure, selective integration, is then advocated.

1. Reduced Integration

Significant improvements in the solution accuracy are attained in thin shells by employing reduced integration on the transverse shear strains, as well as on the membrane strains. On the other hand, uniform application of reduced integration tends to diminish the rank of the stiffness matrix resulting in the occurrence of mechanisms in addition to the rigid body modes. The number of such zero energy modes for the quadratic isoparametric elements are [53],

	8-node S	Heterosis	9-node L
No. of membrane mech.	1(*)	1(*)	2 + 1(*)
No. of bending mech.	1(*)	2 + 1(*)	3 + 1(*)

((*) - not communicable in a mesh of two or more elements)

The 8-node Serendipity element has only two non-communicable mechanisms so that, the element may be viewed as having correct rank. Unfortunately, in the thin-shell limit this element may present bizarre divergence characteristics. The 9-node Lagrangian and Heterosis element do not present this drawback, but they have 7 and 4 mechanisms respectively, two of which are not communicable. Although Lagrange shape functions may yield better membrane behaviour than the Serendipity pattern in highly distorted elements, the Heterosis is preferred for thin-shell applications in view of its lower number of mechanisms.

2. Selective Integration

In the selective integration technique, bending (or bending and membrane) energies are integrated using the normal rule, while the shear and membrane (or just shear) terms are computed using the reduced integration rule. The process involves the following three steps, which are particularized for quadratic elements:

a) Shear membrane (or shear) strains, i.e. the corresponding terms in the strain matrix $\underline{B}$, are computed at the four Gauss points I, II, III, IV (Fig. 5) of the 2x2 reduced integration rule.

b) Using the shape functions $\bar{N}_r(\xi,\eta)$, the terms of the B-matrix previous computed are extrapolated to the nine Gauss

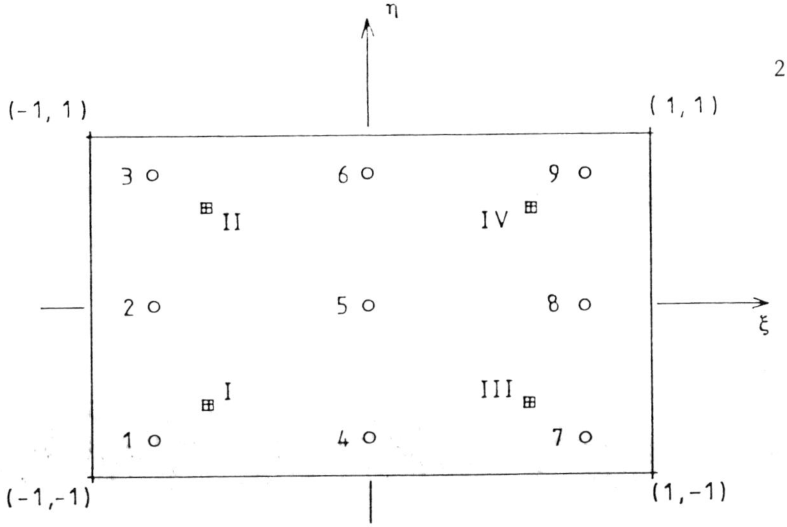

2 × 2 GAUSS RULE

$\xi_r, \eta_r = \pm 0{\cdot}57735$

$(r = 1, 4)$

3 × 3 GAUSS RULE

$\xi_n, \eta_n = 0{\cdot}0, \pm 0{\cdot}77460$

$(n = 1, 9)$

Fig. 5 Four and nine Gauss point positions for selective integration.

Table 1 Constants for the shape functions used in the selective integration rule.

POINT NUMBER	SHAPE FUNCTION	$\bar{\xi}_r$	$\bar{\eta}_r$
I	$\bar{N}_1$	-1.0/0.57735	-1.0/0.57735
II	$\bar{N}_2$	-1.0/0.57735	+1.0/0.57735
III	$\bar{N}_3$	+1.0/0.57735	-1.0/0.57735
IV	$\bar{N}_4$	+1.0/0.57735	+1.0/0.57735

points (Fig. 5) of the 3x3 (normal) integration rule.

$$\underline{B}(\xi,\eta)_n = \sum_{r=1}^{4} \bar{N}_r(\xi,\eta)_n \; B(\xi,\eta)_r \tag{43}$$

where $(\xi,\eta)_n$ and $(\xi,\eta)_r$ correspond to Gauss point positions in the normal and reduced integration rules respectively.
The four shape functions $\bar{N}_r(\xi,\eta)$ have been defined by,

$$\bar{N}_r(\xi,\eta) = \frac{1}{4}\,(1 + \xi\bar{\xi}_r)\;(1 + \eta\bar{\eta}_r) \tag{44}$$

where $\bar{\xi}_r$ and $\bar{\eta}_r$ (see Table 1) are determined so that, each of the $\bar{N}_r$ shape functions takes an unit value at the corresponding Gauss point (2x2 rule) and is null at the other three positions.

c) The bending (or bending and membrane) strains (corresponding $\underline{B}$-matrix terms) are computed at the nine Gauss points of the normal (3x3) integration rule.

The sampling points for all components of strains and stresses are the nine positions of the normal integration rule.

Using the selective integration technique, the number of mechanisms becomes,

	Heterosis	9-node L
No. of membrane mechanisms	1(*)	2 + 1(*)
No. of bending mechanisms	--	1

((*) - not-communicable in a mesh of two or more elements)

Therefore the Heterosis element can be viewed as having correct rank.

2.10 Layered Model

If the shell is built up from a series of layers of different materials, such that the material properties (and stress) are discontinuous functions of ζ, an appropriate integration through the thickness has to be carried out. In nonlinear material problems the stress profile in the thickness direction should be known at a reasonable number of points where the constitutive laws are satisfied. In reinforced concrete shell problems, the nonlinear behaviour of compressive concrete, concrete cracking and reinforcement response need a convenient representation across the structure thickness. In the present finite element code all these cases are accommodated in a simple and effective manner by using a layered approach, wherein a mid-point rule integration scheme is adopted for each layer.

Layers are numbered sequentially, starting at the bottom surface of the shell. Each layer contains stress points on its mid-surface. The stress components of the layer are computed at these stress points and are assumed to be constant over the thickness of each layer, so that the actual stress distribution of the shell is modelled by a piecewise constant approximation (Fig. 6). Layers of different thickness can be employed, as well as different number of layers per element. The specification of the layer thickness in terms of the curvilinear coordinate, ζ, permits the variation of the layer thickness as the shell thickness varies.

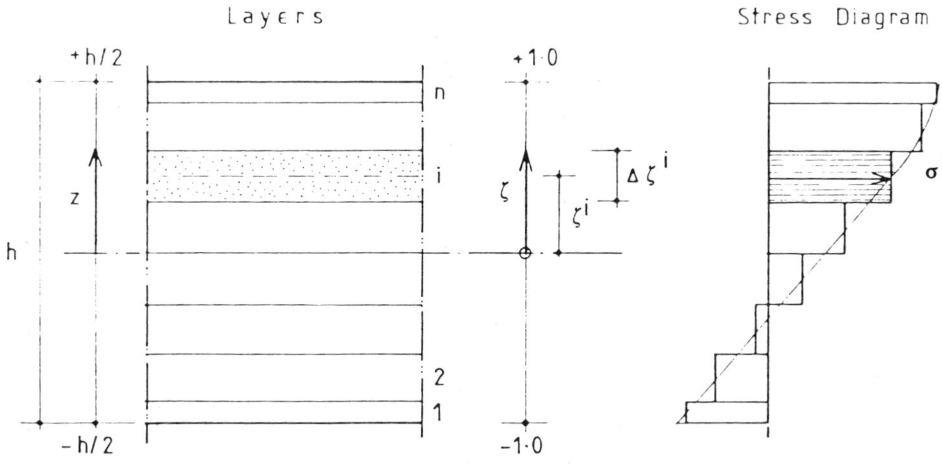

Fig. 6 Layered model and the corresponding stress representation.

The stress resultants are obtained by integrating the corresponding stress components with respect to the thickness coordinate:

- normal forces,

$$N_{x\,(y)} = \int_{-h/2}^{h/2} \sigma_{x\,(y)}\, dz = \frac{h}{2} \sum_{i=1}^{n} \sigma^{i}_{x\,(y)}\, \Delta\zeta^{i} \tag{45}$$

- bending moments

$$M_{x\,(y)\,(xy)} = - \int_{-h/2}^{h/2} \sigma_{x\,(y)\,(\tau_{xy})}\, z\, dz = - \frac{h^2}{4} \sum_{i=1}^{n} \sigma^{i}_{x\,(y)\,(\tau_{xy})}\, \zeta^{i} \Delta\zeta^{i} \tag{46}$$

- shear forces

$$Q_{x\,(y)} = \int_{-h/2}^{h/2} \tau_{xz\,(yz)}\, dz = \frac{h}{2} \sum_{i=1}^{n} \tau^{i}_{xz\,(yz)}\, \Delta\zeta^{i} \tag{47}$$

where n (i=1,n) is the number of layers.
The sign convention for the stress resultants is illustrated in Figure 7.

In the present code the strain matrix $\underline{B}$ is calculated at the mid-surface of each layer; the element stiffness matrix $\underline{K}^e$ and the internal force vector $\underline{f}^e$ are thus defined as follows:

$$\underline{K}^e = \iint_{-1}^{+1} \underline{B}^T \underline{D}\, \underline{B}\, J\, d\zeta\, dA \tag{48}$$

$$\underline{f}^e = \iint_{-1}^{+1} \underline{B}^T \underline{\sigma}\, J\, d\zeta\, dA \tag{49}$$

where

$$\int \ldots dA = \int_{-1}^{+1} \int_{-1}^{+1} \ldots d\xi d\eta$$ (integration on layer mid-surface)

and J is the determinant of the Jacobian matrix (see (20)).
This process of integration in the thickness direction is computationally more expensive, but is more appropriate for thick and variable thickness problems (general shells) in which the variation of the local system of axes (see Fig. 2), as well as the variation of the Jacobian matrix through the shell

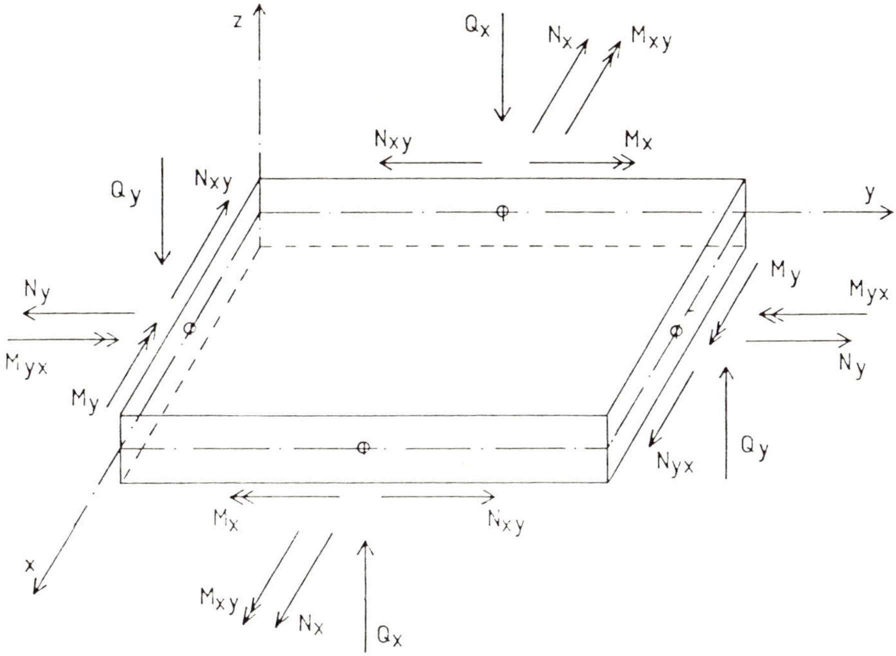

Fig. 7 Sign convention for the stress resultants.

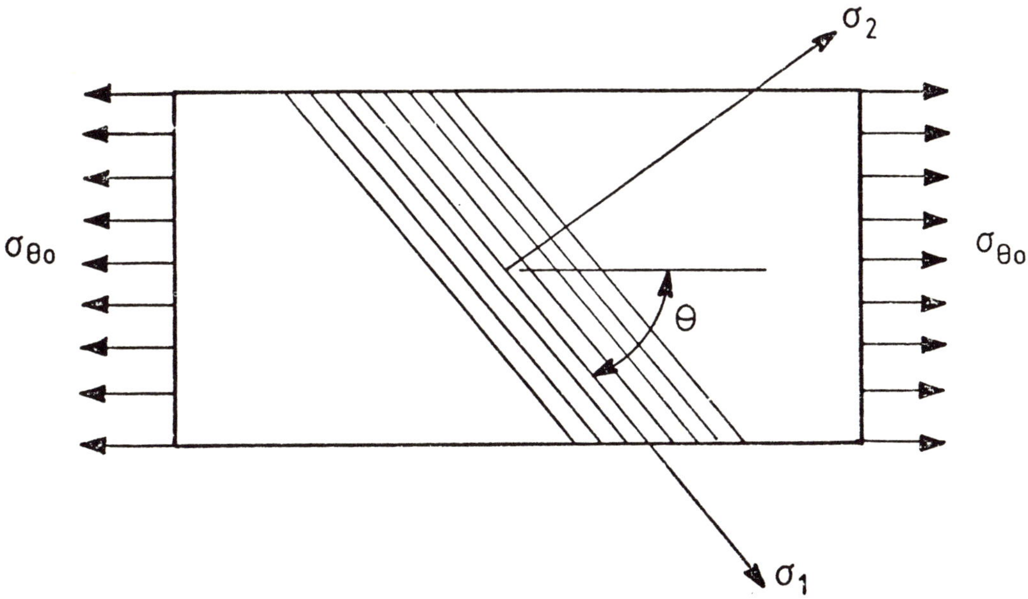

Fig. 8 Inclination of the principal axes of anisotropy to the test specimen axis.

thickness must be taken into account.

The number of layers to be employed for discretizing and integrating in the thickness direction depends on the particular problem. To represent nonlinear material behaviour in moderately thin structures a number of layers between 6 and 10 has proved to be acceptable.

3. THE INCLUSION OF NONLINEAR BEHAVIOUR

In this work two types of nonlinear behaviour are accounted for. Firstly material nonlinearity due to an elasto-plastic material response is considered and anisotropy effects are included in the yielding behaviour. In the next chapter the more complicated nonlinear behaviour of concrete associated with tensile cracking, compressive crushing, etc., is dealt with in detail. The second source of nonlinearity considered is that brought about by changes in the geometrical configuration of the structure.

3.1 General Numerical Procedure for Nonlinear Analysis

During the general stage of the incremental/iterative solution of a finite element elasto-plastic problem, the equilibrium equations will not be exactly satisfied and a system of residual forces $\underline{\psi}$ will exist such that

$$\underline{\psi}_i^n = \underline{f}^n - \underline{p}_i^n = \underline{f}^n - \int_v \underline{B}^T \underline{\sigma}_i^n \, dv \neq 0 \qquad (50)$$

in which $\underline{f}^n$ and $\underline{p}^n$ are respectively the external applied force and internal equivalent force vectors, $\underline{B}$ is the strain/displacement matrix (constant for infinitesimal deformation problems), $\underline{\sigma}^n$ is the current stress field satisfying the yield condition, v denotes the volume of the solid, the superscript n signifies the load increment number, and subscript i the iteration cycle number within that increment.

An iteration sequence must be performed for each load increment in order to obtain a displacement field, $\underline{a}_i^n$ which provides a stress field $\underline{\sigma}_i^n$ in (50) such that the residuals $\underline{\psi}_i^n$

vanish. In particular, the displacements are updated each iteration according to

$$\underline{a}_i^n = \underline{a}_{i-1}^n + \Delta\underline{a}_i^n \qquad (51)$$

where $\Delta\underline{a}_i^n$ denotes the displacement change occurring during the iteration. Several options exist for the choice of the displacement search direction. If the tangential stiffness approach is employed the iterative displacement change is evaluated according to

$$\Delta\underline{a}_i^n = - [\underline{K}_{i-1}^n]^{-1} \underline{\psi}_{i-1}^n \qquad (52)$$

in which $\underline{K}_{i-1}^n$ is the tangential stiffness matrix of the structure evaluated at the beginning of the ith iteration. The updated displacements $\underline{a}_i^n$ obtained from (51) are used to evaluate the current stresses $\underline{\sigma}_i^n$ and hence the residual forces $\underline{\psi}_i^n$ from (50). The iteration process is repeated until these residual forces are deemed to be sufficiently close to zero.

It should be noted that assembly and inversion of the full equation system is required for each iteration. A variant on the above algorithm is offered by the initial stiffness scheme in which the original structural stiffness matrix $\underline{K}_o^o$ is employed at each stage of the iteration process. This reduces the computational costs per iteration but unfortunately also reduces the rate of convergence of the process. In practice the optimum algorithm is generally provided by updating the stiffnesses at selected iterative intervals only. In this work two possibilities are considered: (a) the structural stiffness matrix is updated at the beginning of a load increment and maintained constant during iteration to equilibrium, so that $\underline{K}_{i-1}^n$ in (52) is replaced by $\underline{K}_o^n$. (b) the stiffnesses are updated after the first iteration of each load increment only (i.e. $\underline{K}_1^n$ is used in (52)). The choice of solution algorithm is left as an user option in the computer code described later in Section 4.

3.2 Flow Theory of Plasticity

In order to formulate a theory which models elasto-plastic material deformation three requirements have to be met [48]: (a) the elastic constitutive relation must be formulated (b) A yield criterion indicating the stress level at which plastic flow commences must be postulated, (c) A relationship between stress and strain must be developed for post-yield behaviour.

Before the onset of plastic yielding the relationship between stress and strain is given by expression (31).

*The Yield Criterion

For the anisotropic materials to be considered in this work the yield criterion employed will be a generalisation of the Huber-Mises law. The yield criterion can be written in the general form

$$F(\sigma,\chi) = f(\sigma) - Y(\chi) = 0 \tag{53}$$

in which f is some function of the deviatoric stress invariants and the yield level Y can be a function of a hardening parameter, χ.

Defining the plastic potential, or effective stress, $\bar{\sigma}$, in a similar manner to the Huber-Mises yield function for isotropic materials, we can write

$$f = \bar{\sigma} = [\alpha_{12}(\sigma_{11}-\sigma_{22})^2 + \alpha_{23}(\sigma_{22}-\sigma_{33})^2 + \alpha_{13}(\sigma_{33}-\sigma_{11})^2 + 3\alpha_{44}\tau_{12}^2 + 3\alpha_{55}\tau_{13}^2 + 3\alpha_{66}\tau_{23}^2]^{\frac{1}{2}} \tag{54}$$

where the σ's (τ's) are stress components, the α's are parameters of anisotropy, and the subscripts 1, 2, 3 refer to the directions of the three principal axes of anisotropy. Developing this expression and assuming $\sigma_{33} \equiv \sigma_z' = 0$, we obtain

$$\bar{\sigma}^2 = a_1\sigma_1^2 + 2a_{12}\sigma_1\sigma_2 + a_2\sigma_2^2 + a_3\tau_{12}^2 + a_4\tau_{13}^2 + a_5\tau_{23}^2 \tag{55}$$

where $\sigma_1 \equiv \sigma_{11}$, $\sigma_2 \equiv \sigma_{22}$, τ_{12}, τ_{13}, τ_{23} are the non-zero stress components and a_1, a_{12}, a_2, a_3, a_4 and a_5 are anisotropic parameters which can be determined experimentally.

If the principal axes of anisotropy 1, 2 do not coincide with the reference axes x', y' but are rotated by a certain angle θ, then the anisotropic parameters for the new system are changed according to the stress transformation. In matrix form (55) can be written as

$$\bar{\sigma}^2 = \underline{\sigma}_{1,2,3}^T \, \underline{\bar{A}} \, \underline{\sigma}_{1,2,3} \tag{56}$$

where

$$\underline{\bar{A}} = \begin{bmatrix} \bar{a}_1 & \bar{a}_{12} & 0 & 0 & 0 \\ \bar{a}_{12} & \bar{a}_2 & 0 & 0 & 0 \\ 0 & 0 & \bar{a}_3 & 0 & 0 \\ 0 & 0 & 0 & \bar{a}_4 & 0 \\ 0 & 0 & 0 & 0 & \bar{a}_5 \end{bmatrix}$$

The equation of stress transformation is

$$\underline{\sigma}_{1,2,3} = \underline{T} \, \underline{\sigma}$$

where $\underline{T}$ is the transformation matrix defined previously in (28). The effective stress, expressed in the reference system x', y', z', is then

$$\bar{\sigma}^2 = \underline{\sigma}^T \, \underline{A} \, \underline{\sigma} \tag{57}$$

in which $\underline{A}$ is the matrix of the new anisotropic parameters given by

$$\underline{A} = \underline{T}^T \, \underline{\bar{A}T} = \begin{bmatrix} a_1 & a_{12} & a_{13} & 0 & 0 \\ a_{12} & a_2 & a_{23} & 0 & 0 \\ a_{13} & a_{23} & a_3 & 0 & 0 \\ 0 & 0 & 0 & a_4 & a_{45} \\ 0 & 0 & 0 & a_{45} & a_5 \end{bmatrix}$$

The parameters of anisotropy can be determined by six independent yield tests. The initial parameters (no hardening effects) are obtained by successively allowing all stress components to be zero in the yield function (55) except the one under consideration. For a tensile test in the 1 direction, we have

$$a_{1o} = \bar{\sigma}_o^2/\sigma_{1o}^2 \tag{58}$$

where $\bar{\sigma}_o$ is the uniaxial yield stress in the reference direction, σ_{1o} is the uniaxial yield stress in the 1 direction, and the subscripts 'o' denote values before hardening commences. Taking the 1-direction as the reference direction, then $a_{1o}=1.0$. Similarly

$$a_{2o} = \bar{\sigma}_o^2/\sigma_{2o}^2, \quad a_{3o} = \bar{\sigma}_o^2/\tau_{12o}^2$$
$$a_{4o} = \bar{\sigma}_o^2/\tau_{13o}^2, \quad a_{5o} = \bar{\sigma}_o^2/\tau_{23o}^2 \tag{59}$$

To obtain a_{12o} another uniaxial tensile test is required, in which the specimen is taken from the $\overline{12}$ material plane [42]. If the specimen axis is rotated by an angle θ to the 1 axis (Figure 8), and assuming $\sigma_{\theta o}$ as the uniaxial yield stress obtained by test, we have

$$\begin{aligned} \sigma_1 &= \sigma_{\theta o} \cos^2 \theta \\ \sigma_2 &= \sigma_{\theta o} \sin^2 \theta \\ \tau_{12} &= \sigma_{\theta o} \sin \theta \cos \theta \\ \tau_{13} &= \tau_{23} = 0 \end{aligned} \tag{60}$$

Substituting these stress components into (55) and assuming, for example, θ = 45 degrees, we obtain the remaining parameter,

$$a_{12o} = 2(\bar{\sigma}_o/\sigma_{\theta o})^2 - \tfrac{1}{2}(1.0 + a_{2o} + a_{3o}) \tag{61}$$

For an isotropic material, $\sigma_{\theta o} = \bar{\sigma}_o$, $a_{2o} = 1$, and $a_{3o} = 3$ and therefore a_{12o} takes the value -0.5.

These parameters are functions of the current yield stresses; therefore, they must vary for a work-hardening material. Their subsequent values [43] are obtained by introducing the current yield stress values into (59) and (61).

Elasto-plastic stress/strain relation

The total strain increment $d\underline{\varepsilon}$ is the sum of the elastic and plastic components, so that

$$d\underline{\varepsilon} = d\underline{\varepsilon}^e + d\underline{\varepsilon}^p \tag{62}$$

The elastic strain increment $d\underline{\varepsilon}^e$ is given by the incremental form of (31) and the plastic strain increment by the flow rule

$$d\underline{\varepsilon}^p = d\lambda \frac{\partial Q}{\partial \underline{\sigma}} \tag{63}$$

where Q is termed the plastic potential and dλ is a proportional constant. The assumption Q ≡ f gives rise to an associated plasticity theory, in which case expression (63) is termed the normality condition; since $\partial f/\partial \underline{\sigma}$ is a vector directed normal to yield surface in a stress space geometrical interpretation.

The differential form of (53) can be written

$$\underline{a}^T d\underline{\sigma} - A\, d\lambda = 0 \tag{64}$$

in which the flow vector $\underline{a}$ is defined by

$$\underline{a}^T = \frac{\partial F}{\partial \underline{\sigma}} \tag{65}$$

and

$$A = -\frac{1}{d\lambda} \frac{\partial F}{\partial \chi} d\chi \tag{66}$$

Manipulation of (62)-(66) leads to the following elasto-plastic incremental stress-strain relationship, [48]

$$d\underline{\sigma} = \underline{D}_{ep}\, d\underline{\varepsilon} \tag{67}$$

with

$$\underline{D}_{ep} = \underline{D} - \frac{\underline{D}\,\underline{a}\,\underline{a}^T\,\underline{D}}{A + \underline{a}^T\,\underline{D}\,\underline{a}} \tag{68}$$

The hardening parameter A can be deduced from uniaxial conditions to be

$$A = H' = \frac{d\bar{\sigma}}{d\bar{\varepsilon}_p} \tag{69}$$

Thus A is the tangent to the effective stress - plastic strain curve and is a function of the accumulated effective plastic strain $\bar{\varepsilon}_p$.

The tangential stiffness matrix of the material for use in (52) can be obtained on use of (50) and (67) to be

$$\underline{K} = \frac{d\underline{p}}{d\underline{a}} = \int_v \underline{B}^T \frac{d\underline{\sigma}}{d\underline{\varepsilon}} \frac{d\underline{\varepsilon}}{d\underline{a}}\, dv = \int_v \underline{B}^T \underline{D}_{ep}\, \underline{B}\, dv \tag{70}$$

3.3 Geometric Nonlinearity

Two basically different formulations can be employed for the description of large deformation problems: (a) A Total Lagrangian approach in which the current (2nd Piola Kirchhoff) stress and (Green-Lagrange) strain fields are referred to the original geometric configuration and the displacement field gives the current configuration of the system in relation to its initial position. (b) Updated Lagrangian approach in which the current configuration of the system is used to define the current (Cauchy) stress and (Almansi) strain state. The geometry of the structure is successively updated during the incremental process and the stress and strain fields are referred to the last evaluated configuration. The stress and strain measures employed above are defined in such a way as to be energy conjugates. Both the Green-Lagrange and Almansi strains reduce to the usual definition of engineering strains for infinitesimal deformation problems. The Cauchy stresses have the usual meaning (i.e. force/unit area in the deformed configuration) but the physical meaning of the Piola-Kirchhoff stresses may be lost if significant changes in the original geometry occurs. Transformation of all quantities from a Total Lagrangian formulation to an Updated Lagrangian formulation, and vice-versa, can be performed by considering the displacement and stretching of two natural systems of axes [54,55].

The most appropriate formulation for numerical solution depends on the type of analysis being considered. For the degenerate shell elements employed in this work a specific (and appropriate) total Lagrangian formulation is adopted in which large deflections and moderate rotations (in the sense of the Von Karman hypotheses) are accounted for. Reference of the problem variables to the original configuration is advantageous for qudratic degenerate shell elements, since the computationally expensive transfer of quantities between local and global axes need then be performed only once. The strain-displacement matrix is calculated once during the nonlinear process and its nonlinear part is updated using the current displacements by a simple matrix product. The constitutive relations defined previously in terms of engineering stresses and strains are considered valid for the new stress/strain quantities measured in the original configuration.

Taking the variation of (50) with respect to a displacement variation $\underline{da}$ gives the tangential stiffness matrix for a geometrically nonlinear problem to be

$$\underline{K}\ \underline{da} = \underline{dp} = \int_V \underline{B}^T\ \underline{d\sigma}\ dv + \int_V \underline{dB}^T\ \underline{\sigma}\ dv \tag{71}$$

The strain-displacement matrix $\underline{B}$ may be separated into the usual infinitesimal part $\underline{B}_o$ and nonlinear contribution $\underline{B}_L$ so that

$$\underline{B} = \underline{B}_o + \underline{B}_L \tag{72}$$

Consequently $\underline{dB}^T = \underline{dB}_L^T$. Defining the initial stress or geometric stiffness matrix, $\underline{K}_\sigma$, as

$$\underline{K}_\sigma\ \underline{da} = \int_V d\ \underline{B}_L^T\ \underline{\sigma}\ dv \tag{73}$$

then use of (31) and (73) in (71) results in

$$\underline{K} = \underline{\bar{K}} + \underline{K}_\sigma \tag{74}$$

in which $\underline{\bar{K}}$ is given by the usual expression

$$\underline{\bar{K}} = \int_V \underline{B}^T\ \underline{D}\ \underline{B}\ dv \tag{75}$$

For elasto plastic problems, use of (67) in place of (31) results in $\underline{D}$ being replaced by $\underline{D}_{ep}$ in (75).

Introducing the Von Karman assumptions, which imply that derivatives of u' and v' with respect to x', y' and z' are small, and noting that the variation of w' with z' may be neglected, the Green-Lagrange strains may be expressed in local coordinates as

$$\underline{\varepsilon} = \begin{bmatrix} \varepsilon_{x'} \\ \varepsilon_{y'} \\ \gamma_{x'y'} \\ \gamma_{x'z'} \\ \gamma_{y'z'} \end{bmatrix} = \begin{bmatrix} \partial u'/\partial x' \\ \partial v'/\partial y' \\ \dfrac{\partial u'}{\partial y'} + \dfrac{\partial v'}{\partial x'} \\ \dfrac{\partial u'}{\partial z'} + \dfrac{\partial w'}{\partial x'} \\ \dfrac{\partial v'}{\partial z'} + \dfrac{\partial w'}{\partial y'} \end{bmatrix} + \begin{bmatrix} \frac{1}{2}\left(\dfrac{\partial w'}{\partial x'}\right)^2 \\ \frac{1}{2}\left(\dfrac{\partial w'}{\partial y'}\right)^2 \\ \dfrac{\partial w'}{\partial x'}\,\dfrac{\partial w'}{\partial y'} \\ 0 \\ 0 \end{bmatrix} = \underline{\varepsilon}_o + \underline{\varepsilon}_L \quad (76)$$

The components of the Piola-Kirchhoff stress vector are again given by relation (22) but in which the current strains are now taken to be the Green-Lagrange strains.

From (76), the nonlinear contribution to the strain vector can be written

$$\underline{\varepsilon}_L = \tfrac{1}{2}\,\underline{A}\,\underline{R} \quad (77)$$

where

$$\underline{S}^T = \begin{bmatrix} \dfrac{\partial w'}{\partial x'} & 0 & \dfrac{\partial w'}{\partial x'} & 0 & 0 \\ 0 & \dfrac{\partial w'}{\partial y'} & \dfrac{\partial w'}{\partial y'} & 0 & 0 \end{bmatrix} \quad (78)$$

and

$$\underline{R} = \begin{bmatrix} \dfrac{\partial w'}{\partial x'} \\ \dfrac{\partial w'}{\partial y'} \end{bmatrix} = \underline{G}\;\underline{a} \quad (79)$$

The term $\underline{G}$ is a matrix with two rows and a number of columns equal to the total number of element nodal variables. The first row contains the contribution of each nodal variable to the local derivative $\frac{\partial w'}{\partial x'}$ (corresponding shape function derivatives) and the second row contains similar contributions for $\partial w'/\partial y'$.

Taking the variation of (77) gives

$$d\underline{\varepsilon}_L = \tfrac{1}{2}\,d\underline{S}\,\underline{R} + \tfrac{1}{2}\,\underline{S}\,d\underline{R} = \underline{S}\,d\underline{R} = \underline{S}\,\underline{G}\,d\underline{a} \quad (80)$$

and then by definition we have

$$\underline{B}_L = \underline{S}\ \underline{G} \tag{81}$$

In the present finite element code, matrices $\underline{B}_o$ and $\underline{G}$ are initially calculated and stored. Using the current displacements and (79) matrix $\underline{S}$ is formed and then $\underline{B}_L$ is obtained from (81). The incrmental strain-displacement matrix $\underline{B}$ is evaluated by summing $\underline{B}_o$ to this current value of $\underline{B}_L$.

In order to determine the tangential stiffness matrix given by (74), for use in the nonlinear solution algorithm, it only now remains to explicitly evaluate the geometric matrix, $\underline{K}_\sigma$. Substituting from (81) in (73) results in

$$\underline{K}_\sigma\ d\underline{a} = \int_v \underline{G}^T\ d\underline{S}^T\ \underline{\sigma}\ dv \tag{82}$$

The term $d\underline{S}^T\underline{\sigma}$ can be written, with the aid of (78) and (79), as

$$d\underline{S}^T\underline{\sigma} = [\underline{\sigma}]\ \underline{G}\ d\underline{a} \tag{83}$$

in which

$$[\underline{\sigma}] = \begin{bmatrix} \sigma_{x'} & \tau_{x'y'} \\ \tau_{x'y'} & \sigma_{y'} \end{bmatrix} \tag{84}$$

Substituting (83) into (82) gives the geometric stiffness matrix to be a symmetric matrix of the form,

$$\underline{K}_\sigma = \int_v \underline{G}^T\ [\underline{\sigma}]\ \underline{G}\ dv \tag{85}$$

Matrix $\underline{G}$ was previously defined in (79) and $[\underline{\sigma}]$ is seen from (84) to be composed of components of the current Piola-Kirchhoff stress vector.

4. FINITE ELEMENT COMPUTER CODE

In this section a computer code is presented which undertakes the elasto-plastic solution of both thick and thin plates and shells. The layered approach described in Section 2.10 is adopted and geometric nonlinear effects may also be included.

Computer subroutines, written in FORTRAN, which undertake the various operations required for nonlinear finite element shell analysis are presented and these are later linked to form

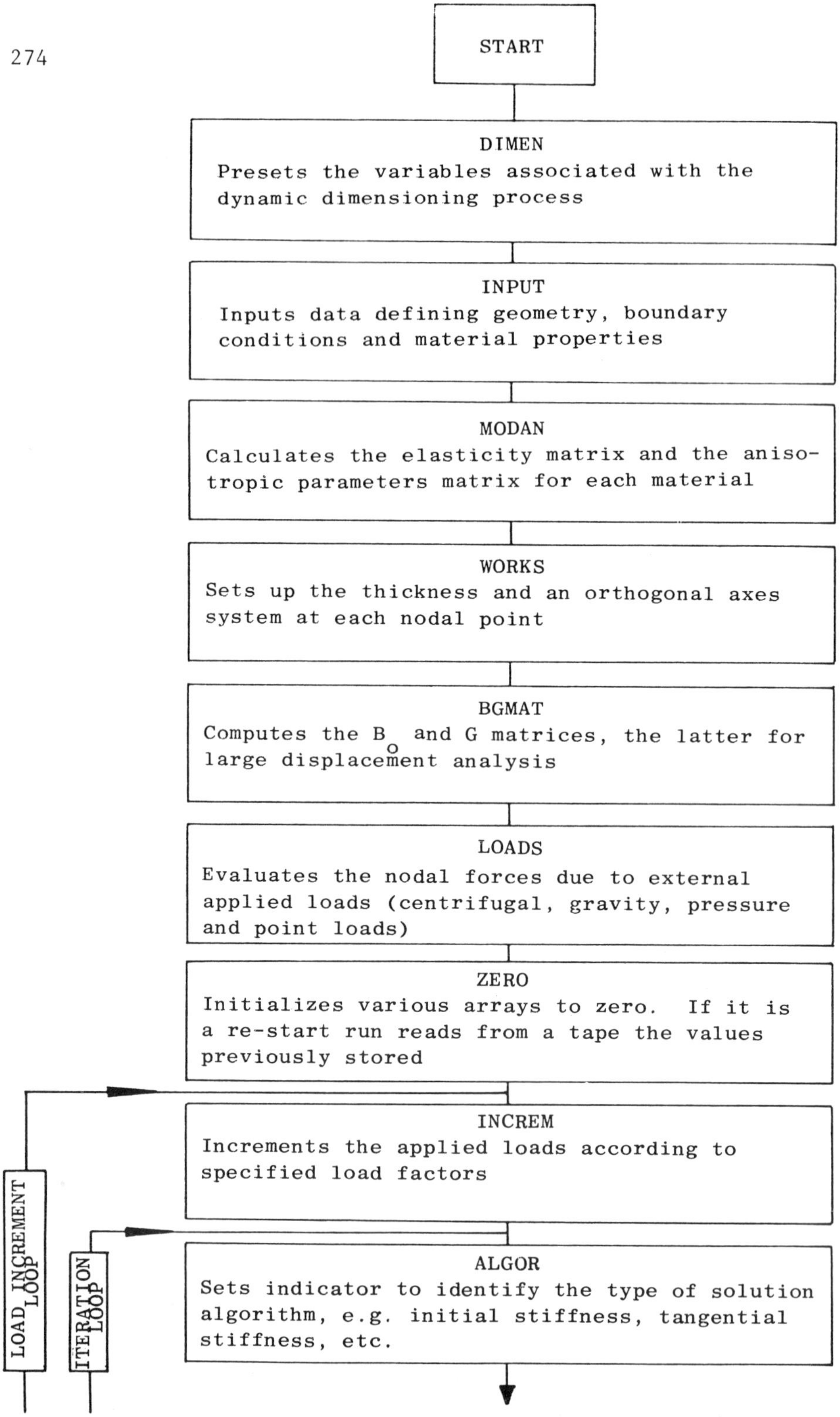

Fig. 9 Overall structure of Program PLASTOSHELL

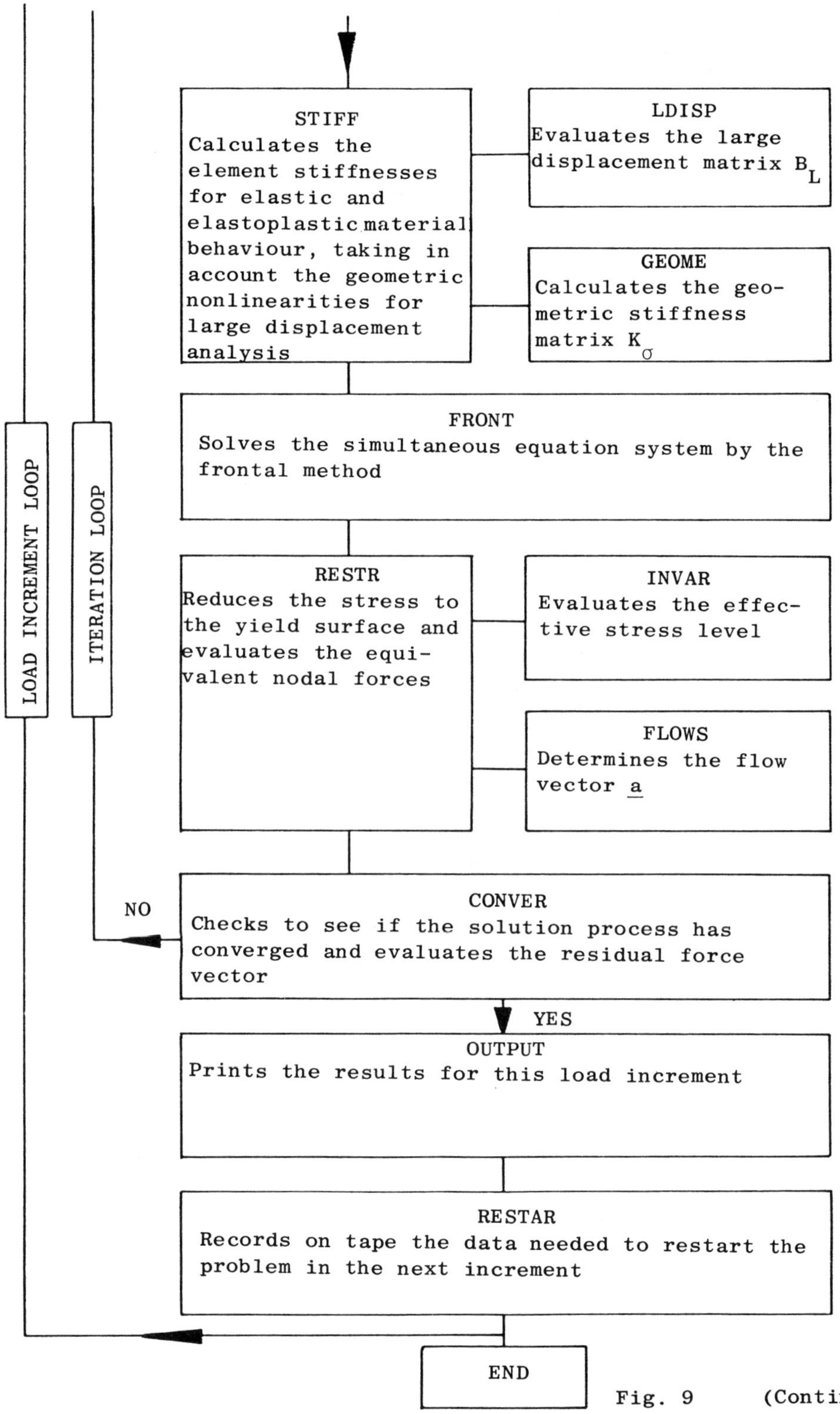

Fig. 9 (Continued)

a working program. Both the logic and the computer coding employed for most of the subroutines utilised here has been described in considerable detail in Ref. 48. User instructions for the program are provided in the appendix following the next chapter.

The structure of the program is shown schematically in Fig. 9 where, in particular, the order in which the various subroutines are called by the controlling or master segment is indicated. The function of each subroutine is now explained and a FORTRAN listing provided.

4.1 Subroutine ALGOR

The function of this subroutine is to set the value of KRESL which controls the nonlinear solution process. A value of KRESL=1 indicates reformulation of the element stiffnesses accompanied by a full equation solution and KRESL=2 indicates that the element stiffnesses are not to be modified and consequently only equation resolution takes place.

```
      SUBROUTINE ALGOR(FIXED,KITER,IITER,KRESL,MTOTV,NALGO,
                       NTOTV,KUNLO,KINCS)
C**********************************************************************
C
C**** THIS SUBROUTINE SETS EQUATION RESOLUTION INDEX,KRESL
C
C**********************************************************************
      DIMENSION FIXED(MTOTV)
      KRESL=2
      IF(NALGO.EQ.1.AND.KITER.EQ.2) KRESL=1
      IF(NALGO.EQ.2) KRESL=1
      IF(NALGO.EQ.3.AND.IITER.EQ.1) KRESL=1
      IF(NALGO.EQ.4.AND.KITER.EQ.2) KRESL=1
      IF(NALGO.EQ.4.AND.IITER.EQ.2) KRESL=1
      IF(NALGO.EQ.4.AND.KUNLO.GT.0) KRESL=1
      IF(KITER.EQ.(KINCS+1+1)) KRESL=1
      IF(IITER.EQ.1) RETURN
      DO 100 ITOTV = 1,NTOTV
      FIXED(ITOTV)=0.0
  100 CONTINUE
      RETURN
      END
```

4.2 Subroutine BGMAT

This subroutine forms the $\underline{B}_o$ matrix and the $\underline{G}$ matrix (for large deformation problems) given by (79). These matrices are stored on tape 8 for subsequent use. Reduced, selective or normal integration rules can be employed in setting up the terms of these matrices. In the present selective integration option, the reduced rule is used for transverse shear terms only.

```
      SUBROUTINE BGMAT(COORD,DICOS,LNODS,MATNO,MELEM,
     .                 MLAYR,MMATS,MPOIN,M3POI,NELEM,
     .                 NEVAB,NGAUS,NGAUZ,NLAYR,NNODE,NPROP,
     .                 POSGP,PROPS,THICK,WEIGP)
C***********************************************************************
C
C*** THIS SUBROUTINE COMPUTES BMATX AND GMATX(THE LATTER FOR LARGE DISPL
C    ANALYSIS).THESE MATRICES ARE STORED ON TAPE 8 FOR LATER USAGE
C    SELECTIVE INTEGRATION(TRANSVERSE SHEAR TERMS) CAN BE ACCOUNTED FOR
C
C***********************************************************************
      COMMON WORMX(3,24),QVALU,DJACB
      DIMENSION BMATX(5,45),BDUMY(8,45),COORD(MPOIN,8),
     .          DICOS(3,M3POI),FUNCT(4),GMATX(2,45),
     .          LNODS(MELEM,9),MATNO(MELEM,MLAYR),POSGP(5),
     .          PROPS(MMATS,NPROP),THICK(MPOIN),SHAPE(3,9),
     .          WEIGP(5)
      REWIND 8
      LGAUS=NGAUS-NGAUZ
C**  LGAUS=0 FOR NORMAL OR REDUCED INTEGRATION RULE.LGAUS=1 FOR
C    SELECTIVE INTEGRATION RULE.
      DO 100 IELEM=1,NELEM
      IF(LGAUS.EQ.0) GO TO 25
      NBORP=0
C
C*** REDUCED INTEGRATION IS USED TO SET UP THE TRANSVERSE SHEAR TERMS
C    OF BMATX.FIRSTLY THESE TERMS ARE STORED IN BDUMY MATRIX
C
      CONS1=1.0/POSGP(4)
      CONS2=-CONS1
      ZETSP=0.0
      KGAUZ=-1
      DO 20 IGAUZ=1,NGAUZ
      DO 20 JGAUZ=1,NGAUZ
      KGAUZ=KGAUZ+1
      EXISP=POSGP(3+IGAUZ)
      ETASP=POSGP(3+JGAUZ)
      CALL      SFR1(SHAPE,EXISP,ETASP)
      CALL      FUNC(BMATX,SHAPE,THICK,NBORP,NNODE,ZETSP,MELEM,
     .               COORD,DICOS,LNODS,IELEM,MPOIN,M3POI,GMATX)
      DO 15 IEVAB=1,NEVAB
      DO 15 IPOSI=1,2
      JPOSI=2*KGAUZ+IPOSI
   15 BDUMY(JPOSI,IEVAB)=GMATX(IPOSI,IEVAB)
   20 CONTINUE
C
C*** SET UP BMATX AND GMATX FOR NORMAL OR REDUCED INTEGRATION RULE
C
   25 NBORP=1
      DO 50 IGAUS=1,NGAUS
      DO 50 JGAUS=1,NGAUS
      EXISP=POSGP(IGAUS)
      ETASP=POSGP(JGAUS)
      CALL      SFR1(SHAPE,EXISP,ETASP)
      ZETSP=-1.0
      DO 45 ILAYR=1,NLAYR
      LPROP=MATNO(IELEM,ILAYR)
      DZETA=PROPS(LPROP,3)
      ZETSP=ZETSP+DZETA/2.0
      CALL      FUNC(BMATX,SHAPE,THICK,NBORP,NNODE,ZETSP,MELEM,
     .               COORD,DICOS,LNODS,IELEM,MPOIN,M3POI,GMATX)
      DVOLU=DJACB*WEIGP(IGAUS)*WEIGP(JGAUS)*DZETA
      IF(LGAUS.EQ.0) GO TO 40
C
C*** SET UP THE INTEPOLATION FUNCTIONS TO PERFORM THE SELECTIVE INTEG.
C
      FUNCT(1)=0.25*(1.0+CONS1*EXISP)*(1.0+CONS1*ETASP)
      FUNCT(2)=0.25*(1.0+CONS1*EXISP)*(1.0+CONS2*ETASP)
      FUNCT(3)=0.25*(1.0+CONS2*EXISP)*(1.0+CONS1*ETASP)
      FUNCT(4)=0.25*(1.0+CONS2*EXISP)*(1.0+CONS2*ETASP)
C
C*** INTEPRPOLATE THE TANSVERSE SHEAR TERMS OF BMATX FROM 4 TO 9 G.P.
C
```

```
      DO 30 IEVAB=1,NEVAB
      DO 30 IDOFN=4,5
      BMATX(IDOFN,IEVAB)=0.0
      DO 30 INTPO=1,4
      IGASH=2*INTPO+IDOFN-5
   30 BMATX(IDOFN,IEVAB)=BMATX(IDOFN,IEVAB)+FUNCT(INTPO)*
     .                      BDUMY(IGASH,IEVAB)
   40 CONTINUE
      WRITE(8) BMATX,GMATX,DVOLU
      ZETSP=ZETSP+DZETA/2.0
   45 CONTINUE
   50 CONTINUE
  100 CONTINUE
      RETURN
      END
```

4.3 Subroutine CHECK1

The function of this subroutine is to scrutinise the problem control parameters for possible errors. Any errors detected are signalled by means of printed error numbers whose interpretation are given in Table 2.

```
      SUBROUTINE CHECK1(NDOFN,NELEM,NGAUS,NMATS,NNODE,NPOIN,
     .                  MMATS,NVFIX,NGAUZ,NLAYR)
C*********************************************************************
C
C**** THIS SUBROUTINE CHECKS THE MAIN CONTROL DATA
C
C*********************************************************************
      DIMENSION NEROR(20)
      DO 10 IEROR=1,8
   10 NEROR(IEROR)=0
C
C*** CREATE THE DIAGNOSTIC MESSAGES
C
      IF(NPOIN.LE.0) NEROR(1)=1
      IF(NELEM*NNODE.LT.NPOIN) NEROR(2)=1
      IF(NVFIX.LT.2.OR.NVFIX.GT.NPOIN) NEROR(3)=1
      IF(NNODE.LT.8.OR.NNODE.GT.9) NEROR(4)=1
      IF(NDOFN.NE.5.OR.NLAYR.GT.10) NEROR(5)=1
      IF(NMATS.LT.1.OR.NMATS.GT.MMATS) NEROR(6)=1
      IF(NGAUS.LT.2.OR.NGAUS.GT.3) NEROR(7)=1
      IF(NGAUZ.LT.2.OR.NGAUZ.GT.3) NEROR(8)=1
C
C*** EITHER RETURN,OR ELSE PRINT THE ERRORS DIAGNOSED
C
      KEROR=0
      DO 20 IEROR=1,8
      IF(NEROR(IEROR).EQ.0) GO TO 20
      KEROR=1
      WRITE(6,900) IEROR
  900 FORMAT(//31H *** DIAGNOSIS BY CHECK1, ERROR,I3)
   20 CONTINUE
      IF(KEROR.EQ.0) RETURN
C
C*** OTHERWISE ECHO ALL THE REMAINING DATA WITHOUT FURTHER COMMENT
C
      CALL ECHO
      END
```

Table 2 Errors diagnosed by Subroutine CHECK1

Error label	Interpretation
1	The specified total number of node points, NPOIN, in the structure is less than or equal to zero
2	The possible maximum total number of node points in the structure is less than the specified total, NPOIN
3	The number of restrained nodal points is less than 2 or greater than NPOIN (for shell problems at least 2 points must be restrained to eliminate rigid body motions)
4	The number of nodes/element is not equal to 8 or 9
5	The number of degrees of freedom per node is not equal to 5 or the number of layers is greater than 10
6	The total number of different materials is less than or equal to zero or greater than the maximum number of different materials specified in Subroutine DIMEN
7	The number of Gaussian integration points in each direction is not equal to either 2 or 3
8	The number of Gaussian integration points for transverse shear is not equal to either 2 or 3

4.4 Subroutine CHECK2

If the problem control parameters have passed the scrutiny of subroutine CHECK1, the geometric data, boundary conditions and material data are then checked by Subroutine CHECK2 where error types 9 to 20 listed in Table 3 are checked for.

This subroutine also ensures that the maximum front-width of the problem does not exceed the dimensions specified in subroutine FRONT.

```
      SUBROUTINE CHECK2(COORD,IFFIX,LNODS,MATNO,MELEM,MFRON,MPOIN,MTOTV,CEK2
     .                  MVFIX,NDFRO,NDOFN,NELEM,NMATS,NNODE,NOFIX,NPOIN,CEK2
     .                  NVFIX,NLAYR)                                    CEK2
C***********************************************************************CEK2
C                                                                       CEK2
C**** THIS SUBROUTINE CHECKS THE REMAINDER OF THE INPUT DATA            CEK2
C                                                                       CEK2
C***********************************************************************CEK2
      DIMENSION COORD(MPOIN,8),IFFIX(MTOTV),LNODS(MELEM,12),            CEK2
     .          MATNO(MELEM,NLAYR),NDFRO(MELEM),NEROR(20),NOFIX(MVFIX)  CEK2
C                                                                       CEK2
C*** CHECK AGAINST TWO IDENTICAL NONZERO NODAL TOP-COORDINATES          CEK2
C                                                                       CEK2
      DO 5 IEROR=9,20                                                   CEK2
    5 NEROR(IEROR)=0                                                    CEK2
      DO 10 IELEM=1,NELEM                                               CEK2
   10 NDFRO(IELEM)=0                                                    CEK2
      DO 50 IPOIN=2,NPOIN                                               CEK2
      KPOIN=IPOIN-1                                                     CEK2
      DO 30 JPOIN=1,KPOIN                                               CEK2
      DO 20 IDIME=1,3                                                   CEK2
      IF(COORD(IPOIN,IDIME).NE.COORD(JPOIN,IDIME)) GO TO 30             CEK2
   20 CONTINUE                                                          CEK2
```

```
      NEROR(9)=NEROR(9)+1
   30 CONTINUE
   40 CONTINUE
C
C*** CHECK THE LIST OF ELEMENT PROPERTY NUMBERS
C
      DO 50 IELEM=1,NELEM
      DO 50 ILAYR=1,NLAYR
   50 IF(MATNO(IELEM,ILAYR).GT.NMATS) NEROR(10)=NEROR(10)
     . +1
C
C*** CHECK FOR IMPOSSIBLE NODE NUMBERS
C
      DO 70 IELEM=1,NELEM
      DO 60 INODE=1,NNODE
      IF(LNODS(IELEM,INODE).EQ.0) NEROR(11)=NEROR(11)+1
   60 IF(LNODS(IELEM,INODE).LT.0.OR.LNODS(IELEM,INODE).GT.NPOIN) NEROR(
     . 12)=NEROR(12)+1
   70 CONTINUE
C
C*** CHECK FOR ANY REPETITION OF A NODE NUMBER WITHIN AN ELEMENT
C
      DO 140 IPOIN=1,NPOIN
      KSTAR=0
      DO 100 IELEM=1,NELEM
      KZERO=0
      DO 90 INODE=1,NNODE
      IF(LNODS(IELEM,INODE).NE.IPOIN) GO TO 90
      KZERO=KZERO+1
      IF(KZERO.GT.1) NEROR(13)=NEROR(13)+1
C
C*** SEEK FIRST,LAST AND INTERMEDIATE APPEARANCES OF NODE IPOIN
C
      IF(KSTAR.NE.0) GO TO 80
      KSTAR=IELEM
C
C*** CALCULATE INCREASE OR DECREASE IN FRONTWIDTH AT EACH ELEMENT STAGE
C
      NDFRO(IELEM)=NDFRO(IELEM)+NDOFN
   80 CONTINUE
C
C*** AND CHANGE THE SIGN OF THE LAST APPEARANCE OF EACH NODE
C
      KLAST=IELEM
      NLAST=INODE
   90 CONTINUE
  100 CONTINUE
      IF(KSTAR.EQ.0) GO TO 110
      IF(KLAST.LT.NELEM) NDFRO(KLAST+1)=NDFRO(KLAST+1)-NDOFN
      LNODS(KLAST,NLAST)=-IPOIN
      GO TO 140
C
C*** CHECK THAT COORDINATES FOR AN UNUSED NODE HAVE NOT BEEN SPECIFIED
C
  110 WRITE(6,900) IPOIN
  900 FORMAT(/15H CHECK WHY NODE,I4,14H NEVER APPEARS)
      NEROR(14)=NEROR(14)+1
      SIGMA=0.0
      DO 120 IDIME=1,3
  120 SIGMA=SIGMA+ABS(COORD(IPOIN,IDIME))
      IF(SIGMA.NE.0.0) NEROR(15)=NEROR(15)+1
C
C*** CHECK THAT AN UNUSED NODE NUMBER IS NOT A RESTRAINED NODE
C
      DO 130 IVFIX=1,NVFIX
  130 IF(NOFIX(IVFIX).EQ.IPOIN) NEROR(16)=NEROR(16)+1
  140 CONTINUE
C
C*** CALCULATE THE LARGEST FRONTWIDTH
C
      NFRON=0
      KFRON=0
      DO 150 IELEM=1,NELEM
      NFRON=NFRON+NDFRO(IELEM)
  150 IF(NFRON.GT.KFRON) KFRON=NFRON
      WRITE(6,905) KFRON
  905 FORMAT(//33H MAXIMUM FRONTWIDTH ENCOUNTERED =,I5//)
      IF(KFRON.GT.MFRON) NEROR(17)=1
```

```
C
C*** CONTINUE CHECKING THE DATA FOR THE FIXED VALUES
C
      DO 170 IVFIX=1,NVFIX
      IF(NOFIX(IVFIX).LE.0.OR.NOFIX(IVFIX).GT.NPOIN) NEROR(18)=NEROR(18)
     . +1
      KOUNT=0
      NLOCA=(NOFIX(IVFIX)-1)*NDOFN
      DO 160 IDOFN=1,NDOFN
      NLOCA=NLOCA+1
  160 IF(IFFIX(NLOCA).GT.0) KOUNT=1
      IF(KOUNT.EQ.0) NEROR(19)=NEROR(19)+1
      KVFIX=IVFIX-1
      DO 170 JVFIX=1,KVFIX
  170 IF(IVFIX.NE.1.AND.NOFIX(IVFIX).EQ.NOFIX(JVFIX)) NEROR(20)=NEROR(20
     . )+1
      KEROR=0
      DO 180 IEROR=9,20
      IF(NEROR(IEROR).EQ.0) GO TO 180
      KEROR=1
      WRITE(6,910) IEROR,NEROR(IEROR)
  910 FORMAT(//31H *** DIAGNOSIS BY CHECK2, ERROR,I3,6X,18H ASSOCIATED N
     .UMBER,I5)
  180 CONTINUE
      IF(KEROR.NE.0) GO TO 200
C
C*** RETURN ALL NODAL CONNECTION NUMBERS TO POSITIVE VALUES
C
      DO 190 IELEM=1,NELEM
      DO 190 INODE=1,NNODE
  190 LNODS(IELEM,INODE)=IABS(LNODS(IELEM,INODE))
      RETURN
  200 CALL ECHO
      END
```

Table 3 Errors diagnosed by Subroutine CHECK2

Error Label	Interpretation
9	A total of x identical nodal coordinates have been detected, i.e. x nodal points have coordinates which are identical to those of one or more of the remaining nodes
10	A total of x material identification numbers are greater than the total number of different materials
11	A total of x nodal connection numbers have zero value
12	A total of x nodal connection numbers are negative or greater than the specified maximum value, NPOIN
13	A total of x repetitions of node numbers within individual elements have been detected
14	A total of x nodes exist in the list of nodal points which do not appear anywhere in the list of element nodal connection numbers
15	Non-zero coordinates have been specified for a total of x nodes which do not appear in the list of element nodal connection numbers
16	A total of x node numbers which do not appear in the element nodal connections list have been specified as restrained nodal points

17	The largest frontwidth encountered in the problem has exceeded the maximum value specified in subroutine FRONT of the program
18	A total of x restrained nodal points have numbers less than or equal to zero or greater than the specified maximum value, NPOIN
19	A total of x restrained nodal points at which the fixity code is less than or equal to zero have been detected
20	A total of x repetitions in the list of restrained nodal points have been detected

4.5 Subroutine CONVER

This subroutine monitors convergence of the nonlinear solution iteration process. For convergence, the norm of the residual forces, $\underline{\psi}$ is required to be less than a specified percentage, TOLER, of the norm of the applied loads.

```
      SUBROUTINE CONVER(ELOAD,IITER,LNODS,MELEM,MEVAB,MTOTV,NCHEK,
     .                  NDOFN,NELEM,NEVAB,NNODE,NTOTV,STFOR,
     .                  TLOAD,TOFOR,TOLER)
C***********************************************************************
C
C**** THIS SUBROUTINE CHECKS FOR CONVERGENCE OF THE ITERATION PROCESS
C
C***********************************************************************
      DIMENSION ELOAD(MELEM,MEVAB),LNODS(MELEM,12) ,STFOR(MTOTV),
     .          TOFOR(MTOTV),TLOAD(MELEM,MEVAB)
      NCHEK=0
      RESID=0.0
      RETOT=0.0
      REMAX=0.0
      DO 5 ITOTV=1,NTOTV
      STFOR(ITOTV)=0.0
      TOFOR(ITOTV)=0.0
    5 CONTINUE
      DO 40 IELEM=1,NELEM
      KEVAB=0
      DO 40 INODE=1,NNODE
      LOCNO=IABS(LNODS(IELEM,INODE))
      DO 40 IDOFN=1,NDOFN
      KEVAB=KEVAB+1
      NPOSI=(LOCNO-1)*NDOFN+IDOFN
      STFOR(NPOSI)=STFOR(NPOSI)+ELOAD(IELEM,KEVAB)
   40 TOFOR(NPOSI)=TOFOR(NPOSI)+TLOAD(IELEM,KEVAB)
      DO 50 ITOTV=1,NTOTV
      REFOR=TOFOR(ITOTV)-STFOR(ITOTV)
      RESID=RESID+REFOR*REFOR
      RETOT=RETOT+TOFOR(ITOTV)*TOFOR(ITOTV)
      AGASH=ABS(REFOR)
   50 IF(AGASH.GT.REMAX) REMAX=AGASH
      DO 10 IELEM=1,NELEM
      DO 10 IEVAB=1,NEVAB
   10 ELOAD(IELEM,IEVAB)=TLOAD(IELEM,IEVAB)-ELOAD(IELEM,IEVAB)
      RESID=SQRT(RESID)
      RETOT=SQRT(RETOT)
      RATIO=100.0*RESID/RETOT
      IF(RATIO.GT.TOLER) NCHEK=1
      IF(IITER.EQ.1) GO TO 20
      IF(RATIO.GT.PVALU) NCHEK=999
   20 PVALU=RATIO
      WRITE(6,30) NCHEK,RATIO,REMAX
   30 FORMAT(1H0,3X,18HCONVERGENCE CODE =,I4,3X,28HNORM OF RESIDUAL SUM
     .RATIO =,E14.6,3X,18HMAXIMUM RESIDUAL =,E14.6)
      RETURN
      END
```

4.6 Subroutine DIMEN

The function of the subroutine is to preset the values of variables employed in the program.

```
      SUBROUTINE DIMEN(MBUFA,MELEM,MEVAB,MFRON,MMATS,MPOIN,MSTIF,MTOTG,
     .                 MTOTV,MVFIX,NDOFN,NPROP,NSTRE,M3POI,MLAYR)
C***********************************************************************
C
C**** THIS SUBROUTINE PRESETS VARIABLES ASSOCIATED WITH DYNAMIC
C     DIMENSIONING
C
C***********************************************************************
      MBUFA = 10
      MELEM = 20
      MFRON = 75
      MLAYR = 10
      MMATS = 5
      MPOIN = 100
      M3POI = 3*MPOIN
      NDOFN = 5
      NSTRE = 5
      MEVAB = NDOFN*9
      MSTIF = (MFRON+1)*MFRON/2
      MTOTG = MELEM*MLAYR*9
      MTOTV = MPOIN*NDOFN
      MVFIX = 36
      NPROP = 17
      RETURN
      END
```

4.7 Subroutine ECHO

The function of this subroutine is to list all the remaining data cards after at least one error has been detected by either Subroutine CHECK1 or CHECK2.

```
      SUBROUTINE ECHO
C***********************************************************************
C
C**** IF DATA ERRORS HAVE BEEN DETECTED BY SUBROUTINES CHECK1 OR
C     CHECK2,THIS SUBROUTINE READS AND WRITES THE REMAINING DATA CARDS
C
C***********************************************************************
       DIMENSION NTITL(80)
      WRITE(6,900)
  900 FORMAT(//50H NOW FOLLOWS A LISTING OF POST-DISASTER DATA CARDS/)
   10 READ(5,905) NTITL
  905 FORMAT(80A1)
      WRITE(6,910) NTITL
  910 FORMAT(20X,80A1)
      GO TO 10
      END
```

4.8 Subroutine FLOWS

The function of this subroutine is to evaluate the flow vector $\underline{a}$ in (65) as well as the denominator of the last term

```
      SUBROUTINE FLOWS(ABETA,AVECT,DVECT,LPROP,
     .                 MMATS,NPROP,PROPS,SG,A,DMATT)
C***********************************************************************
C
C*** THIS SUBROUTINE CALCULATES THE FLOW VECTOR -AVECT- AND
C    COMPUTES  -DVECT- AND -ABETA-
C
C***********************************************************************
      DIMENSION AVECT(5),DMATT(5,5,MMATS),DVECT(5),
     .          PROPS(MMATS,NPROP),SG(5),A(9,MMATS)
C
C*** SET UP MATERIAL PROPERTIES
C
      HARDS=PROPS(LPROP,7)
C
C*** COMPUTES THE VECTOR  AVECT
C
      L=LPROP
      AFUNC=(A(1,L)*SG(1)*SG(1)+2.0*A(2,L)*SG(1)*SG(2)+2.0*A(3,L)*
     .      SG(1)*SG(3)+A(4,L)*SG(2)*SG(2)+2.0*A(5,L)*SG(2)*SG(3)+
     .      A(6,L)*SG(3)*SG(3)+A(7,L)*SG(4)*SG(4)+2.0*A(8,L)*SG(4)*
     .      SG(5)+A(9,L)*SG(5)*SG(5))**0.5
      AVECT(1)=(A(1,L)*SG(1)+A(2,L)*SG(2)+A(3,L)*SG(3))/AFUNC
      AVECT(2)=(A(2,L)*SG(1)+A(4,L)*SG(2)+A(5,L)*SG(3))/AFUNC
      AVECT(3)=(A(3,L)*SG(1)+A(5,L)*SG(2)+A(6,L)*SG(3))/AFUNC
      AVECT(4)=(A(7,L)*SG(4)+A(8,L)*SG(5))/AFUNC
      AVECT(5)=(A(8,L)*SG(4)+A(9,L)*SG(5))/AFUNC
C     WRITE(6,910) AVECT
C 910 FORMAT(8H AVECT =,5E15.6)
C
C*** COMPUTE  DVECT=DMATX*AVECT
C
      DO 10 I=1,5
      DVECT(I)=0.0
      DO 10 J=1,5
   10 DVECT(I)=DVECT(I)+DMATT(I,J,LPROP)*AVECT(J)
C     WRITE(6,920) DVECT
C 920 FORMAT(8H DVECT =,5E15.6)
C
C*** COMPUTE  ABETA=1.0/(H+ATDA)
C
      DENOM=HARDS
      DO 20 ISTRE=1,5
   20 DENOM=DENOM+AVECT(ISTRE)*DVECT(ISTRE)
      ABETA=1.0/DENOM
C     WRITE(6,930) ABETA
C 930 FORMAT(8H ABETA =,E15.6)
      RETURN
      END
```

4.9 Subroutine FRAME

This subroutine forms part of a matrix handling scheme which performs the various tasks indicated in the listing.

```
      SUBROUTINE FRAME(N1,N2,N3,NOPN)
C***********************************************************************
C
C     MULTIPLE VECTOR AND/OR MATRIX MANIPULATIONS
C
C     NOPN = 1, CREATE UNIQUE ORTHOGONAL AXES IN MATRIX N1 INCLUDING VEC
C     NOPN = 2, SCISSORS ON OTHER TWO VECTORS IN N1, THEN N2 MADE  ORTHO
C     NOPN = 3, BEST ORTHOGONAL APPROXIMATION TO GIVEN NON-CARTESIAN FRA
C     NOPN = 4, N2 BECOMES N1T*N2*N1 USING N3 = GASH
C     NOPN = 5, N2 BECOMES N1*N2*N1T USING N3 = GASH
C
C***********************************************************************
```

```
      COMMON WORMX(3,24),QVALU,DJACB
      M3 = N1 + 2
      I2 = N2 - 1
      IF (I2.GE.N1) GO TO 10
      I2 = I2 + 3
   10 I1 = N1 + N1 + N1 + 3 - N2 - I2
      GO TO (1,2,3,4,5  ),NOPN
    1 WORMX(1,I1)=WORMX(3,N2)
      WORMX(2,I1)=0.0
      WORMX(3,I1)=-WORMX(1,N2)
      IF(WORMX(1,I1).EQ.0.0.AND.WORMX(3,I1).EQ.0.0)
     .WORMX(1,I1)=-WORMX(2,N2)
      CALL VECT (N2,I1,I2,4)
      GO TO 14
    2 CALL MATM (I1,I2,0,7)
      CALL VECT (I1,I2,N2,4)
   14 CALL MATM (N1,N1,0,6)
      RETURN
    3 I1 = N1 + 1
      I2 = M3
      DO 11 I = 1,50
      DO 11 N = N1, M3
      CALL MATM (I1,I2,0,7)
      I1 = I2
   11 I2 = N
      RETURN
    4 CALL MATM (N1,N2,N3,2)
      CALL MATM (N3,N1,N2,3)
      RETURN
    5 CALL MATM (N1,N2,N3,3)
      CALL SINGOP (N3,3)
      CALL MATM (N1,N3,N2,3)
      RETURN
      END
```

4.10 Subroutine FRONT

The function of this subroutine is to assemble the contributions from each element to form the global stiffness matrix and global load vector and to solve the resulting set of simultaneous equations by Gaussian direct elimination.

```
      SUBROUTINE FRONT(ASDIS,ELOAD,EQRHS,EQUAT,ESTIF,FIXED,
     .                 GLOAD,GSTIF,IFFIX,IINCS,IITER,KRESL,
     .                 LOCEL,LNODS,MBUFA,MELEM,MEVAB,MFRON,
     .                 MSTIF,MTOTV,MVFIX,NACVA,NAMEV,NDEST,
     .                 NDOFN,NELEM,NEVAB,NNODE,NOFIX,NPIVO,
     .                 NPOIN,NTOTV,TDISP,TLOAD,TREAC,VECRV)
C**********************************************************************
C
C**** THIS SUBROUTINE UNDERTAKES EQUATION SOLUTION BY THE FRONTAL
C     METHOD
C
C**********************************************************************
      DIMENSION ASDIS(MTOTV),ELOAD(MELEM,MEVAB),EQRHS(MBUFA),
     .          EQUAT(MFRON,MBUFA),ESTIF(MEVAB,MEVAB),FIXED(MTOTV),
     .          GLOAD(MFRON),GSTIF(MSTIF),IFFIX(MTOTV),
     .          LNODS(MELEM,9),LOCEL(MEVAB),NACVA(MFRON),
     .          NAMEV(MBUFA),NDEST(MEVAB),NOFIX(MVFIX),
     .          NPIVO(MBUFA),TDISP(MTOTV),TLOAD(MELEM,MEVAB),
     .          TREAC(MVFIX,NDOFN),VECRV(MFRON)
      NFUNC(I,J)=(J*J-J)/2+I
C
C*** CHANGE THE SIGN OF THE LAST APPEARANCE OF EACH NODE
C
      IF(IINCS.GT.1.OR.IITER.GT.1) GO TO 455
      DO 140 IPOIN=1,NPOIN
      KLAST=0
```

```
      DO 130 IELEM=1,NELEM
      DO 120 INODE=1,NNODE
      IF(LNODS(IELEM,INODE).NE.IPOIN) GO TO 120
      KLAST=IELEM
      NLAST=INODE
  120 CONTINUE
  130 CONTINUE
      IF(KLAST.NE.0) LNODS(KLAST,NLAST)=-IPOIN
  140 CONTINUE
  455 CONTINUE
C
C*** START BY INITIALIZING EVERYTHING THAT MATTERS TO ZERO
C
      DO 145 IBUFA=1,MBUFA
  145 EQRHS(IBUFA)=0.0
      DO 150 ISTIF=1,MSTIF
  150 GSTIF(ISTIF)=0.0
      DO 160 IFRON=1,MFRON
      GLOAD(IFRON)=0.0
      VECRV(IFRON)=0.0
      NACVA(IFRON)=0
      DO 160 IBUFA=1,MBUFA
  160 EQUAT(IFRON,IBUFA)=0.0
C
C*** AND PREPARE FOR DISC READING AND WRITING OPERATIONS
C
      NBUFA=0
      IF(KRESL.GT.1) NBUFA=MBUFA
      REWIND 1
      REWIND 2
      REWIND 4
      REWIND 7
C
C*** ENTER MAIN ELEMENT ASSEMBLY-REDUCTION LOOP
C
      NFRON=0
      KELVA=0
      DO 320 IELEM=1,NELEM
      IF(KRESL.GT.1) GO TO 400
      KEVAB=0
      READ(1) ESTIF
      DO 170 INODE=1,NNODE
      DO 170 IDOFN=1,NDOFN
      NPOSI=(INODE-1)*NDOFN+IDOFN
      LOCNO=LNODS(IELEM,INODE)
      IF(LOCNO.GT.0) LOCEL(NPOSI)=(LOCNO-1)*NDOFN+IDOFN
      IF(LOCNO.LT.0) LOCEL(NPOSI)=(LOCNO+1)*NDOFN-IDOFN
  170 CONTINUE
C
C*** START BY LOOKING FOR EXISTING DESTINATIONS
C
      DO 210 IEVAB=1,NEVAB
      NIKNO=IABS(LOCEL(IEVAB))
      KEXIS=0
      DO 180 IFRON=1,NFRON
      IF(NIKNO.NE.NACVA(IFRON)) GO TO 180
      KEVAB=KEVAB+1
      KEXIS=1
      NDEST(KEVAB)=IFRON
  180 CONTINUE
      IF(KEXIS.NE.0) GO TO 210
C
C*** WE NOW SEEK NEW EMPTY PLACES FOR DESTINATION VECTOR
C
      DO 190 IFRON=1,MFRON
      IF(NACVA(IFRON).NE.0) GO TO 190
      NACVA(IFRON)=NIKNO
      KEVAB=KEVAB+1
      NDEST(KEVAB)=IFRON
      GO TO 200
  190 CONTINUE
C
C*** THE NEW PLACES MAY DEMAND AN INCREASE IN CURRENT FRONTWIDTH
C
  200 IF(NDEST(KEVAB).GT.NFRON) NFRON=NDEST(KEVAB)
  210 CONTINUE
      WRITE(7) LOCEL,NDEST,NACVA,NFRON
  400 IF(KRESL.GT.1) READ(7) LOCEL,NDEST,NACVA,NFRON
```

```
C
C*** ASSEMBLE ELEMENT LOADS
C
      DO 220 IEVAB=1,NEVAB
      IDEST=NDEST(IEVAB)
      GLOAD(IDEST)=GLOAD(IDEST)+ELOAD(IELEM,IEVAB)
C
C*** ASSEMBLE THE ELEMENT STIFFNESSES-BUT NOT IN RESOLUTION
C
      IF(KRESL.GT.1) GO TO 402
      DO 222 JEVAB=1,IEVAB
      JDEST=NDEST(JEVAB)
      NGASH=NFUNC(IDEST,JDEST)
      NGISH=NFUNC(JDEST,IDEST)
      IF(JDEST.GE.IDEST) GSTIF(NGASH)=GSTIF(NGASH)+ESTIF(IEVAB,JEVAB)
      IF(JDEST.LT.IDEST) GSTIF(NGISH)=GSTIF(NGISH)+ESTIF(IEVAB,JEVAB)
  222 CONTINUE
  402 CONTINUE
  220 CONTINUE
C
C*** RE-EXAMINE EACH ELEMENT NODE, TO ENQUIRE WHICH CAN BE ELIMINATED
C
      DO 310 IEVAB=1,NEVAB
      NIKNO=-LOCEL(IEVAB)
      IF(NIKNO.LE.0) GO TO 310
C
C*** FIND POSITIONS OF VARIABLES READY FOR ELIMINATION
C
      DO 300 IFRON=1,NFRON
      IF(NACVA(IFRON).NE.NIKNO) GO TO 300
      NBUFA=NBUFA+1
C
C*** WRITE EQUATIONS TO DISC OR TO TAPE
C
      IF(NBUFA.LE.MBUFA) GO TO 406
      NBUFA=1
      IF(KRESL.GT.1) GO TO 408
      WRITE(2) EQUAT,EQRHS,NPIVO,NAMEV
      GO TO 406
  408 WRITE(4) EQRHS
      READ(2) EQUAT,EQRHS,NPIVO,NAMEV
  406 CONTINUE
C
C*** EXTRACT THE COEFFICIENTS OF THE NEW EQUATION FOR ELIMINATION
C
      IF(KRESL.GT.1) GO TO 404
      DO 230 JFRON=1,MFRON
      IF(IFRON.LT.JFRON) NLOCA=NFUNC(IFRON,JFRON)
      IF(IFRON.GE.JFRON) NLOCA=NFUNC(JFRON,IFRON)
      EQUAT(JFRON,NBUFA)=GSTIF(NLOCA)
  230 GSTIF(NLOCA)=0.0
  404 CONTINUE
C
C*** AND EXTRACT THE CORRESPONDING RIGHT HAND SIDES
C
      EQRHS(NBUFA)=GLOAD(IFRON)
      GLOAD(IFRON)=0.0
      KELVA=KELVA+1
      NAMEV(NBUFA)=NIKNO
      NPIVO(NBUFA)=IFRON
C
C*** DEAL WITH PIVOT
C
      PIVOT=EQUAT(IFRON,NBUFA)
      IF(PIVOT.GT.0.0) GO TO 235
      WRITE(6,900) NIKNO,PIVOT
  900 FORMAT(1H0,3X,52HNEGATIVE OR ZERO PIVOT ENCOUNTERED FOR VARIABLE N
     .O. ,I4,10H OF VALUE ,E17.6)
      STOP
  235 CONTINUE
      EQUAT(IFRON,NBUFA)=0.0
C
C*** ENQUIRE WHETHER PRESENT VARIABLE IS FREE OR PRESCRIBED
C
      IF(IFFIX(NIKNO).EQ.0) GO TO 250
C
C*** DEAL WITH A PRESCRIBED DEFLECTION
C
      DO 240 JFRON=1,NFRON
  240 GLOAD(JFRON)=GLOAD(JFRON)-FIXED(NIKNO)*EQUAT(JFRON,NBUFA)
      GO TO 280
```

```
C
C*** ELIMINATE A FREE VARIABLE - DEAL WITH THE RIGHT HAND SIDE FIRST
C
  250 DO 270 JFRON=1,NFRON
      GLOAD(JFRON)=GLOAD(JFRON)-EQUAT(JFRON,NBUFA)*EQRHS(NBUFA)/PIVOT
C
C*** NOW DEAL WITH THE COEFFICIENTS IN CORE
C
      IF(KRESL.GT.1) GO TO 418
      IF(EQUAT(JFRON,NBUFA).EQ.0.0) GO TO 270
      NLOCA=NFUNC(0,JFRON)
      CUREQ=EQUAT(JFRON,NBUFA)
      DO 260 LFRON=1,JFRON
      NGASH=LFRON+NLOCA
  260 GSTIF(NGASH)=GSTIF(NGASH)-CUREQ*EQUAT(LFRON,NBUFA)
     . /PIVOT
  418 CONTINUE
  270 CONTINUE
  280 EQUAT(IFRON,NBUFA)=PIVOT
C
C*** RECORD THE NEW VACANT SPACE, AND REDUCE FRONTWIDTH IF POSSIBLE
C
      NACVA(IFRON)=0
      GO TO 290
C
C*** COMPLETE THE ELEMENT LOOP IN THE FORWARD ELIMINATION
C
  300 CONTINUE
  290 IF(NACVA(NFRON).NE.0) GO TO 310
      NFRON=NFRON-1
      IF(NFRON.GT.0) GO TO 290
  310 CONTINUE
  320 CONTINUE
      IF(KRESL.EQ.1) WRITE(2) EQUAT,EQRHS,NPIVO,NAMEV
      BACKSPACE 2
C
C*** ENTER BACK-SUBSTITUTION PHASE. LOOP BACKWARDS THROUGH VARIABLES
C
      DO 340 IELVA=1,KELVA
C
C***READ A NEW BLOCK OF EQUATIONS - IF NEEDED
C
      IF(NBUFA.NE.0) GO TO 412
      BACKSPACE 2
      READ(2) EQUAT,EQRHS,NPIVO,NAMEV
      BACKSPACE 2
      NBUFA=MBUFA
      IF(KRESL.EQ.1) GO TO 412
      BACKSPACE 4
      READ(4) EQRHS
      BACKSPACE 4
  412 CONTINUE
C
C*** PREPARE TO BACK-SUBSTITUTE FROM THE CURRENT EQUATION
C
      IFRON=NPIVO(NBUFA)
      NIKNO=NAMEV(NBUFA)
      PIVOT=EQUAT(IFRON,NBUFA)
      IF(IFFIX(NIKNO).NE.0) VECRV(IFRON)=FIXED(NIKNO)
      IF(IFFIX(NIKNO).EQ.0) EQUAT(IFRON,NBUFA)=0.0
C
C*** BACK-SUBSTITUTE IN THE CURRENT EQUATION
C
      DO 330 JFRON=1,MFRON
  330 EQRHS(NBUFA)=EQRHS(NBUFA)-VECRV(JFRON)*EQUAT(JFRON,NBUFA)
C
C*** PUT THE FINAL VALUES WHERE THEY BELONG
C
      IF(IFFIX(NIKNO).EQ.0) VECRV(IFRON)=EQRHS(NBUFA)/PIVOT
      IF(IFFIX(NIKNO).NE.0) FIXED(NIKNO)=-EQRHS(NBUFA)
      NBUFA=NBUFA-1
      ASDIS(NIKNO)=VECRV(IFRON)
  340 CONTINUE
C
C*** ADD DISPLACEMENTS TO PREVIOUS TOTAL VALUES
C
      DO 345 ITOTV=1,NTOTV
  345 TDISP(ITOTV)=TDISP(ITOTV)+ASDIS(ITOTV)
C
C*** STORE REACTIONS FOR PRINTING LATER
C
```

```
      KBOUN=1
      DO 370 IPOIN=1,NPOIN
      NLOCA=(IPOIN-1)*NDOFN
      DO 350 IDOFN=1,NDOFN
      NGUSH=NLOCA+IDOFN
      IF(IFFIX(NGUSH).GT.0) GO TO 360
  350 CONTINUE
      GO TO 370
  360 DO 510 IDOFN=1,NDOFN
      NGASH=NLOCA+IDOFN
  510 TREAC(KBOUN,IDOFN)=TREAC(KBOUN,IDOFN)+FIXED(NGASH)
      KBOUN=KBOUN+1
  370 CONTINUE
C
C*** ADD REACTIONS INTO THE TOTAL LOAD ARRAY
C
      DO 700 IPOIN=1,NPOIN
      DO 710 IELEM=1,NELEM
      DO 710 INODE=1,NNODE
      NLOCA=IABS(LNODS(IELEM,INODE))
  710 IF(IPOIN.EQ.NLOCA) GO TO 720
  720 DO 730 IDOFN=1,NDOFN
      NGASH=(INODE-1)*NDOFN+IDOFN
      MGASH=(IPOIN-1)*NDOFN+IDOFN
  730 TLOAD(IELEM,NGASH)=TLOAD(IELEM,NGASH)+FIXED(MGASH)
  700 CONTINUE
      RETURN
      END
```

4.11 Subroutine FUNC

This subroutine evaluates the components of the B matrix and the Jacobian matrix for quadratic degenerate shell elements.

```
      SUBROUTINE FUNC(BMATX,SHAPE,THICK,NBORP,NNODE,ZETA,MELEM,
     .                COORD,DICOS,LNODS,IELEM,MPOIN,M3POI,GMATX)
C*************************************************************************
C
C*** SETS UP THE B MATRIX AND JACOBIAN,BEING THE MOST CHARACTERISTIC
C    SUBROUTINE OF THIS ELEMENT
C
C*************************************************************************
      COMMON WORMX(3,24),QVALU,DJACB
      DIMENSION BMATX(5,45),SHAPE(3,9),THICK(MPOIN),GMATX(2,45),
     .          COORD(MPOIN,8),DICOS(3,M3POI),LNODS(MELEM,9)
C
C*** THIS CREATES X,Y,Z IN COLUMN 1 AND J-TRANSPOSE IN COLUMN 2 - 4
C
      DO 20 I = 1,3
      DO 20 J = 1,4
   20 WORMX (I,J) = 0.0
C**  THE ELEMENT GEOMETRY IS DEFINED BY THE 8-NODE SERENDIPITY
      DO 24 INODE = 1,8
      IPOIN=IABS(LNODS(IELEM,INODE))
      DO 24 K = 1,3
      GTOP = COORD(IPOIN,K)
      GBOT = COORD(IPOIN,K+4)
      GOSH = ((1.0+ZETA)*GTOP + (1.0-ZETA)*GBOT)/2.0
      DO 22 J = 1,3
   22 WORMX (K,J) = WORMX(K,J) + GOSH*SHAPE(J,INODE)
   24 WORMX (K,4) = WORMX (K,4) + SHAPE(1,INODE)*(GTOP-GBOT)/2.0
C**  THIS CREATES J INVERSE IN COLUMNS 5-7 OF`WORMX`
      CALL MATM (2,5,0,1)
      DJACB = 1.0/QVALU
C**  EXIT FOR A CASE WHEN J, J-INVERSE AND DET.J ONLY ARE REQUIRED
      IF (NBORP.EQ.2) GO TO 50
C**  THIS CREATES DIRECTION ZETA NORMAL TO XI AND ETA
      CALL VECT (2,3,10,4)
C**  THIS CRATES A LOCAL CARTESIAN SET
      CALL FRAME (8,10,0,1)
      IF(NBORP.EQ.4) GO TO 50
```

```
C
C**  CREATES THE TERMS OF STRAIN/DISPLACEMENT MATRIX
C
      DO 40 INODE = 1,NNODE
      DO 40 J = 1,5
C**  5 SHAPE FUNCTIONS CREATED AT A NORMAL
      NSHAP = 5*(INODE-1) + J
      DO 26 I = 11,17
      DO 26 K = 1,3
   26 WORMX(K,I) = 0.0
      IF (J.GE.4) GO TO 30
      DO 29 K = 1,3
C**  SHAPE AND XI AND ETA DERIVATIVES
   29 WORMX (J,K+10) = SHAPE(K,INODE)
      GO TO 36
   30 JPOSI = J-3
C**  JPOSI IS 1,2 FOR LOCAL X,Y DEFLECTIONS OF ENDS OF NORMAL
      DO 35 M = 1,3
C**  X,Y,Z COMPONENTS OF LOCAL X,Y DEFLECTIONS
      IPOIN=IABS(LNODS(IELEM,INODE))
      IPOSI=(IPOIN-1)*3
      GASH=DICOS(M,IPOSI+JPOSI)
      IF (JPOSI.NE.2) GO TO 32
      GASH = -GASH
   32 DO 34 K = 1,4
C** SHAPE AND ITS XI,ETA DERIVATIVES - K=4 GIVES SPECIAL ZETA DERIVATIVE
      IF (K.EQ.4) GO TO 31
      WORMX (M,K+10) = ZETA*SHAPE(K,INODE)*GASH*(THICK(IPOIN)/2.)
C**  U,V,W ARE NOW IN COL 11
      GO TO 34
   31 WORMX(M,K+10) = SHAPE(1,INODE)*GASH*(THICK(IPOIN)/2.)
   34 CONTINUE
   35 CONTINUE
C**  THIS TRANSPOSES XI, ETA AND ZETA DERIVATIVES OF U, V, W
   36 CALL SINGOP (12,3)
C**  MULTIPLIES BY J-INVERSE TO FORM X,Y,Z DERIVATIVES OF U,V,W
      CALL MATM (5,12,15,3)
C**  TRANSFERS X,Y,Z DERIVATIVES OF U,V,W FROM COLUMN 15 TO COL. 18
      CALL MATM(15,18,0,8)
C**  THIS CONVERTS TO LOCAL AXES AT INTEGRATING POINT
      CALL FRAME (8,18,21,4)
C
C*** SETS UP THE STRAIN MATRIX TERMS
C
      IF(NBORP.EQ.0) GO TO 39
      BMATX(1,NSHAP) = WORMX(1,18)
      BMATX(2,NSHAP) = WORMX(2,19)
      BMATX(3,NSHAP) = WORMX(2,18) + WORMX(1,19)
      BMATX(4,NSHAP) = WORMX(1,20) + WORMX(3,18)
      BMATX(5,NSHAP) = WORMX(2,20) + WORMX(3,19)
C**  LOCAL STRAINS GO IN BMATX IN THE ORDER X,Y,XY,XZ,YZ
      GMATX(1,NSHAP)=WORMX(1,20)
      GMATX(2,NSHAP)=WORMX(2,20)
C**  LOCAL DERIVATIVES GO IN -GMATX- IN THE ORDER  DW/DX , DW/DY
      GO TO 40
   39 CONTINUE
C**  TRANSVERSE SHEAR TERMS FOR SELECTIVE INTEGRATION
      GMATX(1,NSHAP)=WORMX(1,20)+WORMX(3,18)
      GMATX(2,NSHAP)=WORMX(2,20)+WORMX(3,19)
   40 CONTINUE
   50 RETURN
      END
```

4.12 Subroutine GAUSSQ

The function of this subroutine is to set up the sampling point positions and weighting factors for numerical integration.

```
      SUBROUTINE GAUSSQ(NGAUS,POSGP,WEIGP)
C***********************************************************************
C
C**** THIS SUBROUTINE SETS UP THE GAUSS-LEGENDRE INTEGRATION CONSTANTS
C
C***********************************************************************
      DIMENSION POSGP(5),WEIGP(5)
      DO 2 IGASH=1,5
      POSGP(IGASH)=0.0
    2 WEIGP(IGASH)=0.0
      IF(NGAUS.GT.2) GO TO 4
      POSGP(1)=-0.577350269189626
      WEIGP(1)=1.0
      GO TO 8
    4 POSGP(1)=-0.774596669241483
      POSGP(2)=0.0
      WEIGP(1)=0.555555555555556
      WEIGP(2)=0.888888888888889
    8 KGAUS=NGAUS/2
      DO 10 IGASH=1,KGAUS
      JGASH=NGAUS+1-IGASH
      POSGP(JGASH)=-POSGP(IGASH)
      WEIGP(JGASH)=WEIGP(IGASH)
   10 CONTINUE
C
C*** EXTRA POSITION FOR TWO GAUSS POINT RULE(SELECTIVE INTEGRATION)
C
      POSGP(4)=-0.577350269189626
      WEIGP(4)=1.0
      POSGP(5)=-POSGP(4)
      WEIGP(5)=WEIGP(4)
      RETURN
      END
```

4.13 Subroutine GEOME

This subroutine evaluates the geometric stiffness matrix, $\underline{K}_\sigma$, defined by (85) for each element in turn.

```
      SUBROUTINE GEOME(ESTIF,GMATX,STRSG,MEVAB,NEVAB,MTOTG,
     .                 KGAUS,DVOLU)
C************************************************************************
C
C*** THIS SUBROUTINE CALCULATES THE GEOMETRIC MATRIX -GEMTX-
C
C************************************************************************
      DIMENSION ESTIF(MEVAB,MEVAB),GMATX(2,45),STRSG(5,MTOTG),
     .          GEMTX(45,45),GDUMM(2,45),STDUM(2,2)
      DO 5 IEVAB=1,NEVAB
      DO 5 JEVAB=1,NEVAB
    5 GEMTX(IEVAB,JEVAB)=0.0
C**   SET UP -STDUM- MATRIX WITH THE ACTUAL IN-PLANE STRESSES
      STDUM(1,1)=STRSG(1,KGAUS)
      STDUM(1,2)=STRSG(3,KGAUS)
      STDUM(2,1)=STRSG(3,KGAUS)
      STDUM(2,2)=STRSG(2,KGAUS)
C**   EVALUATE THE PRODUCT OF STDUM*GMATX
      DO 10 I=1,2
      DO 10 IEVAB=1,NEVAB
      GDUMM(I,IEVAB)=0.0
      DO 10 J=1,2
   10 GDUMM(I,IEVAB)=GDUMM(I,IEVAB)+STDUM(I,J)*GMATX(J,IEVAB)
C**   CALCULATE THE GEOMETRIC MATRIX
      DO 20 IEVAB=1,NEVAB
      DO 20 JEVAB=IEVAB,NEVAB
      DO 20 I=1,2
   20 GEMTX(IEVAB,JEVAB)=GEMTX(IEVAB,JEVAB)+GMATX(I,IEVAB)*
     .                   GDUMM(I,JEVAB)*DVOLU
C**   EVALUATE THE NEW STIFFNESS MATRIX ADDING -GEMTX-
      DO 30 IEVAB=1,NEVAB
      DO 30 JEVAB=IEVAB,NEVAB
   30 ESTIF(IEVAB,JEVAB)=ESTIF(IEVAB,JEVAB)+GEMTX(IEVAB,JEVAB)
      RETURN
      END
```

4.14 Subroutine INCREM

The role of subroutine INCREM is to increment the applied loading according to the load factors specified as input. This subroutine is accessed on the first iteration of each load increment.

```
      SUBROUTINE INCREM(ELOAD,FIXED,IINCS,MELEM,MEVAB,MITER,
     .                  MTOTV,MVFIX,NDOFN,NELEM,NEVAB,NOUTP,
     .                  NOFIX,NTOTV,NVFIX,PRESC,RLOAD,TFACT,
     .                  TLOAD,TOLER,LNODS,IFFIX,NNODE,NCOLA,
     .                  NREST,KINCS)
C***********************************************************************
C
C**** THIS SUBROUTINE INCREMENTS THE APPLIED LOADING
C
C***********************************************************************
      DIMENSION ELOAD(MELEM,MEVAB),FIXED(MTOTV),
     .          NOUTP(2),NOFIX(MVFIX),PRESC(MVFIX,NDOFN),
     .          RLOAD(MELEM,MEVAB),TLOAD(MELEM,MEVAB),
     .          LNODS(MELEM,9),IFFIX(MTOTV)
      IF(NREST.EQ.0) GO TO 20
      IF(IINCS.GT.KINCS) GO TO 20
      DO 10 IINCS=1,KINCS
   10 READ(5,950) FACTO
   20 WRITE(6,900) IINCS
  900 FORMAT(///,5X,22H***  INCREMENT NUMBER ,I5)
      READ(5,950) FACTO,TOLER,MITER,NOUTP(1),NOUTP(2)
  950 FORMAT(2F10.5,3I5)
      TFACT=TFACT+FACTO
      WRITE(6,960)TFACT,TOLER,MITER,NOUTP(1),NOUTP(2)
  960 FORMAT(1H0,5X,13HLOAD FACTOR =,F10.5,5X,
     .24H CONVERGENCE TOLERANCE =,F10.5,5X,24HMAX. NO. OF ITERATIONS =,
     . I5,//32H       INITIAL OUTPUT PARAMETER =,I5,5X,24HFINAL OUTPUT PA
     .RAMETER =,I5)
      DO 80 IELEM=1,NELEM
      DO 80 IEVAB=1,NEVAB
      ELOAD(IELEM,IEVAB)=ELOAD(IELEM,IEVAB)+RLOAD(IELEM,IEVAB)*FACTO
   80 TLOAD(IELEM,IEVAB)=TLOAD(IELEM,IEVAB)+RLOAD(IELEM,IEVAB)*FACTO
C
C*** INTERPRET FIXITY DATA IN VECTOR FORM
C
      DO 100 ITOTV=1,NTOTV
  100 FIXED(ITOTV)=0.0
      DO 110 IVFIX=1,NVFIX
      NLOCA=(NOFIX(IVFIX)-1)*NDOFN
      DO 110 IDOFN=1,NDOFN
      NGASH=NLOCA+IDOFN
      FIXED(NGASH)=PRESC(IVFIX,IDOFN)*FACTO
  110 CONTINUE
C
C*** ADDITIONAL CONSTRAINTS FOR THE HETEROSIS ELEMENT
C
      IF(NNODE.EQ.8.OR.NCOLA.NE.1) GO TO 130
      DO 120 IELEM=1,NELEM
      LNOD9=LNODS(IELEM,9)
      NLOCA=(LNOD9-1)*NDOFN
C     IDOFN=3
      DO 120 IDOFN=1,3
      NGASH=NLOCA+IDOFN
  120 IFFIX(NGASH)=-1
  130 CONTINUE
      RETURN
      END
```

4.15 Subroutine INPUT

The role of this subroutine is to accept the following input data

*Control data

*Mesh topology data

*Bounday conditions

*Material properties

```
      SUBROUTINE INPUT(ANVEL,COORD,GRAVI,IFFIX,LNODS,
     .                 MATNO,MFRON,MELEM,MMATS,MPOIN,
     .                 MTOTV,MVFIX,NDFRO,NDOFN,NELEM,NCOLA,
     .                 NEVAB,NGAUS,NGAUZ,NMATS,NNODE,MLAYR,
     .                 NOFIX,NPOIN,NPROP,NTOTG,NLAYR,NREST,
     .                 NTOTV,NVFIX,POSGP,PRESC,PROPS,WEIGP,
     .                 NALGO,NINCS,LARGE)
C**********************************************************************
C
C**** THIS SUBROUTINE ACCEPTS MOST OF THE INPUT DATA
C
C**********************************************************************
      DIMENSION COORD(MPOIN,8),IFFIX(MTOTV),LNODS(MELEM,9),
     .          MATNO(MELEM,MLAYR),NDFRO(MELEM),GRAVI(3),
     .          NOFIX(MVFIX),POSGP(5),PRESC(MVFIX,NDOFN),
     .          PROPS(MMATS,NPROP),TITLE(12),WEIGP(5)
      READ(5,920)  TITLE
      WRITE(6,920) TITLE
  920 FORMAT(12A6)
C
C*** READ THE FIRST DATA CARD, AND ECHO IT IMMEDIATELY.
C
      READ(5,900) NPOIN,NELEM,NVFIX,NNODE,NMATS,NGAUS,NGAUZ,NCOLA,
     .NALGO,NINCS,NLAYR,LARGE,NREST
  900 FORMAT(16I5,/5X,15I5)
      NEVAB=NDOFN*NNODE
      NTOTV=NPOIN*NDOFN
      NGAU2=NGAUS*NGAUS
      NTOTG=NELEM*NGAU2*NLAYR
      WRITE(6,901)NPOIN,NELEM,NVFIX,NNODE,NMATS,NGAUS,NGAUZ,
     .NEVAB,NCOLA,NALGO,NINCS,NLAYR,LARGE,NREST
  901 FORMAT(///,5X,8H NPOIN =,I5/5X,8H NELEM =,I5/5X,8H NVFIX =,I5/
     .       5X,8H NNODE =,I5/5X,8H NMATS =,I5/5X,8H NGAUS =,I5/
     .       5X,8H NGAUZ =,I5/5X,8H NEVAB =,I5/5X,8H NCOLA =,I5/5X,
     .       8H NALGO =,I5/5X,8H NINCS =,I5,/,
     .       5X,8H NLAYR =,I5/5X,8H LARGE =,I5/5X,8H NREST =,I5)
      WRITE (6,912)
      READ (5,913) GRAVI(1),GRAVI(2),GRAVI(3),ANVEL
      WRITE (6,913) GRAVI(1),GRAVI(2),GRAVI(3),ANVEL
  912 FORMAT(//41H X-GRAVITY  Y-GRAVITY  Z-GRAVITY  ANG VEL/)
  913 FORMAT(4F10.5)
      CALL       CHECK1(NDOFN,NELEM,NGAUS,NMATS,NNODE,NPOIN,
     .                  MMATS,NVFIX,NGAUZ,NLAYR)
C
C*** READ THE ELEMENT NODAL CONNECTIONS, AND THE PROPERTY NUMBERS.
C
      WRITE(6,902)
  902 FORMAT(//8H ELEMENT,5X,15H PROPERTY/LAYER,35X,12HNODE NUMBERS)
      DO 2 IELEM=1,NELEM
      READ(5,900) NUMEL,(MATNO(NUMEL,ILAYR),ILAYR=1,NLAYR),
     .            (LNODS(NUMEL,INODE),INODE=1,NNODE)
      WRITE(6,903) NUMEL,(MATNO(NUMEL,ILAYR),ILAYR=1,NLAYR)
    2 WRITE(6,940) (LNODS(NUMEL,INODE),INODE=1,NNODE)
  903 FORMAT(1X,I5,4X,10I5)
  940 FORMAT(60X,12I5)
C
C*** ZERO ALL THE NODAL COORDINATES, PRIOR TO READING SOME OF THEM.
```

```
C
      DO 4 IPOIN=1,NPOIN
      DO 4 IDIME=1,8
    4 COORD(IPOIN,IDIME)=0.0
C
C*** READ SOME NODAL COORDINATES, FINISHING WITH THE LAST NODE OF ALL.
C
      WRITE(6,904)
  904 FORMAT(//5H NODE,8X,1HX,14X,1HY,14X,1HZ,13X,5HPRESS)
    6 READ(5,905) IPOIN,(COORD(IPOIN,IDIME),IDIME=1,8)
  905 FORMAT(I5,4F15.10/5X,4F15.10)
      IF(IPOIN.NE.NPOIN) GO TO 6
C
C*** INTERPOLATE COORDINATES OF MID-SIDE NODES
C
      CALL       NODEX(COORD,LNODS,MELEM,MPOIN,NELEM,NNODE)
      DO 10 IPOIN=1,NPOIN
   10 WRITE(6,906) IPOIN,(COORD(IPOIN,IDIME),IDIME=1,8)
  906 FORMAT(I5,4F15.10/5X,4F15.10)
C
C*** READ THE FIXED VALUES.
C
      WRITE(6,907)
  907 FORMAT(//5H NODE,6X,4HCODE,15X,12HFIXED VALUES)
      DO 8 IVFIX=1,NVFIX
      READ(5,908) NOFIX(IVFIX),IFPRE,(PRESC(IVFIX,IDOFN),IDOFN=1,NDOFN)
      WRITE(6,908) NOFIX(IVFIX),IFPRE,(PRESC(IVFIX,IDOFN),IDOFN=1,NDOFN)
      NLOCA=(NOFIX(IVFIX)-1)*NDOFN
      IFDOF=10**(NDOFN-1)
      DO 8 IDOFN=1,NDOFN
      NGASH=NLOCA+IDOFN
      IF(IFPRE.LT.IFDOF) GO TO 8
      IFFIX(NGASH)=1
      IFPRE=IFPRE-IFDOF
    8 IFDOF=IFDOF/10
  908 FORMAT(1X,I4,5X,I5,5X,5F10.6)
C
C*** READ THE AVAILABLE SELECTION OF ELEMENT PROPERTIES.
C
      WRITE(6,910)
  910 FORMAT(//7H  NUMAT,10X,18HELEMENT PROPERTIES)
      DO 18 IMATS=1,NMATS
      READ(5,900) NUMAT
      READ(5,930) (PROPS(NUMAT,IPROP),IPROP=1,NPROP)
  930 FORMAT(7F10.5/7F10.5/7F10.5)
   18 WRITE(6,911) NUMAT,(PROPS(NUMAT,IPROP),IPROP=1,NPROP)
  911 FORMAT(1X,I4,3X,7E15.5/,8X,7E15.5/,8X,7E15.5)
C
C*** SET UP GAUSSIAN INTEGRATION CONSTANTS
C
      CALL       GAUSSQ(NGAUS,POSGP,WEIGP)
      CALL       CHECK2(COORD,IFFIX,LNODS,MATNO,MELEM,MFRON,MPOIN,MTOTV,
     .                  MVFIX,NDFRO,NDOFN,NELEM,NMATS,NNODE,NOFIX,NPOIN,
     .                  NVFIX,NLAYR)
      RETURN
      END
```

4.16 Subroutine INVAR

This subroutine merely evaluates the current value of the yield function given by (57).

```
      SUBROUTINE INVAR(A,ST,LPROP,MMATS,YIELD)
C***********************************************************************
C
C***THIS SUBROUTINE EVALUATES THE CURRENT VALUE OF THE
C    YIELD FUNCTION
C
C***********************************************************************
```

```
      DIMENSION ST(5),A(9,MMATS)
      L=LPROP
      GASH=A(1,L)*ST(1)*ST(1)+2.0*A(2,L)*ST(1)*ST(2)+2.0*A(3,L)*
     .     ST(1)*ST(3)+A(4,L)*ST(2)*ST(2)+2.0*A(5,L)*ST(2)*ST(3)+
     .     A(6,L)*ST(3)*ST(3)+A(7,L)*ST(4)*ST(4)+2.0*A(8,L)*ST(4)*
     .     ST(5)+A(9,L)*ST(5)*ST(5)
      YIELD=SQRT(GASH)
      RETURN
      END
```

4.17 Subroutine LDISP

This subroutine evaluates matrix $\underline{B}_L$ defined by (81) and adds it to $\underline{B}_o$ as indicated in (72)

```
      SUBROUTINE LDISP(BMATX,GMATX,ETDIS,NEVAB)
C*********************************************************************
C
C*** THIS SUBROUTINE EVALUATES THE INITIAL DISPLACEMENT MATRIX -BLARG-
C    AND ADDS IT UP TO BMATX
C
C*********************************************************************
      DIMENSION BMATX(5,45),GMATX(2,45),ETDIS(45),ADUMM(3,2),
     .          BLARG(3,45)
C**  CALCULATE THE ACTUAL -X- AND -Y- DERIVATIVES OF -W- DISPLACEMENT
      DWDXX=0.0
      DWDYY=0.0
      DO 10 IEVAB=1,NEVAB
      DWDXX=DWDXX+GMATX(1,IEVAB)*ETDIS(IEVAB)
   10 DWDYY=DWDYY+GMATX(2,IEVAB)*ETDIS(IEVAB)
C**  SET UP THE -ADUMM- MATRIX
      ADUMM(1,1)=DWDXX
      ADUMM(1,2)=0.0
      ADUMM(2,1)=0.0
      ADUMM(2,2)=DWDYY
      ADUMM(3,1)=DWDYY
      ADUMM(3,2)=DWDXX
C**  NOW CALCULATE -BLARG- MATRIX
      DO 20 IEVAB=1,NEVAB
      DO 20 I=1,3
      BLARG(I,IEVAB)=0.0
      DO 20 J=1,2
   20 BLARG(I,IEVAB)=BLARG(I,IEVAB)+ADUMM(I,J)*GMATX(J,IEVAB)
C**  THE NEW -BMATX- IS EQUAL TO -BMATX+BLARG-
      DO 30 IEVAB=1,NEVAB
      DO 30 I=1,3
   30 BMATX(I,IEVAB)=BMATX(I,IEVAB)+BLARG(I,IEVAB)
      RETURN
      END
```

4.18 Subroutine LOADS

The function of this subroutine is to evaluate the equivalent nodal force due to pressure, centrifugal, gravity or point loads.

```
      SUBROUTINE LOADS(ANVEL,COORD,ELOAD,GRAVI,LNODS,
     .                 MATNO,MELEM,MEVAB,MMATS,MPOIN,DICOS,
     .                 NELEM,NEVAB,NGAUS,THICK,
     .                 NNODE,NPROP,NSTRE,POSGP,M3POI,
     .                 PROPS,WEIGP,MLAYR,NLAYR)
```

```
C***********************************************************************
C
C*** THIS SUBROUTINE EVALUATES THE NODAL FORCES DUE TO EXTERNAL
C    APPLIED LOADS(CENTRIFUGAL,GRAVITY,PRESSURE AND POINT LOADS)
C
C***********************************************************************
      COMMON WORMX(3,24),QVALU,DJACB
      DIMENSION BMATX(5,45),COORD(MPOIN,8),
     .          ELOAD(MELEM,MEVAB),GRAVI(3),LNODS(MELEM,9),
     .          MATNO(MELEM,MLAYR),POSGP(5),DICOS(3,M3POI),
     .          PROPS(MMATS,NPROP),SHAPE(3,9),STREN(5),
     .          THICK(MPOIN),WEIGP(5),
     .          GMATX(2,45)
      REWIND 8
C
C*** LOOP OVER EACH ELEMENT
C
      DO 150 IELEM=1,NELEM
C
C*** READ THE CHARACTERISTICS OF THE  APPLIED LOADS
C
      READ(5,900) NPRES,NUCLO,NBODY
      WRITE(6,901) NPRES,NUCLO,NBODY
  900 FORMAT(5I5)
  901 FORMAT(8H NPRES =,I5,5X,8H NUCLO =,I5,5X,8H NBODY =,I5)
      IF(NPRES.EQ.0) GO TO 3
      READ(5,902) KPRES,CFACE,PREVA,SURFA
      WRITE(6,902) KPRES,CFACE,PREVA,SURFA
  902 FORMAT(I5,F5.1,2F15.5)
C**  CFACE IS +1.0 OR -1.0 ACCORDING AS PRESSURE IS ON TOP OR BOTTOM SUR
    3 CONTINUE
C
C*** INITIALIZE THE LOAD MATRIX ELOAD  ONE COLUMN AT EACH TIME
C
      DO 4 IEVAB=1,NEVAB
    4 ELOAD(IELEM,IEVAB)=0.0
C
C*** ENTER LOOPS OVER GAUSS POINTS FOR NUMERICAL INTEGRATION
C
      DO 145 IGAUS=1,NGAUS
      DO 145 JGAUS=1,NGAUS
      EXISP=POSGP(IGAUS)
      ETASP=POSGP(JGAUS)
      CALL       SFR1(SHAPE,EXISP,ETASP)
      IF(NBODY.EQ.0) GO TO 141
      ZETSP=-1.0
      DO 140 ILAYR=1,NLAYR
      LPROP=MATNO(IELEM,ILAYR)
      DZETA=PROPS(LPROP,3)
      ZETSP=ZETSP+DZETA/2.0
      READ(8) BMATX,GMATX,DVOLU
C
C*** CALCULATE THE CENTRIFUGAL,GRAVITATIONAL PRESSURE AND POINT LOADS
C
C
C*** CENTRIFUGAL FORCE
C
      IF(ANVEL.EQ.0.0) GO TO 70
      NBORP = 2
      CALL       FUNC(BMATX,SHAPE,THICK,NBORP,NNODE,ZETSP,
     .                MELEM,COORD,DICOS,LNODS,IELEM,MPOIN,
     .                M3POI,GMATX)
      GASH=PROPS(LPROP,4)*ANVEL*ANVEL*DVOLU
      DO 45 I=1,2
   45 STREN(I)=GASH*WORMX(I,1)
      STREN(3)=0.0
      DO 65 INODE=1,NNODE
C**  FIND THE POSITION OF THE  V-1 AND  V-2 VECTORS
      IPOIN=IABS(LNODS(IELEM,INODE))
      JPOSI=(IPOIN-1)*3
      DO 65 ISTRE=1,NSTRE
      IEVAB=(INODE-1)*5+ISTRE
      IF(ISTRE.GT.3) GO TO 50
      ELOAD(IELEM,IEVAB)=ELOAD(IELEM,IEVAB)+STREN(ISTRE)*
     .                            SHAPE(1,INODE)
      GO TO 65
   50 JPOSI=JPOSI+1
      GASH=SHAPE(1,INODE)*(THICK(IPOIN)/2.0)*ZETSP
      IF(ISTRE.NE.5) GO TO 55
      GASH=-GASH
```

```
   55 DO 60 I=1,2
   60 ELOAD(IELEM,IEVAB)=ELOAD(IELEM,IEVAB)+STREN(I)*
     .                        DICOS(I,JPOSI)*GASH
   65 CONTINUE
   70 CONTINUE
C
C*** GRAVITY
C
      GASH=PROPS(LPROP,4)*DVOLU
      DO 75 I=1,3
   75 STREN(I)=GRAVI(I)*GASH
      DO 95 INODE=1,NNODE
      IPOIN=IABS(LNODS(IELEM,INODE))
      JPOSI=(IPOIN-1)*3
      DO 95 ISTRE=1,NSTRE
      IEVAB=(INODE-1)*5+ISTRE
      IF(ISTRE.GT.3) GO TO 80
      ELOAD(IELEM,IEVAB)=ELOAD(IELEM,IEVAB)+STREN(ISTRE)*
     .                        SHAPE(1,INODE)
      GO TO 95
   80 JPOSI=JPOSI+1
      GASH=SHAPE(1,INODE)*(THICK(IPOIN)/2.0)*ZETSP
      IF(ISTRE.NE.5) GO TO 85
      GASH=-GASH
   85 DO 90 I=1,3
   90 ELOAD(IELEM,IEVAB)=ELOAD(IELEM,IEVAB)+STREN(I)*
     .                        DICOS(I,JPOSI)*GASH
   95 CONTINUE
      ZETSP=ZETSP+DZETA/2.0
  140 CONTINUE
C
C*** CALCULATE THE NODAL LOADS DUE TO PRESSURE
C
  141 IF(NPRES.EQ.0) GO TO 142
      CALL        PRES(BMATX,COORD,ELOAD,LNODS,POSGP,SHAPE,THICK,
     .                 WEIGP,IELEM,IGAUS,JGAUS,MELEM,MPOIN,NNODE,
     .                 NEVAB,KPRES,CFACE,PREVA,SURFA,DICOS,M3POI)
  142 CONTINUE
  145 CONTINUE
C
C*** POINT LOADS
C
      IF(NUCLO.EQ.0) GO TO 150
C
C*** IS THE PRESENT ELEMENT A LOADED ELEMENT. IF IT IS
C    READ AND ACCUMULATE THE LOADS IN ELOAD
C
      DO 120 IGASP=1,NUCLO
      READ(5,950) LPOIN,LDOFN,CARGA
      WRITE(6,960) LPOIN,LDOFN,CARGA
      IEVAB=(LPOIN-1)*5+LDOFN
  120 ELOAD(IELEM,IEVAB)=ELOAD(IELEM,IEVAB)+CARGA
  150 CONTINUE
  950 FORMAT(2I5,F10.5)
  960 FORMAT(//8H LPOIN =,I5,8H LDOFN =,I5,7H LOAD =,F16.8/)
      RETURN
      END
```

4.19 Subroutine MATM

This subroutine forms part of a matrix handling scheme and performs the tasks indicated in the listing.

```
      SUBROUTINE MATM(N1,N2,N3,NOPN)
C***********************************************************************
C
C     MATRIX MANIPULATIONS
C
C     NOPN = 1,TRANSPOSE-INVERT N1 INTO N2,DJACB = 1/QVALU
C     NOPN = 2, TRANSPOSE-MULT., A(K,I)*B(K,J) = C(I,J)
C     (I.E. A IS TRANSPOSED)
```

```
C       NOPN = 3, TRUE MULTIPLY, A(I,K)*B(K,J) = C(I,J)
C       NOPN = 4, MATRIX(TRANSPOSED)*VECTOR
C       NOPN = 5, TRANSPOSE MATRIX N1 INTO N2
C       NOPN = 6, NORMALISE N1 INTO N2, IN COLUMNS
C       NOPN = 7, N1 AND N2 OPEN SCISSORS-FASHION TO BE ORTHOGONAL
C       NOPN = 8, TRANSFER MATRIX N1 INTO N2
C       NOPN = 9, MATRIX N1*VECTOR N2 = VECTOR N3
C
C***********************************************************************
      COMMON WORMX(3,24),QVALU,DJACB
      GO TO (1,2,3,4,5,6,7,8,9),NOPN
    1 K = 2
      DO 10 I = 1,3
      J = 4 - I - K
      M1 = N1 + J
      M2 = N1 + K
      M3 = N2 + I - 1
      M4 = N1 + I - 1
      CALL VECT(M1,M2,M3,4)
      CALL VECT(M4,M3,0,1)
      IF (QVALU.NE.0.0) GO TO 22
      WRITE (6,21)
   21 FORMAT (17H ZERO DETERMINANT)
      STOP
C**  EXECUTION IS TERMINATED WHEN THE DETERMINANT IS ZERO
   22 QVALU = 1.0/QVALU
      CALL VECT(M3,M3,0,3)
   10 K = I - 1
      RETURN
    2 DO 11 I = 1,3
      M1 = N1 + I - 1
      DO 11 J = 1,3
      M2 = N2 + J - 1
      M3 = N3 + J - 1
      CALL VECT(M1,M2,0,1)
   11 WORMX(I,M3) = QVALU
      RETURN
    3 DO 13 I = 1,3
      DO 13 K = 1,3
      M2 = N2 + K - 1
      M3 = N3 + K - 1
      GASH = 0.0
      DO 12 L = 1,3
      M1 = N1 + L - 1
   12 GASH = GASH + WORMX(I,M1)*WORMX(L,M2)
   13 WORMX(I,M3) = GASH
      RETURN
    4 DO 14 I = 1,3
      M1 = N1 + I - 1
      CALL VECT(M1,N2,0,1)
   14 WORMX(I,N3) = QVALU
      RETURN
    5 DO 15 I = 1,3
      N1I = N1 + I- 1
      N2I = N2 + I- 1
      DO 15 J = I,3
      N1J = N1 + J- 1
      N2J = N2 + J- 1
      GASH = WORMX(J ,N1I)
      WORMX(J ,N2I) = WORMX(I ,N1J)
   15 WORMX(I ,N2J) = GASH
      RETURN
    6 DO 16 I = 1,3
      I1 = N1 + I - 1
      I2 = N2 + I - 1
   16 CALL VECT (I1,I2,0,2)
      RETURN
    7 CALL SINGOP (N1,2)
      CALL SINGOP (N2,2)
      CALL VECT (N1,N2,0,1)
      GASH = -QVALU/(1.0 + SQRT(1.0-QVALU*QVALU))
      DO 17 I = 1,3
      GISH = WORMX(I,N1)
      GOSH = WORMX(I,N2)
      WORMX (I,N1) = GISH + GASH*GOSH
   17 WORMX (I,N2) = GOSH + GASH*GISH
      RETURN
```

```
  8 DO 18 J = 1,3
    N1J = N1 + J - 1
    N2J = N2 + J - 1
    DO 18 I = 1,3
 18 WORMX(I,N2J) = WORMX(I,N1J)
    RETURN
  9 DO 20 I = 1,3
    GASH = 0.0
    DO 19 J = 1,3
    N1J = N1 + J - 1
 19 GASH = GASH + WORMX(I,N1J)*WORMX(J,N2)
 20 WORMX(I,N3) = GASH
    RETURN
    END
```

4.20 Subroutine MODAN

This subroutine evaluates the elasticity matrix $\underline{D}$ defined by (31) and also forms the matrix of anisotropic parameters $\underline{A}$ defined in (57).

```
      SUBROUTINE MODAN(AMATX,DMATT,NMATS,NPROP,PROPS,MMATS,
     .                 MATNO,MELEM,MLAYR,NELEM,NLAYR)
C***********************************************************************
C
C*** CALCULATES THE MATRIX OF ELASTICITY -D- AND THE MATRIX OF THE
C    ANISOTROPIC PARAMETERS -AMATX- FOR EACH MATERIAL
C
C***********************************************************************
      DIMENSION AMATX(9,MMATS),DMATT(5,5,MMATS),PROPS(MMATS,NPROP),
     .          APARA(5,5),TRANS(5,5),GASHM(5,5),MATNO(MELEM,MLAYR),
     .          COEFE( 2 )
      DO 15 IMATS=1,NMATS
C
C*** SETS UP THE MATRIX OF THE ANISOTROPIC PARAMETERS
C
      UNIAX=PROPS(IMATS,6)
      DO 5 I=1,9
    5 AMATX(I,IMATS)=0.0
      AMATX(1,IMATS)=1.0
      AMATX(4,IMATS)=(UNIAX/PROPS(IMATS,12))**2.0
      A3Z=(UNIAX/PROPS(IMATS,13))**2.0
      AMATX(6,IMATS)=(UNIAX/PROPS(IMATS,14))**2.0
      AMATX(2,IMATS)=2.0*A3Z-0.5*(1.0+AMATX(4,IMATS)+AMATX(6,IMATS))
      AMATX(7,IMATS)=(UNIAX/PROPS(IMATS,15))**2.0
      AMATX(9,IMATS)=(UNIAX/PROPS(IMATS,16))**2.0
C
C*** SETS UP THE ELASTICITY MATRIX -D-
C
      GASH=1.0-PROPS(IMATS,2)**2.0*PROPS(IMATS,8)/PROPS(IMATS,1)
      DO 10 I=1,5
      DO 10 J=1,5
   10 DMATT(I,J,IMATS)=0.0
      DMATT(1,1,IMATS)=PROPS(IMATS,1)/GASH
      DMATT(2,2,IMATS)=PROPS(IMATS,8)/GASH
      DMATT(1,2,IMATS)=PROPS(IMATS,2)*DMATT(2,2,IMATS)
      DMATT(2,1,IMATS)=DMATT(1,2,IMATS)
      DMATT(3,3,IMATS)=PROPS(IMATS,9)
      DMATT(4,4,IMATS)=PROPS(IMATS,10)
      DMATT(5,5,IMATS)=PROPS(IMATS,11)
   15 CONTINUE
C
C*** CALCULATE THE SHEAR CORRECTION FACTOR
C
      IF(NMATS.NE.1) GO TO 25
   19 DO 20 I=1,2
   20 COEFE(I)=5./6.
      GO TO 27
   25 DO 26 IELEM=1,NELEM
      KOUNT=0
      DO 26 ILAYR=2,NLAYR
      IF(MATNO(IELEM,ILAYR).EQ.MATNO(IELEM,ILAYR-1))GO TO 26
      KOUNT=KOUNT+1
   26 CONTINUE
```

```
      IF(KOUNT.EQ.0)GO TO 19
      CALL        SHEARC(MATNO,MELEM,MLAYR,PROPS,MMATS,NPROP,
     .                   COEFE,NLAYR,DMATT)
   27 DO 28 IMATS=1,NMATS
      DMATT(4,4,IMATS)=DMATT(4,4,IMATS)*COEFE(1)
   28 DMATT(5,5,IMATS)=DMATT(5,5,IMATS)*COEFE(2)
      WRITE(6,900)(COEFE(I),I=1,2)
  900 FORMAT(/` COEFE(1)=`,E15.8,5X,`COEFE(2)=`,E15.8/)
C
      DO 80 IMATS=1,NMATS
C
C*** IF THE REFERENTIAL SYSTEM OF AXES COINCIDES WITH THE
C    PRINCIPAL AXES OF MATERIAL -GOTO 80
C
      THETA=PROPS(IMATS,17)
      IF(ABS(THETA).LT.0.001) GO TO 80
C
C*** SETS UP THE TRANSFORMATION MATRIX -TRANS-
C
      DO 30 I=1,5
      DO 30 J=1,5
   30 TRANS(I,J)=0.0
      C=COS(THETA)
      S=SIN(THETA)
      TRANS(1,1)=C*C
      TRANS(1,2)=S*S
      TRANS(2,1)=TRANS(1,2)
      TRANS(2,2)=TRANS(1,1)
      TRANS(1,3)=C*S
      TRANS(3,1)=-2.0*TRANS(1,3)
      TRANS(2,3)=-TRANS(1,3)
      TRANS(3,2)=-TRANS(3,1)
      TRANS(3,3)=TRANS(1,1)-TRANS(1,2)
      TRANS(4,4)=C
      TRANS(4,5)=S
      TRANS(5,4)=-S
      TRANS(5,5)=C
C
C*** CALCULATE THE PRODUCT OF  D  MATRIX BY  T  MATRIX
C
      DO 35 I=1,5
      DO 35 J=1,5
      GASHM(I,J)=0.0
      DO 35 K=1,5
   35 GASHM(I,J)=GASHM(I,J)+DMATT(I,K,IMATS)*TRANS(K,J)
C
C*** CALCULATED THE TRANSFORMED  D  MATRIX
C
      DO 40 I=1,5
      DO 40 J=I,5
      DMATT(I,J,IMATS)=0.0
      DO 40 K=1,5
   40 DMATT(I,J,IMATS)=DMATT(I,J,IMATS)+TRANS(K,I)*GASHM(K,J)
      DO 45 I=1,5
      DO 45 J=1,5
   45 DMATT(J,I,IMATS)=DMATT(I,J,IMATS)
C
C*** SET UP THE MATRIX OF THE ANISOTROPIC PARAMETERS FOR THIS MATERIAL
C
      DO 50 I=1,5
      DO 50 J=1,5
   50 APARA(I,J)=0.0
      APARA(1,1)=AMATX(1,IMATS)
      APARA(1,2)=AMATX(2,IMATS)
      APARA(2,1)=APARA(1,2)
      APARA(2,2)=AMATX(4,IMATS)
      APARA(3,3)=AMATX(6,IMATS)
      APARA(4,4)=AMATX(7,IMATS)
      APARA(5,5)=AMATX(9,IMATS)
C
C*** SET UP THE NEW TRANSFORMATION MATRIX
C
      TRANS(3,1)=-C*S
      TRANS(2,3)=2.0*TRANS(3,1)
      TRANS(3,2)=-TRANS(3,1)
      TRANS(1,3)=-TRANS(2,3)
C
C*** CALCULATE THE PRODUCT OF  A  MATRIX BY  T  MATRIX
C
```

```
      DO 55 I=1,5
      DO 55 J=1,5
      GASHM(I,J)=0.0
      DO 55 K=1,5
   55 GASHM(I,J)=GASHM(I,J)+APARA(I,K)*TRANS(K,J)
C
C*** CALCULATE THE NEW ANISOTROPIC PARAMETERS
C
      DO 60 I=1,5
      DO 60 J=I,5
      APARA(I,J)=0.0
      DO 60 K=1,5
   60 APARA(I,J)=APARA(I,J)+TRANS(K,I)*GASHM(K,J)
      AMATX(1,IMATS)=APARA(1,1)
      AMATX(2,IMATS)=APARA(1,2)
      AMATX(3,IMATS)=APARA(1,3)
      AMATX(4,IMATS)=APARA(2,2)
      AMATX(5,IMATS)=APARA(2,3)
      AMATX(6,IMATS)=APARA(3,3)
      AMATX(7,IMATS)=APARA(4,4)
      AMATX(8,IMATS)=APARA(4,5)
      AMATX(9,IMATS)=APARA(5,5)
   80 CONTINUE
      RETURN
      END
```

4.21 Subroutine NODEX

This subroutine checks each midside node of quadratic elements. If all coordinates of a midside node are found to be zero, its coordinates are linearly interpolated between the two adjacent corner nodes.

```
      SUBROUTINE NODEX(COORD,LNODS,MELEM,MPOIN,NELEM,NNODE)
C**********************************************************************
C
C**** THIS SUBROUTINE INTERPOLATES THE MIDE SIDE NODES OF STRAIGHT
C     SIDES OF ELEMENTS
C
C**********************************************************************
      DIMENSION COORD(MPOIN,8),LNODS(MELEM,9),ELCOR(8,8)
C*** LOOP OVER EACH ELEMENT
C
      DO 60 IELEM=1,NELEM
C
C*** LOOP OVER EACH ELEMENT EDGE
C
      NNOD1=7
      DO 20 INODE=1,NNOD1,2
C
C*** COMPUTE THE NODE NUMBER OF THE FIRST NODE
C
      NODST=LNODS(IELEM,INODE)
      IGASH=INODE+2
      IF(IGASH.GT.8) IGASH=1
C
C*** COMPUTE THE NODE NUMBER OF THE LAST NODE
C
      NODFN=LNODS(IELEM,IGASH)
      MIDPT=INODE+1
C
C*** COMPUTE THE NODE NUMBER OF THE INTERMEDIATE NODE
C
      NODMD=LNODS(IELEM,MIDPT)
      TOTAL=ABS(COORD(NODMD,1))+ABS(COORD(NODMD,2))+ABS(COORD(NODMD,3))
C
C*** IF THE TOP-COORDINATES OF THE INTERMEDIATE NODE ARE ALL ZERO
C     INTERPOLATE BY A STRAIGHT LINE
C
```

```
      IF(TOTAL.GT.0.0) GO TO 20
      KOUNT=1
   10 COORD(NODMD,KOUNT)=(COORD(NODST,KOUNT)+COORD(NODFN,KOUNT))/2.0
      KOUNT=KOUNT+1
      IF(KOUNT.LE.8) GO TO 10
   20 CONTINUE
      IF(NNODE.EQ.8) GO TO 60
C
C*** SET UP THE CENTRAL POINT COORDINATES
C
      NODCE=LNODS(IELEM,9)
      DO 30 INODE=1,8
      NODEB=LNODS(IELEM,INODE)
      DO 30 IDIME=1,8
   30 ELCOR(IDIME,INODE)=COORD(NODEB,IDIME)
      DO 50 IDIME=1,8
      CENCO=0.0
      DO 35 INODE=1,7,2
   35 CENCO=CENCO+ELCOR(IDIME,INODE)
      CENCO=CENCO*(-0.5)
      DO 40 INODE=2,8,2
   40 CENCO=CENCO+ELCOR(IDIME,INODE)
      CENCO=CENCO*0.5
   50 COORD(NODCE,IDIME)=CENCO
   60 CONTINUE
      RETURN
      END
```

4.22 Subroutine OUTPUT

This subroutine outputs the results at a frequency determined by the output parameters NOUTP(1) and NOUTP(2). This subroutine also calculates the stress resultants given by (45), (46) and (47).

```
      SUBROUTINE OUTPUT(IITER,MTOTG,MTOTV,MVFIX,NCHEK,NELEM,NGAUS,
     .                  NOFIX,NOUTP,NPOIN,NSTRE,
     .                  NVFIX,STRSG,TDISP,TREAC,EPSTN,POSGP,
     .                  EFFST,MATNO,MMATS,PROPS,NPROP,MELEM,THICK,
     .                  MPOIN,LNODS,MLAYR,NLAYR)
C*******************************************************************
C
C**** THIS SUBROUTINE OUTPUTS DISPLACEMENTS,REACTIONS AND STRESSES
C
C*******************************************************************
      DIMENSION NOFIX(MVFIX),NOUTP(2),STRSG(5,MTOTG),STRES(6),
     .          TDISP(MTOTV),TREAC(MVFIX,5),EPSTN(MTOTG),
     .          POSGP(5),SHAPE(3,9),EFFST(MTOTG),
     .          THICK(MPOIN),LNODS(MELEM,9),
     .          FORCE(8),MATNO(MELEM,MLAYR),
     .          PROPS(MMATS,NPROP)
      KOUTP=NOUTP(1)
      IF(IITER.GT.1) KOUTP=NOUTP(2)
      IF(IITER.EQ.1.AND.NCHEK.EQ.0) KOUTP=NOUTP(2)
C
C*** OUTPUT DISPLACEMENTS
C
      IF(KOUTP.LT.1) GO TO 10
      WRITE(6,900)
  900 FORMAT(1H0,5X,13HDISPLACEMENTS)
      WRITE(6,905)
  905 FORMAT(1H0,6X,4HNODE,4X,6HX-DISP,8X,6HY-DISP,8X,6HZ-DISP,
     .       8X,6HAF-ROT,8X,6HBT-ROT)
      DO 20 IPOIN=1,NPOIN
      NGASH=IPOIN*5
      NGISH=NGASH-4
   20 WRITE(6,910) IPOIN,(TDISP(IGASH),IGASH=NGISH,NGASH)
  910 FORMAT(I10,5E14.6)
   10 CONTINUE
```

```
C
C*** OUTPUT REACTIONS
C
      IF(KOUTP.LT.2) GO TO 30
      WRITE(6,920)
  920 FORMAT(1H0,5X,9HREACTIONS)
      WRITE(6,925)
  925 FORMAT(1H0,6X,4HNODE,4X,6HX-REAC,8X,6HY-REAC,8X,6HZ-REAC,
     .        8X,6HAF-MOM,8X,6HBT-MOM)
      DO 40 IVFIX=1,NVFIX
   40 WRITE(6,910) NOFIX(IVFIX),(TREAC(IVFIX,IDOFN),IDOFN=1,5)
   30 CONTINUE
C
C*** OUTPUT STRESSES
C
      IF(KOUTP.LT.3) GO TO 120
      WRITE(6,927)
  927 FORMAT(1H0,5X,8HSTRESSES)
      WRITE(6,926)
  926 FORMAT(1H0,1X,5HKLAYR,5X,6HXX-STR,8X,6HYY-STR,8X,6HXY-STR,
     .         8X,6HXZ-STR,8X,6HYZ-STR,6X,10HEFF.STRESS,3X,
     .         13HEFF.PL.STRAIN)
      KGAUS=0
      DO 110 IELEM=1,NELEM
      KELGS=0
      WRITE(6,940) IELEM
  940 FORMAT(1H0,18H *    ELEMENT NO. =,I5,/)
      DO 105 IGAUS=1,NGAUS
      DO 105 JGAUS=1,NGAUS
      EXISP=POSGP(IGAUS)
      ETASP=POSGP(JGAUS)
C
C*** SET TO ZERO THE STRESS RESULTANT VECTOR
C
      DO 70 JFORC=1,8
   70 FORCE(JFORC)=0.0
      KGASP=0
      KELGS=KELGS+1
      WRITE(6,945) KELGS
  945 FORMAT(11H G.P. NO. =,I5)
      CALL       SFR1(SHAPE,EXISP,ETASP)
C
C*** COMPUTE GAUSS POINT THICKNESS
C
      THIGP=0.0
      DO 65 INODE=1,8
      IPOIN=IABS(LNODS(IELEM,INODE))
   65 THIGP=THIGP+SHAPE(1,INODE)*THICK(IPOIN)
      ZETSP=-1.0
      DO 100 ILAYR=1,NLAYR
      LPROP=MATNO(IELEM,ILAYR)
      DZETA=PROPS(LPROP,3)
      ZETSP=ZETSP+DZETA/2.0
      KGAUS=KGAUS+1
      KGASP=KGASP+1
C
C*** THE FIVE LOCAL STRESSES IN THE ORDER  XX,YY,XY,XZ,YZ
C
      DO 50 ISTRE=1,NSTRE
   50 STRES(ISTRE)=STRSG(ISTRE,KGAUS)
      WRITE(6,950) KGASP,(STRES(ISTRE),ISTRE=1,NSTRE),
     .             EFFST(KGAUS),EPSTN(KGAUS)
C
C*** SET UP THE STRESS RESULTANTS IN THE ORDER NX,NY,NXY,MX,MY,MXY,QX,QY
C
      DO 75 ISTRE=1,3
      FORCE(ISTRE)=FORCE(ISTRE)+STRES(ISTRE)*THIGP/2.0*DZETA
   75 FORCE(ISTRE+3)=FORCE(ISTRE+3)-STRES(ISTRE)*THIGP*THIGP*
     .             ZETSP*DZETA/4.0
      DO 80 ISTRE=4,5
   80 FORCE(ISTRE+3)=FORCE(ISTRE+3)+STRES(ISTRE)*THIGP/2.0*DZETA
  950 FORMAT(I5,2X,8E14.6)
      ZETSP=ZETSP+DZETA/2.0
  100 CONTINUE
      WRITE(6,960) FORCE(1),FORCE(4),FORCE(2),FORCE(5),FORCE(3),
     .             FORCE(6),FORCE(7),FORCE(8)
  960 FORMAT(/,22H STRESS RESULTANTS =  ,6HN-XX =,E12.5,5X,
     .       6HM-XX =,E12.5/22X,6HN-YY =,E12.5,5X,6HM-YY =,
     .       E12.5/22X,6HN-XY =,E12.5,5X,6HM-XY =,E12.5/,
     .       22X,6HQ-XZ =,E12.5,5X,6HQ-YZ =,E12.5)
```

```
 105 CONTINUE
 110 CONTINUE
 120 CONTINUE
     RETURN
     END
```

4.23 Subroutine PRES

This subroutine is called by Subroutine LOADS and evaluates the equivalent nodal forces for pressure loading.

```
      SUBROUTINE PRES(BMATX,COORD,ELOAD,LNODS,POSGP,SHAPE,THICK,
     .                WEIGP,IELEM,IGAUS,JGAUS,MELEM,MPOIN,NNODE,
     .                NEVAB,KPRES,CFACE,PREVA,SURFA,DICOS,M3POI)
C***********************************************************************
C
C*** THIS SUBROUTINE EVALUATES THE NODAL LOADS DUE TO PRESSURE
C
C***********************************************************************
      COMMON WORMX(3,24),QVALU,DJACB
      DIMENSION BMATX(5,45),COORD(MPOIN,8),ELOAD(MELEM,NEVAB),
     .          LNODS(MELEM,9),POSGP(5),PREMX(2,9),SHAPE(3,9),
     .          THICK(MPOIN),WEIGP(5),DICOS(3,M3POI),GMATX(2,45)
      ZETA=CFACE
      NBORP=2
      CALL      FUNC(BMATX,SHAPE,THICK,NBORP,NNODE,ZETA,MELEM,
     .               COORD,DICOS,LNODS,IELEM,MPOIN,M3POI,GMATX)
C
C*** EVALUATE THE PRESSURE AT SAMPLING POINTS   KPRES = 0,1 OR 2 ACCORDI
C    AS PRESSURE IS U.D.,HYDROSTATIC,OR SPECIFIED AS NODAL COORDINATES
C
      IF(KPRES.EQ.0) GO TO 20
      IF(KPRES.EQ.2) GO TO 10
      WORMX(3,1)=WORMX(3,1)-SURFA
      PRESS=PREVA*WORMX(3,1)
      IF(PRESS.GE.0.0) GO TO 25
      PRESS=0.0
      GO TO 25
   10 PREVA=0.0
      DO 15 INODE=1,8
      NGASH=IABS(LNODS(IELEM,INODE))
C
C*** SET UP ARRAY OF NODAL PRESSURE ; ROW 1  TOP , ROW 2  BOTTOM
C
      PREMX(1,INODE)=COORD(NGASH,4)
      PREMX(2,INODE)=COORD(NGASH,8)
      GISH=((1.0+ZETA)*PREMX(1,INODE)+(1.0-ZETA)*PREMX(2,INODE))/2.0
   15 PREVA=PREVA+GISH*SHAPE(1,INODE)
   20 PRESS=PREVA
   25 GMULT=WEIGP(IGAUS)*WEIGP(JGAUS)*CFACE*PRESS
C
C*** CALCULATE CONSISTENT NODAL LOADS
C
      DO 45 INODE=1,NNODE
      IPOIN=IABS(LNODS(IELEM,INODE))
      QVALU=-GMULT*SHAPE(1,INODE)*DJACB
      CALL      VECT(7,21,0,3)
      DO 30 I=1,3
      IPOSI=(INODE-1)*5+I
   30 ELOAD(IELEM,IPOSI)=ELOAD(IELEM,IPOSI)+WORMX(I,21)
      QVALU=ZETA*THICK(IPOIN)/2.0
      CALL      SINGOP(21,1)
      NPOSI=(IPOIN-1)*3
      DO 40 I=1,2
      JPOSI=(INODE-1)*5+(I+3)
      NPOSI=NPOSI+1
      DO 32 K=1,3
   32 WORMX(K,24)=DICOS(K,NPOSI)
```

```
      CALL       VECT(21,24,0,1)
      IF(I.EQ.2) GO TO 35
      QVALU=-QVALU
   35 ELOAD(IELEM,JPOSI)=ELOAD(IELEM,JPOSI)+QVALU
   40 CONTINUE
   45 CONTINUE
      RETURN
      END
```

4.24 Subroutine RESTR

The function of this subroutine is to evaluate the nodal forces which are statically equivalent to the stress field satisfying the yield condition.

```
      SUBROUTINE RESTR(ASDIS,EFFST,ELOAD,LNODS,
     .                 MATNO,MELEM,MMATS,MPOIN,MTOTG,MTOTV,
     .                 NDOFN,NELEM,NEVAB,NGAUS,NNODE,
     .                 NPROP,NSTRE,POSGP,PROPS,STRSG,
     .                 TDISP,WEIGP,EPSTN,KUNLO,AMATX,DMATT,
     .                 THICK,MLAYR,NLAYR,LARGE)
C***********************************************************************
C
C*** THIS SUBROUTINE REDUCES THE STRESSES TO THE YIELD SURFACE AND
C    EVALUATES THE EQUIVALENTE NODAL FORCES
C
C***********************************************************************
      DIMENSION ASDIS(MTOTV),AVECT(5),BMATX(5,45),
     .          DMATT(5,5,MMATS),DVECT(5),EFFST(MTOTG),ELDIS(45),
     .          ELOAD(MELEM,NEVAB),GVECT(5),LNODS(MELEM,9),
     .          MATNO(MELEM,MLAYR),POSGP(5),PROPS(MMATS,NPROP),
     .          DESIG(5),SIGMA(5),SGTOT(5),ETDIS(45),
     .          STRES(5),EPSTN(MTOTG),TDISP(MTOTV),THICK(MPOIN),
     .          STRSG(5,MTOTG),WEIGP(5),
     .          AMATX(9,MMATS),GMATX(2,45)
      REWIND 8
      DO 5 IELEM=1,NELEM
      DO 5 IEVAB=1,NEVAB
    5 ELOAD(IELEM,IEVAB)=0.0
      KUNLO=0
      KGAUS=0
C
C*** LOOP OVER EACH ELEMENT
C
      DO 210 IELEM=1,NELEM
C
C*** IDENTIFY THE DISPLACEMENTS OF THE ELEMENT NODAL POINTS
C
      JPOSI=0
      DO 10 INODE=1,NNODE
      LNODE=IABS(LNODS(IELEM,INODE))
      NPOSN=(LNODE-1)*NDOFN
      DO 10 IDOFN=1,NDOFN
      NPOSN=NPOSN+1
      JPOSI=JPOSI+1
      ELDIS(JPOSI)=ASDIS(NPOSN)
      ETDIS(JPOSI)=TDISP(NPOSN)
   10 CONTINUE
      KELGS=0
C
C*** ENTER LOOPS OVER EACH SAMPLING POINTS
C
      DO 205 IGAUS=1,NGAUS
      DO 205 JGAUS=1,NGAUS
      DO 200 ILAYR=1,NLAYR
      LPROP=MATNO(IELEM,ILAYR)
      UNIAX=PROPS(LPROP,6)
      HARDS=PROPS(LPROP,7)
      KGAUS=KGAUS+1
      KELGS=KELGS+1
      EPSTN(KGAUS)=ABS(EPSTN(KGAUS))
      READ(8) BMATX,GMATX,DVOLU
```

```
C
C*** CALL SUBROUTINE WHICH SETS UP -BMATX- TAKING INTO ACCOUT
C    THE GEOMETRIC NONLINEARITY
C
      IF(LARGE.EQ.1)
     .CALL        LDISP(BMATX,GMATX,ETDIS,NEVAB)
C
C*** NOW PROCEED TO CALCULATE STRESSES FROM STRES=DMATX*BMATX*ELDIS
C    FIRST STORE IN GASH VECTOR GVECT THE PRODUCT BMATX*ELDIS
C
      DO 30 IDOFN=1,NDOFN
      GASH=0.0
      DO 25 IEVAB=1,NEVAB
   25 GASH=GASH+BMATX(IDOFN,IEVAB)*ELDIS(IEVAB)
   30 GVECT(IDOFN)=GASH
C
C*** CALCULATE THE FIVE LOCAL STRESSES IN THE ORDER XX,YY,XY,XZ,YZ
C
      DO 50 ISTRE=1,NSTRE
      GASH=0.0
      DO 45 JSTRE=1,NSTRE
   45 GASH=GASH+DMATT(ISTRE,JSTRE,LPROP)*GVECT(JSTRE)
   50 STRES(ISTRE)=GASH
C
C*** REDUCE STRESSES TO THE YIELD SURFACE FOR YIELDED GAUSS POINTS
C
      PREYS=UNIAX+EPSTN(KGAUS)*HARDS
      DO 150 ISTR1=1,NSTRE
      DESIG(ISTR1)=STRES(ISTR1)
  150 SIGMA(ISTR1)=STRSG(ISTR1,KGAUS)+STRES(ISTR1)
      CALL        INVAR(AMATX,SIGMA,LPROP,MMATS,YIELD)
      ESPRE=EFFST(KGAUS)-PREYS
      IF(ESPRE.GE.0.0) GO TO 55
      ESCUR=YIELD-PREYS
      IF(ESCUR.LE.0.0) GO TO 60
      RFACT=ESCUR/(YIELD-EFFST(KGAUS))
      GO TO 70
   55 ESCUR=YIELD-EFFST(KGAUS)
      IF(ESCUR.LE.0.0) GO TO 60
      RFACT=1.0
   70 MSTEP=ESCUR*8.0/UNIAX+1.0
      ASTEP=MSTEP
      REDUC=1.0-RFACT
      DO 80 ISTR1=1,NSTRE
      SGTOT(ISTR1)=STRSG(ISTR1,KGAUS)+REDUC*STRES(ISTR1)
   80 STRES(ISTR1)=RFACT*STRES(ISTR1)/ASTEP
      DO 90 ISTEP=1,MSTEP
      CALL        INVAR(AMATX,SGTOT,LPROP,MMATS,YIELD)
      CALL        FLOWS(ABETA,AVECT,DVECT,LPROP,
     .                  MMATS,NPROP,PROPS,SGTOT,AMATX,DMATT)
      AGASH=0.0
      DO 100 ISTR1=1,NSTRE
  100 AGASH=AGASH+AVECT(ISTR1)*STRES(ISTR1)
      DLAMD=AGASH*ABETA
      IF(DLAMD.LT.0.0) DLAMD=0.0
      BGASH=0.0
      DO 110 ISTR1=1.NSTRE
      BGASH=BGASH+AVECT(ISTR1)*SGTOT(ISTR1)
  110 SGTOT(ISTR1)=SGTOT(ISTR1)+STRES(ISTR1)-DLAMD*DVECT(ISTR1)
      EPSTN(KGAUS)=EPSTN(KGAUS)+DLAMD*BGASH/YIELD
   90 CONTINUE
      CALL        INVAR(AMATX,SGTOT,LPROP,MMATS,YIELD)
      CURYS=UNIAX+EPSTN(KGAUS)*HARDS
      BRING=1.0
      IF(YIELD.GT.CURYS) BRING=CURYS/YIELD
      DO 130 ISTR1=1,NSTRE
  130 STRSG(ISTR1,KGAUS)=BRING*SGTOT(ISTR1)
      EFFST(KGAUS)=BRING*YIELD
C*** ALTERNATIVE LOCATION OF STRESS REDUCTION LOOP TERMINATION CARD
C  90 CONTINUE
C***
      GO TO 190
   60 DO 180 ISTR1=1,NSTRE
  180 STRSG(ISTR1,KGAUS)=STRSG(ISTR1,KGAUS)+DESIG(ISTR1)
      EFFST(KGAUS)=YIELD
      IF(EPSTN(KGAUS).EQ.0.0.OR.ESCUR.EQ.0.0) GO TO 190
      EPSTN(KGAUS)=-EPSTN(KGAUS)
      KUNLO=KUNLO+1
  190 CONTINUE
```

```
C
C*** CALCULATE THE EQUIVALENT NODAL FORCES AND ASSOCIATE WITH THE
C    ELEMENT NODES
      MGASH=0
      DO 140 INODE=1,NNODE
      DO 140 IDOFN=1,NDOFN
      MGASH=MGASH+1
      DO 140 ISTRE=1,NSTRE
  140 ELOAD(IELEM,MGASH)=ELOAD(IELEM,MGASH)+BMATX(ISTRE,MGASH)*
     .STRSG(ISTRE,KGAUS)*DVOLU
  200 CONTINUE
  205 CONTINUE
  210 CONTINUE
      RETURN
      END
```

4.25 Subroutine SFR1

This subroutine evaluates the element shape functions and their derivatives.

```
      SUBROUTINE SFR1(W,G,H)
C**********************************************************************
C
C     PARABOLIC SHAPE FUNCTIONS AND THEIR FIRST DERIVATIVES FOR
C    8-NODE ELEMENT PLUS THE CENTRAL HIERARCHICAL FUNCTION
C     G AND H DENOTE THE XI AND ETA VALUES AT THE POINT CONSIDERED
C
C**********************************************************************
      DIMENSION W(3,9)
      GG = G*G
      GH = G*H
      HH = H*H
      GGH = GG*H
      GHH = G*HH
      G2 = G*2.
      H2 = H*2.
      GH2 = GH*2.
      W(1,1) = (-1.+GH+GG+HH-GGH-GHH)/4.
      W(1,2) = ( 1.-H-GG+GGH)/2.
      W(1,3) = (-1.-GH+GG+HH-GGH+GHH)/4.
      W(1,4) = ( 1.+G-HH-GHH)/2.
      W(1,5) = (-1.+GH+GG+HH+GGH+GHH)/4.
      W(1,6) = ( 1.+H-GG-GGH)/2.
      W(1,7) = (-1.-GH+GG+HH+GGH-GHH)/4.
      W(1,8) = ( 1.-G-HH+GHH)/2.
      W(1,9) = 1.0-GG-HH+GG*HH
      W(2,1) = ( H+G2-GH2-HH)/4.
      W(2,2) =  -G+GH
      W(2,3) = (-H+G2-GH2+HH)/4.
      W(2,4) = ( 1.-HH)/2.
      W(2,5) = ( H+G2+GH2+HH)/4.
      W(2,6) =  -G-GH
      W(2,7) = (-H+G2+GH2-HH)/4.
      W(2,8) = (-1.+HH)/2.
      W(2,9) = -G2*(1.0-HH)
      W(3,1) = ( G+H2-GG-GH2)/4.
      W(3,2) = (-1.+GG)/2.
      W(3,3) = (-G+H2-GG+GH2)/4.
      W(3,4) =  -H-GH
      W(3,5) = ( G+H2+GG+GH2)/4.
      W(3,6) = ( 1.-GG)/2.
      W(3,7) = (-G+H2+GG-GH2)/4.
      W(3,8) =  -H+GH
      W(3,9) = -H2*(1.0-GG)
      RETURN
      END
```

4.26 Subroutine SINGOP

This subroutine forms part of a matrix handling scheme and undertakes the tasks described in the listing.

```
      SUBROUTINE SINGOP(N1,NOPN)
C***********************************************************************
C
C     VECTOR OR MATRIX MANIPULATIONS INVOLVING SINGLE SPACE
C
C     NOPN = 1, MULT.VECTOR BY QVALU
C     NOPN = 2, NORMALISE VECTOR
C     NOPN = 3, TRANSPOSE MATRIX
C     NOPN = 4, FIND VECTOR SQUARED
C     NOPN = 5, FORM UNIT DIAGONAL MATRIX IN N1
C
C***********************************************************************
      COMMON WORMX(3,24),QVALU,DJACB
      GO TO (1,2,3,4,5),NOPN
    1 CALL VECT(N1,N1,0,3)
      RETURN
    2 CALL VECT(N1,N1,0,2)
      RETURN
    3 CALL MATM(N1,N1,0,5)
      RETURN
    4 CALL VECT(N1,N1,0,1)
      RETURN
    5 N2 = N1 + 2
      II = 0
      DO 12 J = N1,N2
      DO 11 I = 1,3
   11 WORMX(I,J) = 0.0
      II = II + 1
   12 WORMX(II,J) = 1.0
      RETURN
      END
```

4.27 Subroutine STIFF

This subroutine evaluates the stiffness matrix for each element in turn according to (70) for plastified elements or (74) if large deformation effects are to be considered.

```
      SUBROUTINE STIFF(EPSTN,ESTIF,KITER,LNODS,MATNO,
     .                 MELEM,MEVAB,MMATS,MPOIN,MTOTG,NDOFN,
     .                 NELEM,NEVAB,NGAUS,NNODE,NPROP,
     .                 NSTRE,POSGP,PROPS,STRSG,WEIGP,AMATX,
     .                 DMATT,MLAYR,NLAYR,THICK,
     .                 TDISP,MTOTV,LARGE)
C***********************************************************************
C
C     THIS SUBROUTINE EVALUATES THE STIFFNESS MATRIX
C     FOR EACH ELEMENT IN TURN
C
C***********************************************************************
      DIMENSION BMATX(5,45),DBMAT(5,45),DMATX(5,5),
     .          ESTIF(MEVAB,MEVAB),LNODS(MELEM,9),MATNO(MELEM,MLAYR),
     .          POSGP(5),PROPS(MMATS,NPROP),AMATX(9,MMATS),
     .          STRES(5),THICK(MPOIN),WEIGP(5),EPSTN(MTOTG),
     .          STRSG(5,MTOTG),AVECT(5),DVECT(5),DMATT(5,5,MMATS),
     .          TDISP(MTOTV),ETDIS(45),GMATX(2,45)
      REWIND 1
      REWIND 8
      KGAUS=0
C
C*** LOOP OVER EACH ELEMENT
C
```

```
      DO 110 IELEM=1,NELEM
C**   SET UP THE ELEMENT DISPLACEMENT VECTOR -ETDIS-
      JPOSI=0
      DO 10 INODE=1,NNODE
      LNODE=IABS(LNODS(IELEM,INODE))
      NPOSN=(LNODE-1)*NDOFN
      DO 10 IDOFN=1,NDOFN
      NPOSN=NPOSN+1
      JPOSI=JPOSI+1
   10 ETDIS(JPOSI)=TDISP(NPOSN)
C
C*** INITIALIZE THE ELEMENT STIFFNESS MATRIX
C
      DO 20 IEVAB=1,NEVAB
      DO 20 JEVAB=1,NEVAB
   20 ESTIF(IEVAB,JEVAB)=0.0
      KGASP=0
C
C*** ENTER LOOPS OVER GAUSS POINTS FOR NUMERICAL INTEGRATION
C
      DO 105 IGAUS=1,NGAUS
      DO 105 JGAUS=1,NGAUS
      DO 100 ILAYR=1,NLAYR
      LPROP=MATNO(IELEM,ILAYR)
      KGASP=KGASP+1
      KGAUS=KGAUS+1
      READ(8) BMATX,GMATX,DVOLU
C**   CALL SUBROUTINE WHICH SETS UP -BMATX- TAKING INTO ACCOUT
C     THE LARGE DISPLACEMENTS
      IF(LARGE.EQ.1.AND.KITER.GT.2)
     .CALL       LDISP(BMATX,GMATX,ETDIS,NEVAB)
      IF(KITER.EQ.2) GO TO 80
      IF(EPSTN(KGAUS).LE.0.0) GO TO 80
C
C*** CALCULATE THE ELASTO-PLASTIC -D- MATRIX
C
      DO 50 ISTRE=1,NSTRE
   50 STRES(ISTRE)=STRSG(ISTRE,KGAUS)
      CALL       FLOWS(ABETA,AVECT,DVECT,LPROP,
     .                 MMATS,NPROP,PROPS,STRES,AMATX,DMATT)
      DO 70 ISTRE=1,NSTRE
      DO 70 JSTRE=1,NSTRE
   70 DMATX(ISTRE,JSTRE)=DMATT(ISTRE,JSTRE,LPROP)-ABETA*
     .                   DVECT(ISTRE)*DVECT(JSTRE)
C
C*** CALCULATE THE PRODUCT OF  D MATRIX BY  B MATRIX
C
      DO 35 ISTRE=1,NSTRE
      DO 35 IEVAB=1,NEVAB
      DBMAT(ISTRE,IEVAB)=0.0
      DO 35 JSTRE=1,NSTRE
      DBMAT(ISTRE,IEVAB)=DBMAT(ISTRE,IEVAB)+DMATX(ISTRE,JSTRE)*
     .                   BMATX(JSTRE,IEVAB)
   35 CONTINUE
      GO TO 90
   80 CONTINUE
      DO 85 ISTRE=1,NSTRE
      DO 85 IEVAB=1,NEVAB
      DBMAT(ISTRE,IEVAB)=0.0
      DO 85 JSTRE=1,NSTRE
   85 DBMAT(ISTRE,IEVAB)=DBMAT(ISTRE,IEVAB)+
     .     DMATT(ISTRE,JSTRE,LPROP)*BMATX(JSTRE,IEVAB)
C
C*** CALCULATE THE ELEMENT STIFFNESS
C
   90 DO 40 IEVAB=1,NEVAB
      DO 40 JEVAB=IEVAB,NEVAB
      DO 40 ISTRE=1,NSTRE
   40 ESTIF(IEVAB,JEVAB)=ESTIF(IEVAB,JEVAB)+BMATX(ISTRE,IEVAB)*
     .                   DBMAT(ISTRE,JEVAB)*DVOLU
      IF(LARGE.EQ.0.OR.KITER.EQ.2) GO TO 98
C
C*** CALL SUBROUTINE WHICH CALCULATES THE GEOMETRIC MATRIX -GEMTX-
C
      CALL       GEOME(ESTIF,GMATX,STRSG,MEVAB,NEVAB,MTOTG,
     .                 KGAUS,DVOLU)
   98 CONTINUE
  100 CONTINUE
  105 CONTINUE
```

```
C
C*** CONSTRACT THE LOWER TRIANGLE OF THE STIFFNESS MATRIX
C
      DO 60 IEVAB=1,NEVAB
      DO 60 JEVAB=1,NEVAB
      ESTIF(JEVAB,IEVAB)=ESTIF(IEVAB,JEVAB)
   60 CONTINUE
C
C*** STORE THE STIFFNESS MATRIX FOR EACH
C    ELEMENT ON DISC FILE
C
      WRITE(1) ESTIF
  110 CONTINUE
      RETURN
      END
```

4.28 Subroutine VECT

This subroutine forms part of a matrix handling scheme and undertakes the tasks described in the listing.

```
      SUBROUTINE VECT (N1,N2,N3,NOPN)
C*********************************************************************
C
C     VECTOR MANIPULATIONS
C
C     NOPN = 1, QVALU BECOMES SCALAR PRODUCT OF COL. N1 AND N2
C     NOPN = 2, NORMALISE N1 INTO N2
C     NOPN = 3, MULTIPLY N1 BY QVALU, PLACE IN N2
C     NOPN = 4, N3 BECOMES VECTOR PRODUCT OF N1 AND N2
C     NOPN = 5, N3 BECOMES VECTOR N1 + VECTOR N2*QVALU
C
C*********************************************************************
      COMMON WORMX(3,24),QVALU,DJACB
      I1 = N1
      GO TO (1,2,3,4,5),NOPN
    1 I1 = N2
    2 QVALU = 0.0
      DO 10 I = 1,3
   10 QVALU = QVALU + WORMX(I,N1)*WORMX(I,I1)
      GO TO (15,16),NOPN
   16 IF (QVALU.NE.0.0) GO TO 18
      WRITE (6,17)
   17 FORMAT (12H NULL VECTOR)
      STOP
C     EXECUTION IS TERMINATED WHEN A VECTOR IS NULL
   18 QVALU = 1.0/SQRT(QVALU)
    3 DO 12 I = 1,3
   12 WORMX(I,N2) = WORMX(I,N1)*QVALU
      RETURN
    4 K = 3
      DO 13 I = 1,3
      J = 6 - I - K
      WORMX(I,N3) = WORMX(J,N1)*WORMX(K,N2) - WORMX(K,N1)*WORMX(J,N2)
   13 K = I
      RETURN
    5 DO 14 I = 1,3
   14 WORMX(I,N3) = WORMX(I,N1) + QVALU*WORMX(I,N2)
   15 RETURN
      END
```

4.29 Subroutine WORKS

This subroutine sets up the nodal system of coordinate axes, V_{ik}, defined by (1)-(3) and also evaluates the shell thickness at each nodal point.

```
      SUBROUTINE WORKS(COORD,DICOS,LNODS,THICK,MELEM,MPOIN,
     .                 NPOIN,M3POI)
C***********************************************************************
C
C*** THIS SUBROUTINE SETS UP THE THICKNESS AND ORTHOGONAL
C    SYSTEM OF AXES AT EACH NODAL POINT
C
C***********************************************************************
      DIMENSION COORD(MPOIN,8),LNODS(MELEM,9),THICK(MPOIN),
     .          DICOS(3,M3POI)
      COMMON  WORMX(3,24),QVALU,DJACB
C
C*** TOP AND BOTTOM COORDINATES ARE SET UP AT COLUMNS -1- AND -2-
C
      DO 30 IPOIN=1,NPOIN
      DO 10 I=1,3
      WORMX(I,1)=COORD(IPOIN,I)
   10 WORMX(I,2)=COORD(IPOIN,I+4)
      NGASH=3
      NGISH=NGASH+2
      QVALU=-1.0
C*** VECTOR V-3 IN COLUMN NGISH
      CALL       VECT(1,2,NGISH,5)
C*** SETS QVALU EQUAL TO SCALAR PRODUCT OF THE VECTOR (V-3)*(V-3)
      CALL       SINGOP(NGISH,4)
      THICK(IPOIN)=SQRT(QVALU)
C*** CREATES AND NORMALIZES AT EACH NODE THE VECTORS V-1,V-2 AND V-3
      CALL       FRAME(NGASH,NGISH,0,1)
C
C*** SET UP THE DIRECTION COSSINE MATRIX OF THE LOCAL AXES AT EACH POINT
C    IN ORDER  V-1 , V-2 , V-3
C
      NPOSI=(IPOIN-1)*3
      DO 20 I=1,3
      DO 20 J=1,3
      JPOSI=NPOSI+I
   20 DICOS(J,JPOSI)=WORMX(J,I+2)
   30 CONTINUE
      RETURN
      END
```

4.30 Subroutine RESTAR

This subroutine records (on tape 12) the quantities which are necessary to restart the problem. These quantities are updated every load increment after convergence has been achieved.

```
      SUBROUTINE RESTAR(EFFST,ELOAD,EPSTN,MELEM,MEVAB,MTOTG,
     .                  MTOTV,MVFIX,TDISP,TLOAD,TREAC,STRSG,
     .                  TFACT,KINCS)
C***********************************************************************
C
C*** THIS SUBROUTINE RECORDS ONTO TAPE 12 THE DATA NEEDED TO
C    RESTART THE PROBLEM
C
C***********************************************************************
      DIMENSION EFFST(MTOTG),ELOAD(MELEM,MEVAB),EPSTN(MTOTG),
     .          TDISP(MTOTV),TLOAD(MELEM,MEVAB),TREAC(MVFIX,5),
     .          STRSG(5,MTOTG)
      REWIND 12
      WRITE(12) KINCS,TFACT,EFFST,ELOAD,EPSTN
      WRITE(12) TDISP,TLOAD,TREAC,STRSG
      RETURN
      END
```

4.31 Subroutine ZERO

This subroutine initialises the various arrays which are later used to accumulate quantities as the incremental and iterative process continues. Initially these arrays are set to zero and subsequently when the analysis is restarted these arrays contain the last converged values.

```
      SUBROUTINE ZERO(EFFST,ELOAD,EPSTN,MELEM,MEVAB,KINCS,
     .                MTOTG,MTOTV,NDOFN,NELEM,NEVAB,NREST,
     .                NSTRE,NTOTG,NTOTV,NVFIX,MVFIX,STRSG,
     .                TDISP,TFACT,TLOAD,TREAC)
C***********************************************************************
C
C**** THIS SUBROUTINE INITIALISES VARIOUS ARRAYS TO ZERO
C
C***********************************************************************
      DIMENSION ELOAD(MELEM,MEVAB),STRSG(5,MTOTG),TDISP(MTOTV),
     .          TLOAD(MELEM,MEVAB),TREAC(MVFIX,5),EPSTN(MTOTG),
     .          EFFST(MTOTG)
      IF(NREST.EQ.1) GO TO 70
      KINCS=0
      TFACT=0.0
      DO 30 IELEM=1,NELEM
      DO 30 IEVAB=1,NEVAB
      ELOAD(IELEM,IEVAB)=0.0
   30 TLOAD(IELEM,IEVAB)=0.0
      DO 40 ITOTV=1,NTOTV
   40 TDISP(ITOTV)=0.0
      DO 50 IVFIX=1,NVFIX
      DO 50 IDOFN=1,NDOFN
   50 TREAC(IVFIX,IDOFN)=0.0
      DO 60 ITOTG=1,NTOTG
      EPSTN(ITOTG)=0.0
      EFFST(ITOTG)=0.0
      DO 60 ISTR1=1,NSTRE
   60 STRSG(ISTR1,ITOTG)=0.0
      GO TO 80
   70 REWIND 12
      READ(12) KINCS,TFACT,EFFST,ELOAD,EPSTN
      READ(12) TDISP,TLOAD,TREAC,STRSG
   80 CONTINUE
      RETURN
      END
```

4.32 Subroutine SHEARC

This subroutine undertakes the shear stress correction described in Section 2.7.2.

```
      SUBROUTINE SHEARC(MATNO,MELEM,MLAYR,PROPS,MMATS,NPROP,
     .                  COEFE,NLAYR,DMATT)
C***********************************************************************
C
C*** CALCULATES THE SHEAR CORRECTION FACTOR FOR THE CASE OF
C    LAMINATED COMPOSITE STRUCTURES.
C
C***********************************************************************
      DIMENSION RFACT(2),TRLOW(2),UPTER(2),GBARF(2),MATNO(MELEM,MLAYR),
     .          COEFE(2),ZETA1(2),ZETA2(2),DINDX(2),PROPS(MMATS,NPROP),
     .          GINDX(2),DIFF2(2),DIFF3(2),SUMLA(2),DMATT(5,5,MMATS),
     .          DIFF5(2)
C
C
C*** INITIALISE SOME ARRAYS
C
```

```
      DO 10 I= 1,2
      SUMLA(I) =0.0
      RFACT(I) =0.0
      GBARF(I) =0.0
      UPTER(I) =0.0
      TRLOW(I) =0.0
   10 COEFE(I) =0.0
C
C*** CALCULATE THE POSITION OF THE NEUTRAL AXIS
C
      DSUMM = 0.0
      DO 15 ILAYR = 1,NLAYR
      LPROP = MATNO(1,ILAYR)
      DZETA = PROPS(LPROP,3)
      ZHEIG = DSUMM+DZETA/2.
      DO 14 I = 1,2
      DINDX(I) = DMATT(I,I,LPROP)
      UPTER(I) = UPTER(I)+DINDX(I)*ZHEIG*DZETA
   14 TRLOW(I) = TRLOW(I)+DINDX(I)*DZETA
   15 DSUMM = DSUMM+DZETA
      DO 16 I = 1,2
   16 ZETA2(I) = -UPTER(I)/TRLOW(I)
C
C*** CALCULATE THE SHEAR CORRECTION FACTOR
C
      DO 20 ILAYR = 1,NLAYR
      LPROP = MATNO(1,ILAYR)
      DIFF1 = PROPS(LPROP,3)
      INDEX = 10
      DO 20 I = 1,2
      ZETA1(I) = ZETA2(I)
      ZETA2(I) = ZETA1(I)+DIFF1
      DIFF2(I) = ZETA2(I)**2-ZETA1(I)**2
      DIFF3(I) = ZETA2(I)**3-ZETA1(I)**3
      DIFF5(I) = ZETA2(I)**5-ZETA1(I)**5
C
      DINDX(I) = DMATT(I,I,LPROP)
      GINDX(I) = PROPS(LPROP,INDEX)
C
      RFACT(I) = RFACT(I)+DINDX(I)*DIFF3(I)/3.
      GBARF(I) = GBARF(I)+GINDX(I)*DIFF1/2.
      TERM1 = SUMLA(I)*SUMLA(I)*DIFF1
      TERM2 = DINDX(I)*(ZETA1(I)**4)*DIFF1/4.
      TERM3 = DINDX(I)*DIFF5(I)/20.
      TERM4 =-DINDX(I)*ZETA1(I)*ZETA1(I)*DIFF3(I)/6.
      TERM5 = SUMLA(I)*ZETA1(I)*ZETA1(I)*DIFF1
      TERM6 =-SUMLA(I)*DIFF3(I)/3.
      COEFE(I) = COEFE(I)+(TERM1+DINDX(I)*(TERM2+
     .           TERM3+TERM4+TERM5+TERM6))/GINDX(I)
      INDEX = INDEX+1
      SUMLA(I) = SUMLA(I)-DINDX(I)*DIFF2(I)/2.
   20 CONTINUE
C
      DO 30 I= 1,2
   30 COEFE(I) = RFACT(I)*RFACT(I)/(2.*GBARF(I)*COEFE(I))
C
      RETURN
      END
```

4.33 The main, master or controlling segment

This segment controls the calling, in order, of the other subroutines. Its other function is to control the iterative process and also the incrementing of the applied loads.

The following channel numbers are employed by the program: 5(input channel), 6(output channel), 1,2,4,7,8(scratch files), 12(permanent file for problem restart).

```
      PROGRAM PLSHELL(INPUT,OUTPUT,TAPE5=INPUT,TAPE6=OUTPUT,
     .                TAPE1,TAPE2,TAPE4,TAPE7,TAPE8,TAPE12)
C***********************************************************************
C
C*** PROGRAM FOR ELASTO-PLASTIC ANALYSIS OF ANISOTROPIC SHELL STRUCTURES
C    USING QUADRATIC DEGENERATE SHELL ELEMENTS(8-NODE, HETEROSIS AND 9-NODE)
C    AND A LAYERED APPROACH. ACCOUNTING FOR LARGE DISPLACEMENTS AND SELECTIVE
C    INTEGRATION(TRANSVERSE SHEAR TERMS). THE ANISOTROPIC PARAMETERS REMAIN
C    CONSTANT DURING PLASTIC FLOW.RESTART FACILITIES INCLUDED.
C
C***********************************************************************
      DIMENSION ASDIS(500),COORD(100,8),ELOAD(20,45),EQRHS(10),
     .          EQUAT(75,10),EFFST(1800),EPSTN(1800),ESTIF(45,45),
     .          FIXED(500),GLOAD(75),GSTIF(2850),GRAVI(3),
     .          IFFIX(500),LOCEL(45),LNODS(20,9),MATNO(20,10),
     .          NACVA(75),NAMEV(10),NDEST(45),NDFRO(20),
     .          NOFIX(36),NOUTP(2),NPIVO(10),POSGP(5),THICK(100),
     .          PRESC(36,5),PROPS(5,17),RLOAD(20,45),STFOR(500),
     .          STRSG(5,1800),TDISP(500),TLOAD(20,45),TOFOR(500),
     .          TREAC(36,5),VECRV(75),WEIGP(5),DICOS(3,300),
     .          AMATX(9,5),DMATT(5,5,5)
C
C*** PRESET VARIABLES ASSOCIATED WITH DYNAMIC DIMENSIONING
C
      CALL      DIMEN(MBUFA,MELEM,MEVAB,MFRON,MMATS,MPOIN,MSTIF,
     .                MTOTG,MTOTV,MVFIX,NDOFN,NPROP,NSTRE,M3POI,
     .                MLAYR)
C
C*** CALL THE SUBROUTINE WHICH READS MOST OF THE PROBLEM DATA
C
      CALL      INPUT(ANVEL,COORD,GRAVI,IFFIX,LNODS,
     .                MATNO,MFRON,MELEM,MMATS,MPOIN,
     .                MTOTV,MVFIX,NDFRO,NDOFN,NELEM,NCOLA,
     .                NEVAB,NGAUS,NGAUZ,NMATS,NNODE,MLAYR,
     .                NOFIX,NPOIN,NPROP,NTOTG,NLAYR,NREST,
     .                NTOTV,NVFIX,POSGP,PRESC,PROPS,WEIGP,
     .                NALGO,NINCS,LARGE)
C
C*** CALL SUBROUTINE WHICH COMPUTES THE ELASTICITY MATRIX -D- AND
C    THE MATRIX OF THE ANISOTROPIC PARAMETERS
C
      CALL      MODAN(AMATX,DMATT,NMATS,NPROP,PROPS,MMATS,
     .                MATNO,MELEM,MLAYR,NELEM,NLAYR)
C
C*** CREATE THE THICKNESS AND A LOCAL ORTHOGONAL SET AT EACH NODAL POINT
C
      CALL      WORKS(COORD,DICOS,LNODS,THICK,MELEM,MPOIN,
     .                NPOIN,M3POI)
C
C*** CALL SUBROUTINE WHICH COMPUTES BMATX AND GMATX.THESE MATRICES
C    ARE STORED ON TAPE 8 FOR LATER USAGE
C
      CALL      BGMAT(COORD,DICOS,LNODS,MATNO,MELEM,
     .                MLAYR,MMATS,MPOIN,M3POI,NELEM,
     .                NEVAB,NGAUS,NGAUZ,NLAYR,NNODE,NPROP,
     .                POSGP,PROPS,THICK,WEIGP)
C
C*** CALL THE SUBROUTINE WHICH COMPUTES THE APPLIED LOADS
C    AFTER READING SOME LOADING DATA
C
      CALL      LOADS(ANVEL,COORD,RLOAD,GRAVI,LNODS,
     .                MATNO,MELEM,MEVAB,MMATS,MPOIN,DICOS,
     .                NELEM,NEVAB,NGAUS,THICK,
     .                NNODE,NPROP,NSTRE,POSGP,M3POI,
     .                PROPS,WEIGP,MLAYR,NLAYR)
C
C*** INITIALISE CERTAIN ARRAYS
C
      CALL      ZERO(EFFST,ELOAD,EPSTN,MELEM,MEVAB,KINCS,
     .               MTOTG,MTOTV,NDOFN,NELEM,NEVAB,NREST,
     .               NSTRE,NTOTG,NTOTV,NVFIX,MVFIX,STRSG,
     .               TDISP,TFACT,TLOAD,TREAC)
C
C*** LOOP OVER EACH INCREMENT
C
      DO 100 IINCS=1,NINCS
C
C*** READ DATA FOR CORRENT INCREMENT
C
```

```
      CALL        INCREM(ELOAD,FIXED,IINCS,MELEM,MEVAB,MITER,
     .                   MTOTV,MVFIX,NDOFN,NELEM,NEVAB,NOUTP,
     .                   NOFIX,NTOTV,NVFIX,PRESC,RLOAD,TFACT,
     .                   TLOAD,TOLER,LNODS,IFFIX,NNODE,NCOLA,
     .                   NREST,KINCS)
C
C*** LOOP OVER EACH ITERATION
C
      KSTOP=0
      KUNLO=0
      DO 50 IITER=1,MITER
      KITER=IINCS+IITER
      JINCS=IINCS-KINCS
C
C*** CALL SUBROUTINE WHICH SELECTS SULOTION ALGORITHM VARIABLE KRESL
C
      CALL        ALGOR(FIXED,KITER,IITER,KRESL,MTOTV,NALGO,
     .                  NTOTV,KUNLO,KINCS)
C
C*** CHECK WHETHER A NEW EVALUATION OF THE STIFFNESS MATRIX
C    IS REQUIRED
C
      IF(KRESL.EQ.1)
     .CALL        STIFF(EPSTN,ESTIF,KITER,LNODS,MATNO,
     .                  MELEM,MEVAB,MMATS,MPOIN,MTOTG,NDOFN,
     .                  NELEM,NEVAB,NGAUS,NNODE,NPROP,
     .                  NSTRE,POSGP,PROPS,STRSG,WEIGP,AMATX,
     .                  DMATT,MLAYR,NLAYR,THICK,
     .                  TDISP,MTOTV,LARGE)
C
C*** MERGE AND SOLVE THE RESULTING EQUATIONS BY THE FRONTAL SOLVER
C
      CALL        FRONT(ASDIS,ELOAD,EQRHS,EQUAT,ESTIF,FIXED,
     .                  GLOAD,GSTIF,IFFIX,JINCS,IITER,KRESL,
     .                  LOCEL,LNODS,MBUFA,MELEM,MEVAB,MFRON,
     .                  MSTIF,MTOTV,MVFIX,NACVA,NAMEV,NDEST,
     .                  NDOFN,NELEM,NEVAB,NNODE,NOFIX,NPIVO,
     .                  NPOIN,NTOTV,TDISP,TLOAD,TREAC,VECRV)
C
C*** CALCULATE RESIDUAL FORCES
C
      CALL        RESTR(ASDIS,EFFST,ELOAD,LNODS,
     .                  MATNO,MELEM,MMATS,MPOIN,MTOTG,MTOTV,
     .                  NDOFN,NELEM,NEVAB,NGAUS,NNODE,
     .                  NPROP,NSTRE,POSGP,PROPS,STRSG,
     .                  TDISP,WEIGP,EPSTN,KUNLO,AMATX,DMATT,
     .                  THICK,MLAYR,NLAYR,LARGE)
C
C*** CHECK FOR CONVERGENCE
C
      CALL        CONVER(ELOAD,IITER,LNODS,MELEM,MEVAB,MTOTV,
     .                   NCHEK,NDOFN,NELEM,NEVAB,NNODE,NTOTV,
     .                   STFOR,TLOAD,TOFOR,TOLER)
C
C*** OUTPUT RESULTS IF REQUIRED
C
      IF(IITER.EQ.1.AND.NOUTP(1).GT.0)
     .CALL        OUTPUT(IITER,MTOTG,MTOTV,MVFIX,NCHEK,NELEM,
     .                   NGAUS,NOFIX,NOUTP,NPOIN,
     .                   NSTRE,NVFIX,STRSG,TDISP,TREAC,EPSTN,
     .                   POSGP,EFFST,MATNO,MMATS,PROPS,
     .                   NPROP,MELEM,THICK,
     .                   MPOIN,LNODS,MLAYR,NLAYR)
C
C*** IF SOLUTION HAS CONVERGED STOP ITERATING AND OUTPUT RESULTS
C
      IF(IITER.EQ.1.AND.NCHEK.EQ.0) GO TO 100
      IF(NCHEK.EQ.0) GO TO 75
   50 CONTINUE
C
C***
C
      KSTOP=1
   75 CALL        OUTPUT(IITER,MTOTG,MTOTV,MVFIX,NCHEK,NELEM,
     .                   NGAUS,NOFIX,NOUTP,NPOIN,
     .                   NSTRE,NVFIX,STRSG,TDISP,TREAC,EPSTN,
     .                   POSGP,EFFST,MATNO,MMATS,PROPS,
     .                   NPROP,MELEM,THICK,
     .                   MPOIN,LNODS,MLAYR,NLAYR)
      IF(KSTOP.EQ.1) STOP
```

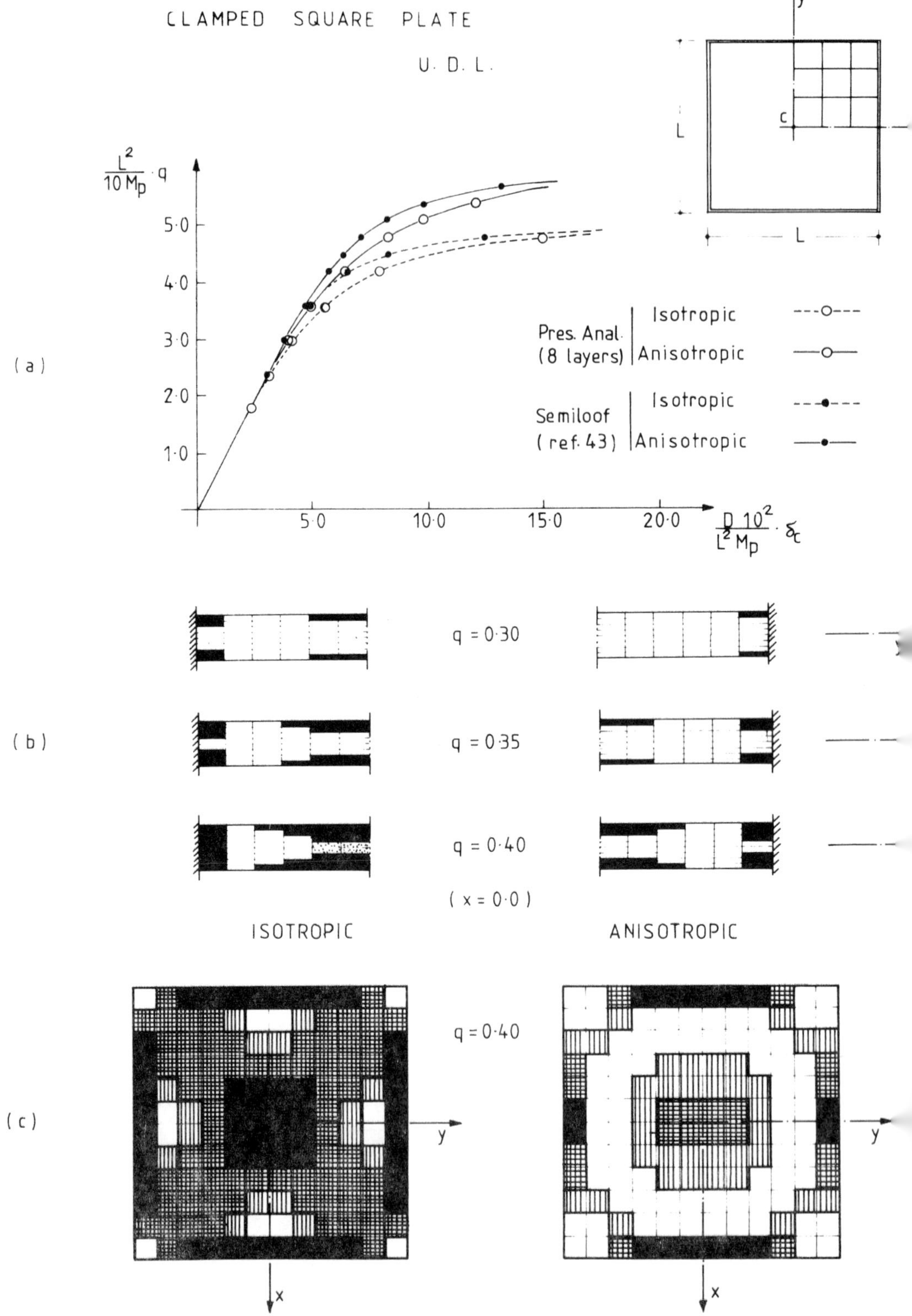

Fig. 10 Elasto-plastic analysis of a uniformly loaded clamped square plate showing (a) load/central deflection characteristics, and (b)-(c) spread of plastic zones

```
C
C*** RECORD ONTO TAPE 12 THE DATA NEEDED FOR RESTART THE PROBLEM
C    IN NEXT INCREMENT
C
      CALL       RESTAR(EFFST,ELOAD,EPSTN,MELEM,MEVAB,MTOTG,
     .                  MTOTV,MVFIX,TDISP,TLOAD,TREAC,STRSG,
     .                  TFACT,IINCS)
  100 CONTINUE
      STOP
      END
```

5. NUMERICAL EXAMPLES

In this section a selection of numerical examples are presented which illustrate the application of the computer code described. The examples presented include nonlinear effects due to either elasto-plastic material behaviour or geometric nonlinearity, but it is worth noting that the program can accommodate linear elastic isotropic shells as a special case.

*Clamped square plate

The first problem considered is the elasto-plastic behaviour of a clamped square plate under uniformly distributed loading. One quadrant of the plate is discretised by 9 heterosis finite elements and 8 equal layers are taken through the thickness. The following material constants are used:

Isotropic: $E_x = E_y = 30000.0; \quad \nu = 0.30;$

$G_{xy} = G_{xz} = G_{yz} = 11540.0; \quad \bar{\sigma}_o = \sigma_{ox} = \sigma_{oy}$

$= \sigma_{045} = 30.0; \quad \tau_{012} = \tau_{013} = \tau_{023} = 17.32;$

$E_p = 300.0; \quad G_p = 100.0$

Anisotropic: $\sigma_{oy} = 40.0; \; \sigma_{045} = 35.0; \; \tau_{012} = 20.2;$

the remaining values are the same as for the isotropic case (units: MN,m)

Geometric characteristics: Thickness, $h = 0.20$; span $L = 6.0$

In Figure 10(a) the vertical displacement of the central point is plotted for different load levels for both isotropic and anisotropic materials. The thick shell layered results are also compared with the non-layered thin shell Semiloof solutions [43]. The spread of plastic zones is shown in Figures 10(b) and (c).

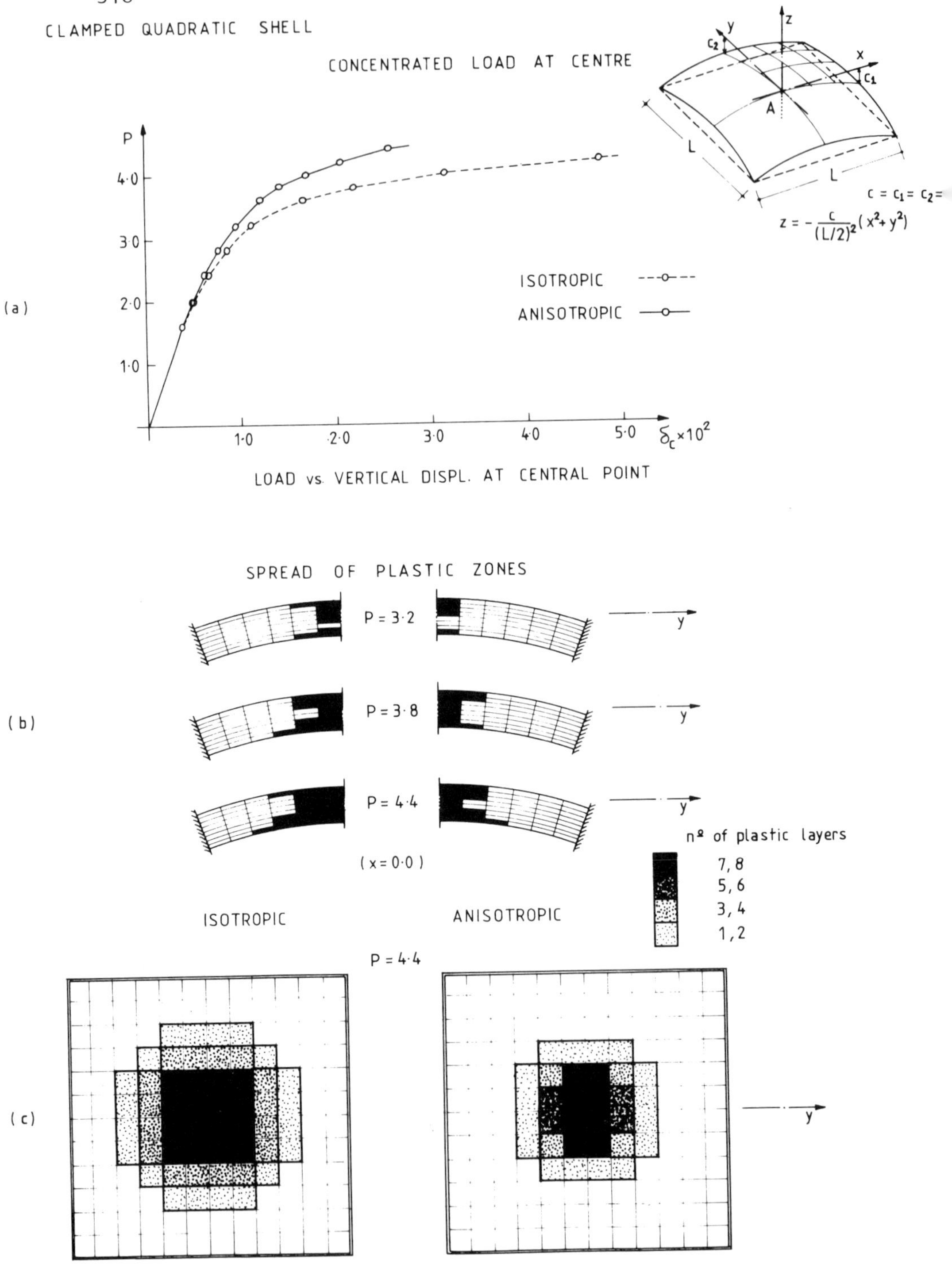

Fig. 11 Elasto-plastic analysis of a centrally point loaded clamped quadratic shell showing (a) load/central deflection characteristics, and (b)-(c) spread of plastic zones.

Clamped quadratic shell

The elasto-plastic analysis of a clamped shell, under a central point load, is undertaken. The material characteristics and finite element discretization are the same as employed in the previous plate example. The shell geometry is defined in Figure 11, in which the thickness h = 0.20 and the span L = 6.0. In figure 11(a) the central displacement is compared for the isotropic and anisotropic cases, together with results obtained using the Semiloof thin shell formulation. The agreement between the two models is good, taking into account that instantaneous plastification of the section is assumed to occur in the Semiloof formulation. In Figure 11(b), the plastic zone distributions are illustrated.

Cylindrical shell

The classical example of the cylindrical shell roof shown in Figure 12, subjected to self-weight loading, is considered. The isotropic case is compared with four anisotropic examples in which one material constant is varied at a time to emphasize its effect. One-quarter of the shell is discretized by 6 elements and, for the layered approach, 6 equal layers are taken through the depth. The following material and geometric characteristics are used in solution:

Isotropic: $E_1 = E_2 = 21{,}000$; $\nu = 0$; $G_{12} = G_{13} = G_{23} = 10{,}500$; $\bar{\sigma}_o = \sigma_1 = \sigma_2 = \sigma_3 = 4.1$; $\tau_{12} = \tau_{13} = \tau_{23} = 2.367$; $E_p = G_p = 0.0$

Anisotropic: one value is changed at a time as specified in Figure 12. L = 7.60; R = 7.60; h = 0.076; α = 40 degrees (units MN,m)

The plots of the vertical displacement of the central point of the free edge A, versus loading, are shown in Figure 12. Results of the thick shell model for both a 'non-layered' and a layered approach are compared with those of the thin shell Semiloof formulation.

Load-deflection results of two anisotropic cases, for which nonlinear geometric effects are included in analysis are illustrated in Figs. 13 and 14. The corresponding curves for isotropic situations are also shown. The influence of the inclusion of geometric nonlinear behaviour in the analysis is immediately apparent.

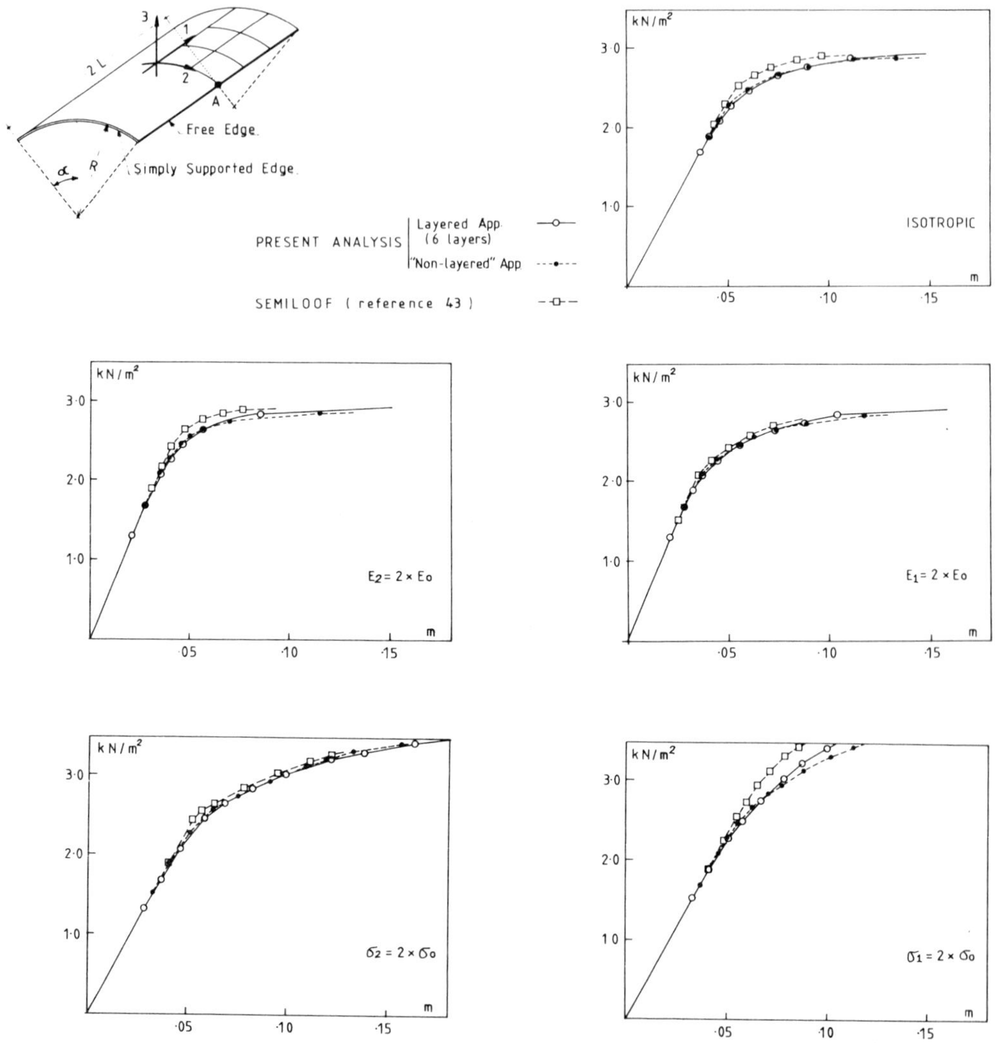

Fig. 12 Elasto-plastic analysis of a cylindrical shell subjected to self-weight loading for various degrees of anisotropy.

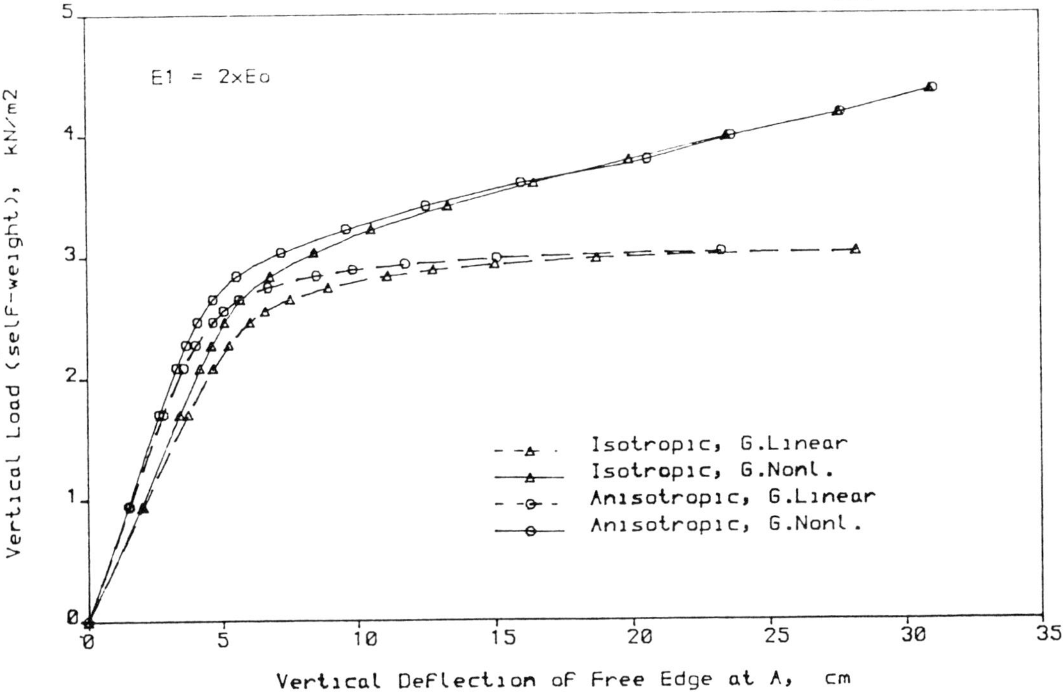

Fig. 13 Comparison of load-free edge deflection curves for the cylindrical shell of Fig. 12 for linear and geometrically nonlinear behaviour.

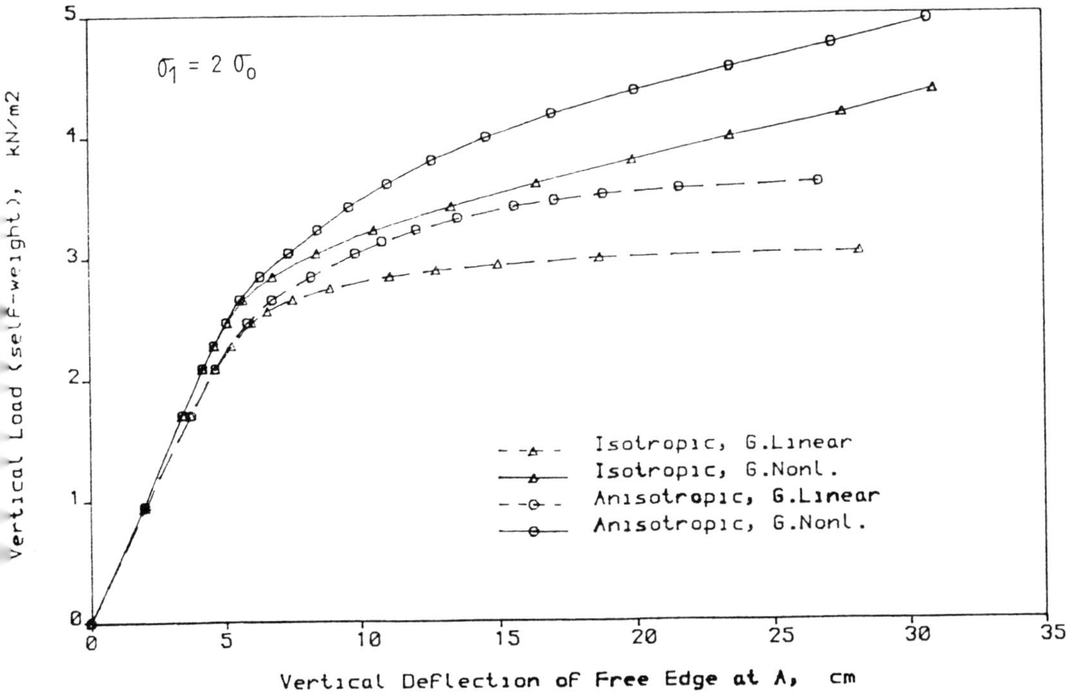

Fig. 14 Comparison of load-free edge deflection curves for the cylindrical shell of Fig. 12 for linear and geometrically nonlinear behaviour.

6. REFERENCES

1. R. H. GALLAGHER
Shell elements, in Proc. of World Congress on Finite Element Methods in Structural Mechanics, (ed. J. Robinson) Bournemouth, Dorset, England, October 1975. pp. E.1-E.35.

2. R. H. GALLAGHER
Finite Element Analysis: Fundamentals, Prentice-Hall, 1975.

3. D. G. ASHWELL and R. H. GALLAGHER (editors)
Finite elements for thin shells and curved members, John Wiley & Sons, London, 1976.

4. O. C. ZIENKIEWICZ
The Finite Element Method (third edition), McGraw-Hill Book Company (UK) Limited, 1977.

5. R. H. GALLAGHER
Problems and progress in thin shell finite element analysis, in Finite Elements for Thin Shells and Curved Members, D. G. Ashwell and R. H. Gallagher (eds.), 1976, pp. 1-14.

6. J. BACKLUND and H. WENNERSTROM
Finite element analysis of elasto-plastic shells, Int. J. of Num. Meth. in Eng., Vol.8, 1974, pp. 415-424.

7. J. H. ARGYRIS, H. BALMER, M. KLEIBER and U. HINDENLANG
Natural description of large inelastic deformations for shells of arbitrary shape - application of Trump element, Comp. Meth. in Appl. Mech. and Eng., Vol.22, 1980, pp. 361-389.

8. K. J. BATHE, E. DVORKIN and L. W. HO
Our discrete-Kirchhoff and isoparametric element for non-linear analysis - an assessment, Computers and Structures, Vol. 16, No. 1-4, 1983, pp. 89-98.

9. M. A. CRISFIELD
Linear and nonlinear finite element analysis of cylindrical shells, TRRL Laboratory Report 987, Transport and Road Research Laboratory, Crowthorne, Berkshire, U.K., 1981.

10. S. IDELSOHN
On the use of deep, shallow or flat shell finite elements for the analysis of thin shell structures, Comp. Meth. in Appl. Mech. and Eng., Vol.26, 1981, pp. 321-330.

11. S. AHMAD, B. M. IRONS and O. C. ZIENKIEWICZ
Analysis of thick and thin shell structures by curved finite elements, in Int. J. Num. Meth. Eng., Vol.2, 1970, pp. 419-451.

12. O. C. ZIENKIEWICZ, R. L. TAYLOR and J. M. TOO
Reduced integration technique in general analysis of plates and shells, in Int. J. Num. Meth. Eng., Vol.3, 1971, pp. 275-290.

13. S. F. PAWSEY and R. W. CLOUGH
Improved numerical integration of thick shell finite elements, in Int. J. Num. Meth. Eng., Vol.3, 1971, pp. 575-586.

14. W. KANOK-NUKULCHAI
A simple and efficient finite element for general shell analysis, in Int. J. Num. Meth. Eng., Vol.14, 1979, pp. 179-200.

15. K. J. BATHE and S. BOLOURCHI
A geometric and material nonlinear plate and shell element, in Computers and Structures, Vol.11, 1980, pp. 23-48.

16. T. J. R. HUGHES nad W. K. LIU
Nonlinear finite element analysis of shells: part I, three-dimensional shells, in Comp. Meth. Appl. Mech. Eng., Vol.26, 1981, pp. 331-362.

17. T. J. R. HUGHES and E. CARNOY
Nonlinear finite element formulation accounting for large membrane strains, in Nonlinear Finite Element Analysis of Plates and Shells (T. J. R. Hughes, ed.) AMD Vol.48, 1981, pp. 193-208.

18. K. S. SURANA
Geometrically nonlinear formulation for the curved shell elements, Int. J. Num. Meth. Eng., Vol.19, 1983, pp. 581-615.

19. I. CORMEAU
Elastoplastic thick shell analysis by viscoplastic solid finite elements, Int. J. Num. Meths. Eng., Vol.12, 1978, pp. 203-227.

20. D. R. J. OWEN and J. A. FIGUEIRAS
Anisotropic elasto-plastic finite element analysis of thick and thin plates and shells, Int. J. Num. Meth. Eng., Vol. 19, 1983, pp. 541-566.

21. E. RAMM and J. M. SATTELE
Elasto-plastic large deformation shell analysis using degenerated elements, in Nonlinear Finite Element Analysis of Plates and Shells (T. J. R. Hughes ed.) AMD-Vol.48, 1981 pp. 265-282.

22. H. PARISCH
Large displacements of shells including material nonlinearities, in Comp. Meths. in Appl. Mech. Eng., Vol.27, 1981, pp. 183-214.

23. D. R. J. OWEN, J. A. FIGUEIRAS and F. DAMJANIC
Finite element analysis of reinforced and prestressed concrete structures including thermal loading. Comp. Meths. in Appl. Mech. Eng. (to be published).

24. E. D. L. PUGH, E. HINTON and O. C. ZIENKIEWICZ
A study of quadrilateral plate bending elements with reduced integration, Int. J. Num. Meth. Eng., Vol.12, 1978, pp. 1059-1079.

25. E. HINTON, E. M. SALONEN and N. BICANIC
A study of locking phenomena in isoparametric elements, in Conf. on the Mathematics of F.E. and Applications II, Brunel University, 1975, pp. 436-447.

26. T. J. R. HUGHES, M. COHEN and M. HAROUN
Reduced and selective integration techniques in the finite element analysis of plates, Nuclear Eng. and Design, Vol. 46, 1978, pp. 203-222.

27. H. PARISCH
A critical survey of the 9-node degenerated shell element with special emphasis on thin shell applications and reduced integration, in Comp. Meths. Appl. Mech. Eng., Vol.20, 1979, pp. 323-350.

28. S. G. LEKHNITSKII
Anisotropic Plates, Gordon and Breach, London, 1968.

29. E. REISSNER and Y. STAVSKY
Bending and stretching of certain types of heterogeneous aelotropic elastic plates, J. Appl. Mech. 28(3), Trans. ASME 83, ser. E.402-408, 1961.

30. J. M. WHITNEY and A. W. LEISSA
Analysis of heterogeneous anisotropic plates, J. Appl. Mech. 36(2), Trans ASME 91, ser. E261-266, 1969.

31. E. REISSNER
The effect of transverse shear deformation on the bending of elastic plates, J. Appl. Mech, 12(2), Trans. ASME 67, 69-77, 1945.

32. R. D. MINDLIN
Influence of rotary inertia and shear on flexural motions of isotropic, elastic plates, J. Appl. Mech. 18(1), Trans. ASME 73, 31-38, 1951.

33. N. J. PAGANO
'Analysis of the flexural test of bidirectional composites' J. Composite Mater. 1, 336, 1967.

34. N. J. PAGANO
'Exact solutions for composite laminates in cylindrical bending', J. Composite Mater. 3, 398-411, 1969.

35. S. SRINIVAS
'A refined analysis of composite laminates', J. Sound Vib. 30(4), 495-507, 1973.

36. P. C. YANG, C. H. NORRIS and Y. STAVSKY
'Elastic wave propagation in heterogeneous plates' Int. J. Solids Struct. 2, 665-684, 1966.

37. J. M. WHITNEY and N. J. PAGANO
'Shear deformation in heterogeneous anisotropic plates', J. Appl. Mech. 37, 1031-1036, 1970.

38. P. C. CHOU and J. CARLEONE
'Transverse shear in laminated plate theories', A.I.A.A.J. 11, 1333-1336, 1973.

39. R. HILL
The Mathematical Theory of Plasticity, Clarendon Press, Oxford, 1950.

40. R. HILL
'A theory of yielding and plastic flow of anisotropic metals' Proc. Roy. Soc. Lond., ser.A. 193, 1948.

41. L. W. HU
'Studies on plastic flow of anisotropic metals', J. Appl. Mech., 1956.

42. B. WHANG
'Elasto-plastic orthotropic plates and shells', Proc. Symp. on Applications of F.E.M. in Civil Engineering, Vanderbilt Univ. Tennessee, 1969.

43. D. R. J. OWEN and J. A. FIGUEIRAS
'Elasto-plastic analysis of anisotropic plates and shells by the Semiloof element', Int. J. Num. Meth. Engng. 19, 521-539, 1983.

44. T. J. R. HUGHES and M. COHEN
The 'heterosis' finite element for plate bending, Computers and Structures, Vol.9, 1978, pp. 445-450.

45. E. ONATE, E. HINTON and N. GLOVER
Techniques for improving the performance of Ahmad shell elements, Report No. C/R/313/78,Univ. of Wales, Swansea.

46. J. M. WHITNEY
'The effect of transverse shear deformation on the bending of laminated plates' J. Composite Mater. 3, 534, 1969.

47. T. S. CHOW
'On the propagation of flexural waves in an orthotropic laminated plate and its response to an impulsive load', J. Composite Mater. 5, 306-319, 1971.

48. D. R. J. OWEN and E. HINTON
Finite Elements in Plasticity: Theory and Practice, Pineridge Press, Swansea, U.K., 1980.

49. T. H. LIN
'On the incremental plastic strain vector of metals and composites' in Recent Research on Mechanical Behaviour of Solids, Univ. of Tokyo Press, 1980, pp. 35-43.

50. S. G. LEKHNITSKII
Theory of Elasticity of an Anisotropic Body, Holder-Day, San Franciso, 1963.

51. A. S. L. WANG and P. C. CHOU
'A comparison of two laminated plate theories', Trans. ASME, ser.E, J. Appl. Mech. 39(2), 611-613, 1972.

52. E. HINTON and D. R. J. OWEN
Finite Element Programming, Academic Press Inc. (London), 1977.

53. E. CARNOY
Shell finite elements, Conference on Structural Dynamics, (Commission of the European Communities) held in Ispra, Italy, 16-20 May, 1983.

54. Y. C. FUNG
Foundations of Solid Mechanics, Prentice-Hall, Inc. Englewood Cliffs, New Jersey, 1965.

55. K. J. BATHE
Finite Element Procedures in Engineering Analysis, Prentice-Hall, Inc. Englewood Cliffs, New Jersey, 1982.

CHAPTER 5

ULTIMATE LOAD ANALYSIS OF REINFORCED CONCRETE PLATES AND SHELLS INCLUDING GEOMETRIC NONLINEAR EFFECTS

D. R. J. Owen   Department of Civil Engineering, University of Wales, Swansea, U.K.

J. A. Figueiras   Department of Civil Engineering, University of Oporto, Portugal

1. INTRODUCTION

In the last chapter, the numerical prediction of the ultimate load capacity of shell structures was considered where the source of material nonlinearity was due to classical elasto-plastic behaviour. In this chapter the techniques and computer code described earlier are extended to the special case of reinforced concrete behaviour. Once again, the effects of geometric nonlinearity are accounted for.

The analysis and design of reinforced and prestressed concrete structures has been based on simple equilibrium conditions and empirical rules for almost a century. These traditional methods generally result in safe designs, but they frequently contain inherent inconsistencies and often do not reflect a clear understanding of the actual composite action of the material. Present-day design codes continue, in many respects, to be based on empirical approaches and rely heavily on the results of a considerable amount of experimental data. This situation is largely attributable to the complex behaviour of reinforced concrete components and structures. Concrete cracking, tension stiffening, nonlinear multiaxial material properties and complex interface behaviours were previously ignored or treated in a very approximate manner. Numerical methods, and particularly the finite element technique, now permit a more rational analysis of these complexities.

The last decade has witnessed rapid advances in the use of finite element methods for the analysis of reinforced concrete structures as reported in several comprehensive review articles [1-5]. Some numerical approaches have been developed to mainly study local behaviours, such as bond effects, cracking, interfaced shear, dowel action [6-9] while other numerical studies have been directed at the analysis and design of components and structures [10-13]. The primary objectives in the latter case is the accurate prediction of the overall deformation characteristics and limit loads. An averaging procedure is generally employed to simulate concrete cracking; with the crack effects being assumed to be distributed within a sampling volume. An average shear modulus which accounts for both crack interface effects and dowel action is employed and a full bond is assumed at the steel-concrete interface.

Extensive experimental studies have been undertaken to characterise the response and ultimate strength of plain concrete under multiaxial stress states [14,15]. Considerable scatter of results has been observed and collaborative studies have been undertaken to identify the principal factors influencing this variation [16]. Several approaches, based on experimental data, have been used to represent the constitutive relationships of concrete under multiaxial stress states and these can be categorised into the four following groups: (a) Linear and nonlinear elasticity theories [17-20], (b) perfect and work-hardening plasticity theories [21-23], (c) endochronic theory of plasticity [24] and (d) plastic fracturing theory [25]. Experimental evidence indicates that the nonlinear deformation in concrete is basically inelastic and therefore the stress-strain behaviour may be separated into recoverable and irrecoverable components. The irrecoverable strain components may either be treated within the classical theory of plasticity or by using a time dependent viscoplastic model.

In the present work both a perfect and strain hardening plasticity approach are employed to model the compressive behaviour of the concrete. A dual criterion for yield and crushing in terms of stresses and strains is considered, which is complemented with a tension cut-off representation. Several discretisation processes have been previously employed in the finite element analysis of reinforced concrete plates and shells. The rigidity of a cracked concrete region has been modelled by a bilinear moment-curvature relationship in Ref. [26] and used in conjunction with simple quadrilateral elements. A layered approach has been used by several investigators to represent the concrete and steel reinforcement and also to simulate progressive cracking through the concrete layers. Such a process has been utilised with triangular elements [10,21], shallow shell parallelogram elements [11,27] and Reissner-Mindlin plate elements [13,28]. In the present study, degenerate thick shell elements employing a layered discretisation through the thickness are adopted. Reduced and selective integration rules are used in order to obtain the best performance from both the 8-node Serendipity and Heterosis elements considered. The concrete characteristics are specified for each layer through the thickness and the reinforcing steel is represented as a smeared layer of equivalent thickness with (anisotropic) uniaxial strength and rigidity properties.

2. MATERIAL MODELLING

In this section the constitutive behaviour of concrete is described in a form suitable for numerical computation. In the present analysis, both a perfect and strain-hardening plasticity approach are employed to model the compressive behaviour. A dual criterion for yielding and crushing in terms of stresses and strains is considered, which is complemented with a tension cut-off representation.

2.1 Compressive Behaviour of Concrete

The following three conditions have to be considered in establishing the nonlinear stress-strain relations, based on the flow theory of plasticity: (a) The yield criterion, (b) the flow and hardening rules and (c) the crushing condition. The yield criterion for concrete under a triaxial stress state is generally assumed to be dependent on the three stress invariants [29-31]. Nevertheless practical formulations have been developed which employ either one [10] or two [22,32] stress invariants only. A dependence of the yield function on the mean normal stress I_1 (or σ_{oct}) and the shear stress invariant J_2 (or τ_{oct}) has proved to be adequate for most situations. The plastic volume dilatation observed near failure, under pressure conditions, can also be modelled by the dependence of the yield function on the first stress invariant.

2.1.1 The yield condition

In the present analysis of thick plates and shells, transverse shear effects are taken into account and therefore a triaxial yield criterion must be employed. This criterion is formulated in terms of the first two stress invariants and only two material parameters are involved in its definition.

$$f(I_1, J_2) = [\beta(3J_2) + \alpha I_1]^{\frac{1}{2}} = \sigma_o \qquad (1)$$

where α and β are material parameters and σ_o is the equivalent effective stress taken as the compressive stress from a uniaxial test. In terms of principal stresses the expression for yielding can be written,

$$\beta[(\sigma_1^2+\sigma_2^2+\sigma_3^2) - (\sigma_1\sigma_2+\sigma_1\sigma_3+\sigma_2\sigma_3)] + \alpha\ (\sigma_1+\sigma_2+\sigma_3) = \sigma_o^2 \qquad (2)$$

The Huber-Mises yield condition is recovered by assuming $\alpha=0$ and $\beta=1.0$.

The stress state in thick plates and shells is usually not far from a biaxial one, since the stress normal to the middle plate ($\sigma_{z'}$) is neglected. Therefore, it is reasonable to obtain the material parameters by fitting biaxial test results. The uniaxial compression test and the biaxial test under equal compression stresses ($\sigma_1 = \sigma_2$) are used to define these constants. For practical purposes a relation can be assumed between the equal biaxial yield stress (or strength), f_{cb}, and the uniaxial yield stress (or strength) f'_c,

$$f_{cb} = 1.16 \text{ (to 1.20) } f'_c \tag{3}$$

In such cases, the yield condition is a function of only one material parameter ($\sigma_o = f'_c$) which is the most reliable constant to characterize the concrete behaviour and is easily obtainable from experimental tests. If Kupfer's results [14] are employed, $f_{cb} = 1.16 f'_c$, then use of (2) gives

$$\begin{aligned} \alpha &= 0.355\sigma_o \\ \beta &= 1.355 \end{aligned} \tag{4}$$

and (1) can be written in terms of the stress components as

$$f(\sigma) = \{1.355[(\sigma_x^2+\sigma_y^2-\sigma_x\sigma_y) + 3(\tau_{xy}^2+\tau_{xz}^2+\tau_{yz}^2)] + $$

$$+ 0.355\sigma_o(\sigma_x+\sigma_y)\}^{\frac{1}{2}} = \sigma_o \tag{5}$$

This expression is compared in Fig. 1 with the experimental results of Kupfer et al. [14] in biaxial stress space. In the perfectly plastic model σ_o is taken as the ultimate stress f'_c obtained from a uniaxial compression test. An elastic response is assumed up to the effective stress value $\sigma_o = f'_c$, after which a perfectly plastic response follows until the crushing surface is reached. In the strain-hardening model the initial yield surface is attained when the effective stress reaches 30 percent of the peak stress f'_c. The subsequent loading surfaces $f(\sigma) = \sigma_o(\chi)$ are functions of the hardening parameter χ defined by the hardening rule expressed in terms of the effective plastic strains as shown in Section 2.1.3. When the effective stress, defined by the yield function, reaches the ultimate stress f'_c, a perfectly plastic response is assumed until the crushing surface is encountered.

Figure 2 illustrates the one dimensional representation of both the perfectly plastic and the strain-hardening model. The tensile behaviour, to be discussed later, is also represented. The two dimensional representation in the principal stress space is shown in Figure 3. For the strain-hardening model, once the initial yield surface has been reached, subsequent loading will produce plastic straining which characterizes the corresponding

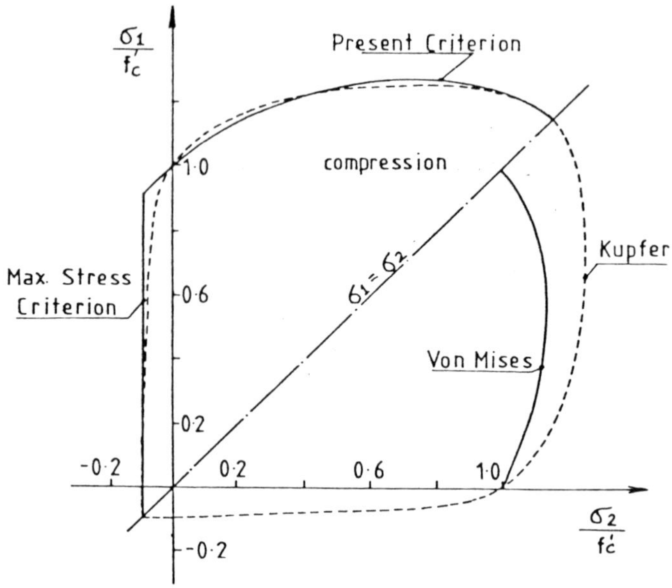

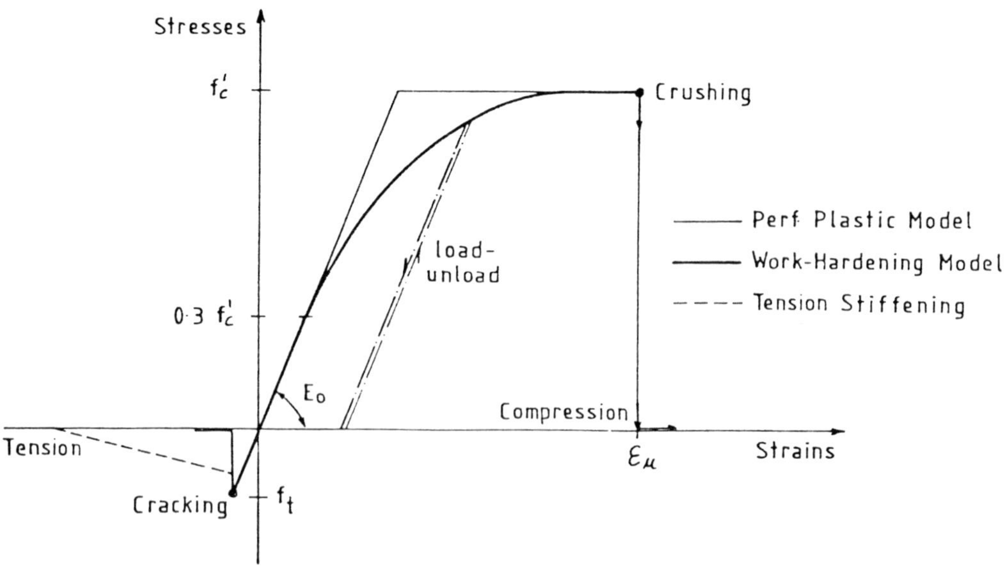

Fig. 2 One-dimensional representation of the concrete constitutive model.

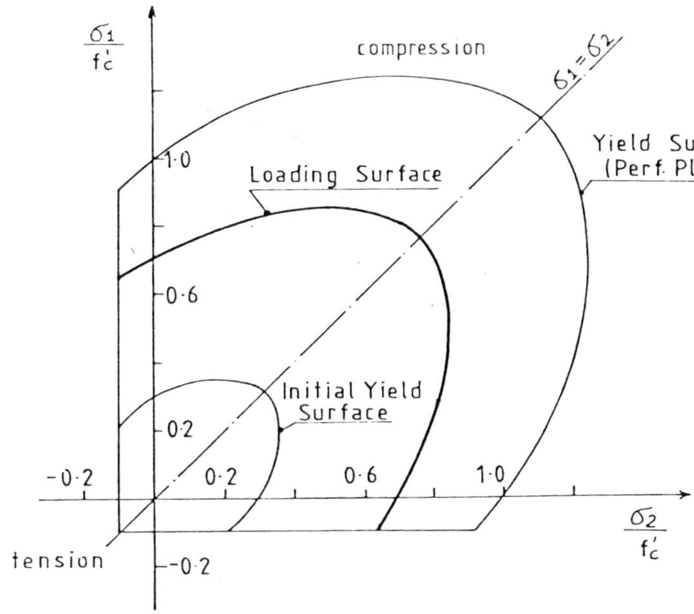

Fig. 3 Two-dimensional stress space representation of the concrete constitutive model.

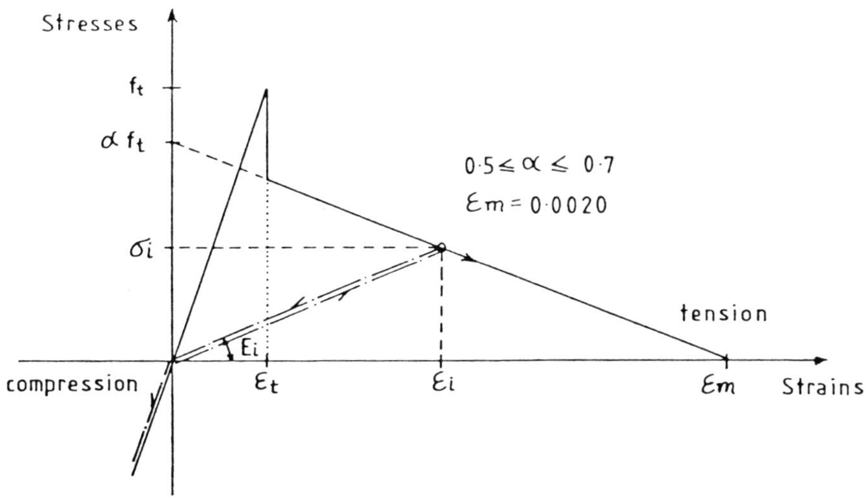

Fig. 4 Loading and unloading behaviour of cracked concrete illustrating tension stiffening behaviour.

level of effective stress or loading surface. Unloading follows the initial elastic modulus E_o and an elastic response occurs for subsequent loading until the corresponding loading surface is reached. Further loading causes an elasto-plastic response with increasing plastic deformation and a corresponding expansion of the loading surface according to the hardening rule.

2.1.2 The flow rule

To construct the stress-strain relationship in the plastic range, the normality of the plasticity deformation rate vector to the yield surface is commonly assumed. This associated flow rule is considered for concrete predominantly for practical reasons, since there is very little supportive experimental evidence available [33]. The plastic strain increment is then defined [34,35] as,

$$d\varepsilon^p_{ij} = d\lambda \frac{\partial f(\sigma)}{\partial \sigma_{ij}} \tag{6}$$

where $d\lambda$ is a proportionality constant which determines the magnitude of the plastic strain increment, and the gradient $\partial f(\sigma)/\partial \sigma_{ij}$ defines its direction to be perpendicular to the yield surface. The current stress function $f(\sigma)$ is the yield condition; or the subsequent loading functions in the strain hardening model.

Formulation of the elasto-plastic constitutive matrix, $\underline{D}_{ep}$ follows an identical procedure to that described in Section 3.2 of the previous chapter resulting in the same expression (68).

The yield function derivatives which define the flow vector $\underline{a}$ take the following explicit expressions for the present yield surface given by (5):

$$\underline{a}^T = [\frac{\partial f}{\partial \sigma_x}, \frac{\partial f}{\partial \sigma_y}, \frac{\partial f}{\partial \tau_{xy}}, \frac{\partial f}{\partial \tau_{xz}}, \frac{\partial f}{\partial \tau_{yz}}] \tag{7}$$

$$a_1 = \frac{\partial f}{\partial \sigma_x} = c + [2(c^2+\beta)\sigma_x + (2c^2-\beta)\sigma_y]/\mathrm{CONST}$$

$$a_2 = \frac{\partial f}{\partial \sigma_y} = c + [2(c^2+\beta)\sigma_y + (2c^2-\beta)\sigma_x]/CONST$$

$$a_3 = \frac{\partial f}{\partial \tau_{xy}} = 6\beta\tau_{xy}/CONST \qquad (8)$$

$$a_4 = \frac{\partial f}{\partial \tau_{xz}} = 6\beta\tau_{xz}/CONST$$

$$a_5 = \frac{\partial f}{\partial \tau_{yz}} = 6\beta\tau_{yz}/CONST$$

where $c = \alpha/(2\sigma_o)$ (equal to 0.1775 for Kupfer's results)

α and β are material constants previously defined, and

$$CONST = 2[(c^2+\beta)\sigma_x^2 + (c^2+\beta)\sigma_y^2 + (2c^2-\beta)\sigma_x\sigma_y + 3\beta(\tau_{xy}^2 + \tau_{xz}^2 + \tau_{yz}^2)]^{\frac{1}{2}} \qquad (9)$$

2.1.3 The Hardening Rule

The hardening rule defines the motion of the subsequent yield surface (the loading surfaces) during plastic deformation. It determines the relation between the loading surfaces $f(\sigma,\chi)$ (or effective stress) and the accumulated plastic strain (or effective plastic strain). The concept of effective stress and effective plastic strain makes it possible to extrapolate from a simple uniaxial test to the multiaxial situation. In the present work the relation between effective stress and effective plastic strain is extrapolated from the uniaxial stress-strain relationship using the conventional "Madrid Parabola".

$$\sigma = E_o\varepsilon - \frac{1}{2}\frac{E_o}{\varepsilon_o}\varepsilon^2 \qquad (10)$$

where E_o is the initial elasticity modulus

ε is the total strain

ε_o is the total strain at peak stress f'_c

Substituting the elastic strain component $\varepsilon_e = \sigma/E_o$ in (10) we obtain

$$\sigma = -E_o\varepsilon_p + \sqrt{2E_o^2\varepsilon_o\varepsilon_p}; \quad 0.3f'_c < \sigma \leq f'_c \qquad (11)$$

where ε_p is the plastic strain component, and ε_o can be taken equal to $2f'_c/E_o$ for normal concrete. Using the current effective plastic strain in expression (11), gives the effective stress level $\sigma=\sigma_o$ defining the current loading surface position.

2.1.4 The crushing condition

The crushing type of concrete fracture is a strain controlled phenomenon. The lack of available experimental data on concrete ultimate-deformation capacity under multiaxial stress states has resulted in the appropriate strain criterion being developed by simply converting the yield criterion described in terms of stresses directly into strains. Thus,

$$\beta(3J'_2) + \alpha I'_1 = \varepsilon_u^2 \tag{12}$$

where I'_1 and J'_2 are strain invariants [32] and ε_u is an ultimate total strain extrapolated from uniaxial test results. Using the material parameters α and β determined from Kupfer's results [14], the crushing condition is expressed in terms of the total strain components as

$$1.355[(\varepsilon_x^2+\varepsilon_y^2-\varepsilon_x\varepsilon_y) + 0.75(\gamma_{xy}^2+\gamma_{xz}^2+\gamma_{yz}^2)] + \\ + 0.355\ \varepsilon_u(\varepsilon_x+\varepsilon_y) = \varepsilon_u^2 \tag{13}$$

When ε_u reaches the value specified as the ultimate strain, the material is assumed to lose all its characteristics of strength and rigidity.

2.2 Tensile Behaviour of Concrete

The response of concrete under tensile stresses is assumed to be linear elastic until the fracture surface is reached and its behaviour is characterised by the isotropic versions of relations (24)-(26) of the previous chapter. The tensile type of fracture, or cracking, is governed by a maximum tensile stress criterion (tension cut-off). Cracks are assumed to form in planes perpendicular to the direction of maximum principal tensile stress as soon as this stress reaches the specified concrete tensile strength f'_t. In order to avoid further complexities, cracks are assumed to form only in planes perpendicular to the structural plane (x'y' plane). After cracking has occurred, the elasticity modulus and Poisson's ratio are reduced to zero in the direction perpendicular to the cracked plane, and a reduced shear modulus is employed. Taking 1 and 2 as the two principal directions in the plane of the structure, the stress-strain relationship for concrete cracked in the 1 direction, is

$$\begin{bmatrix} \sigma_1 \\ \sigma_2 \\ \tau_{12} \\ \tau_{13} \\ \tau_{23} \end{bmatrix} = \begin{bmatrix} 0 & 0 & 0 & 0 & 0 \\ 0 & E & 0 & 0 & 0 \\ 0 & 0 & G^c_{12} & 0 & 0 \\ 0 & 0 & 0 & G^c_{13} & 0 \\ 0 & 0 & 0 & 0 & \frac{5G}{6} \end{bmatrix} \begin{bmatrix} \varepsilon_1 \\ \varepsilon_2 \\ \gamma_{12} \\ \gamma_{13} \\ \gamma_{23} \end{bmatrix} \tag{14}$$

When the tensile stress in the 2 direction reaches the value f'_t, a second cracked plane perpendicular to the first one is assumed to form, and the stress-strain relationship becomes:

$$\begin{bmatrix} \sigma_1 \\ \sigma_2 \\ \tau_{12} \\ \tau_{13} \\ \tau_{23} \end{bmatrix} = \begin{bmatrix} 0 & 0 & 0 & 0 & 0 \\ 0 & 0 & 0 & 0 & 0 \\ 0 & 0 & G^c_{12/2} & 0 & 0 \\ 0 & 0 & 0 & G^c_{13} & 0 \\ 0 & 0 & 0 & 0 & G^c_{23} \end{bmatrix} \begin{bmatrix} \varepsilon_1 \\ \varepsilon_2 \\ \gamma_{12} \\ \gamma_{13} \\ \gamma_{23} \end{bmatrix} \tag{15}$$

The cracked concrete is anisotropic and these relations must be transformed to the reference axes x'y'.

In the analysis and design of reinforced concrete plates and shells we are not directly interested in the tensile strength of the concrete, which is relatively small and unreliable, but in the influence of the cracked concrete zones on the structural behaviour. Therefore a simplified averaging process more convenient for finite element formulation is used. A smeared representation for cracked concrete is assumed, which implies that cracks are not discrete but distributed across a region of the finite element.

2.2.1 Tension stiffening

Due to bond effects, cracked concrete carries between cracks, a certain amount of tensile force normal to the cracked plane. The concrete adheres to the reinforcing bars and contributes to the overall stiffness of the structure [37]. Several approaches based on experimental results have been employed to simulate this tension stiffening behaviour [13,21, 36,38]. A gradual release of the concrete stress component normal to the cracked plane (Fig. 4) is adopted in this work. The process of loading and unloading of cracked concrete is also illustrated in Figure 4. Unloading and reloading of cracked concrete is assumed to follow the linear behaviour shown with a fictitious elasticity modulus E_i given by

$$E_i = \alpha f'_t(1-\varepsilon_i/\varepsilon_m)/\varepsilon_i, \qquad \varepsilon_t \leq \varepsilon_i \leq \varepsilon_m \tag{16}$$

where α, ε_m are tension stiffening parameters (Fig. 4) and ε_i is the maximum value reached by the tensile strain at the point considered. If the crack closes, i.e. if the strain component normal to the crack plane becomes negative, the concrete acquires the uncracked behaviour in the corresponding direction, but the crack direction and the maximum tensile strain continue to be stored. The value ε_i can be readily modified to simulate bond deterioration during reloading. The normal stress σ_1 (and or σ_2) is obtained by the following expression (Fig. 4),

$$\sigma_1 = \alpha f'_t(1-\varepsilon_1/\varepsilon_m), \qquad \varepsilon_t \leq \varepsilon_1 \leq \varepsilon_m \tag{17}$$

or by,

$$\sigma_1 = \sigma_i \varepsilon_1 / \varepsilon_i, \qquad \text{if} \qquad \varepsilon_1 < \varepsilon_i \tag{18}$$

where ε_1 is the current tensile strain in material direction 1.

A layered approach is used to discretize the structure through the thickness and for each layer different material properties can be specified. A modified tension stiffening diagram can be employed for those layers which are relatively distant from reinforcing bars. Particularly, the tension stiffening effect can be neglected ($\alpha = 0$ for $0 < \varepsilon_m < \varepsilon_t$) in layers where the steel cannot supply tensile stresses due to the particular reinforcing arrangement. The value of f'_t should be taken as the modulus of rupture of the concrete. This modulus can be related to the uniaxial compression strength [58] by

$$f_r = 0.62\ (f'_c)^{\frac{1}{2}} \tag{19}$$

where f_r and f'_c are expressed in MPa. However experimental tests have shown large variations in the above coefficient. The tension stiffening constant α in Fig. 4 is taken equal to 0.5, 0.6 or 0.7 depending on the relative percentage of steel in the section. However the change in structural behaviour with this variation is generally small and a constant value of $\alpha = 0.6$ could be used. For the tension stiffening constant, ε_m, a fixed value of 0.002 is employed.

2.2.2 Cracked shear modulus

Experimental results indicate that a considerable amount of shear stress can be transferred across the rough surfaces of cracked concrete [39-41]. Also, the dowel action of steel bars contribute to the shear stiffness across cracks [39]. These tests have shown that the primary variable in the shear transfer mechanism is the crack width, although aggregate size, reinforcement ratio and bar size also have an influence. A common procedure to account for aggregate interlock and dowel action in a smeared cracking model is to attribute an appropriate value to the cracked shear modulus G^c [7,11]. In the present work an approach similar to that used in Ref. [7] is adopted, where the cracked shear modulus is assumed to be a function of the current tensile strain. For concrete cracked in the 1 direction,

$$
\begin{aligned}
G^c_{12} &= 0.25*G(1-\varepsilon_1/0.04); \quad G^c_{12} = 0 \text{ if } \varepsilon_1 \geq 0.004 \\
G^c_{13} &= G^c_{12} \\
G_{23} &= \frac{5G}{6}
\end{aligned}
\tag{20}
$$

where G is the uncracked concrete shear modulus and ε_1 is the tensile strain in the 1 direction. For concrete cracked in both directions,

$$
\begin{aligned}
G^c_{13} &= 0.25*G(1-\varepsilon_1/0.004); \quad G^c_{13} = 0 \text{ if } \varepsilon_1 \geq 0.004 \\
G^c_{23} &= 0.25*G(1-\varepsilon_2/0.004); \quad G^c_{23} = 0 \text{ if } \varepsilon_2 \geq 0.004 \\
G^c_{12} &= 0.5*G^c_{13}; \quad \text{or } G^c_{12} = 0.5*G^c_{23} \text{ if } G^c_{23} < G^c_{13}
\end{aligned}
\tag{21}
$$

If the crack closes the uncracked shear modulus G is again assumed in the corresponding direction.

2.3 Behaviour of Steel in Tension and Compression

The reinforcing bars are considered as steel layers of equivalent thickness in the present model. Each steel layer has an uniaxial behaviour resisting only the axial force in the bar direction. A bilinear or a trilinear idealization can be adopted in order to model the elasto-plastic stress-strain relationships. The basic relationships for uniaxial elasto-plastic behaviour and the corresponding numerical formulation can be found in Ref. [34].

3. FINITE ELEMENT SOLUTION

The same degenerate shell elements described in the previous chapter are again employed in the present analysis of reinforced concrete structures. In brief, the 8-node Serendipity and the Heterosis element with a hierarchical formulation for the ninth node [43] are considered and both reduced and selective integration rules are employed to perform the numerical integration of the element matrices. A layered approach is employed to represent the steel reinforcement and to discretize the concrete behaviour through the thickness. Reinforcing steel is represented as a smeared layer of equivalent thickness with (anisotropic) uniaxial strength and rigidity properties. The progress of cracked zones, as well as the compressive behaviour of the concrete is analysed and monitored for each layer and for each element Gauss point. To simplify the data input and also to deal with variable thickness of the structure the layer thickness is defined in terms of the normalised ζ coordinate.

Geometric nonlinearities may play an important role in the behaviour of reinforced concrete beams, plates and shells [4,56, 57]. Even for relatively small deflections (0.05* thickness) the consideration of nonlinear geometric behaviour may be indispensable for obtaining a correct structural response and collapse load. Disregard of geometric nonlinearity may result either in stiffening or softening of the structure, depending on the particular problem. In this work the Total Lagrangian approach based on the simplified Von Karman strain expressions described in Section 3.3 of the previous chapter is again adopted.

For numerical solution, the general nonlinear algorithm itemised in Section 3.1 of the previous chapter is employed. For the present application the essential steps of the process are detailed below.

At the beginning of the nth load increment the displacements $\underline{a}^{n-1}$ and the stresses $\underline{\sigma}^{n-1}$ are known, as well as the unbalanced nodal forces $\underline{\psi}^{n-1}$ resulting from the previous load increment. The incremental nodal forces are calculated according to:

$$\underline{\psi}_o^n = \underline{\psi}^{n-1} + \Delta\underline{f}^n \tag{22}$$

where $\underline{\psi}^{n-1}$ are the residuals existing at the end of the previous load increment and $\Delta\underline{f}^n$ is the nth load increment. Subsequently, the iterative process is performed with the following steps for a generic iteration, i:

1. The stiffness matrix $\underline{K}$ is updated or not according to the solution algorithm adopted.
2. The incremental displacements $\Delta\underline{a}_i$ are evaluated using the equilibrium equations,

$$\Delta\underline{a}_i = - [K]^{-1} \underline{\psi}_{i-1} \tag{23}$$

where $\underline{\psi}_{i-1}$ are the unbalanced nodal forces resulting from the previous iteration. The total displacement vector $\underline{a}_i$ is then updated,

$$\underline{a}_i = \underline{a}_{i-1} + \Delta\underline{a}_i \tag{24}$$

3. The incremental strains $\Delta\underline{\varepsilon}_i$ and the total strains $\underline{\varepsilon}_i$ are evaluated,

$$\Delta\underline{\varepsilon}_i = \underline{B}\ \Delta\underline{a}_i \tag{25}$$

$$\underline{\varepsilon}_i = \underline{B}'\ \underline{a}_i \tag{26}$$

where $\underline{B}$ and $\underline{B}'$ are updated strain matrices depending on $\underline{a}_i$. Note that $\underline{B}$ and $\underline{B}'$ are equal to the initial strain matrix $\underline{B}_o$ if nonlinear geometrical behaviour is not considered.

4. The incremental stresses $\Delta\underline{\sigma}_i$ and the total stresses $\underline{\sigma}_i$ are calculated,

$$\Delta\underline{\sigma}_i = \underline{D}\ \Delta\underline{\varepsilon}_i \tag{27}$$

$$\underline{\sigma}_i = \underline{\sigma}_{i-1} + \Delta\underline{\sigma}_i \tag{28}$$

where $\underline{D}$ is the elasticity matrix taken as:

- either the elastic matrix of uncracked concrete or the corresponding matrix of cracked concrete - for concrete layers,
- the elastic matrix - for steel layers.

5. The stresses are corrected according to the material constitutive equations:

a) Concrete layers

* Using the total stresses $\underline{\sigma}_i$, the maximum principal stress σ_1, acting in the structural plane, is calculated.

* If $\sigma_1 > f'_t$ or, if the concrete is already cracked, the stresses are updated according to the tensile modelling of Section 2.2.1.

* Using $\underline{\sigma}_i$ or the stresses updated in the previous step, the effective stress $\bar{\sigma}$ is calculated (according to the yield function)

* If $\bar{\sigma}$ is greater than the initial yield stress or if the layer has already yielded, the stresses are corrected according to the elasto-plastic behaviour.

b) Steel layers

* Using the total stresses $\underline{\sigma}_i$, the stresses in the reinforcement direction σ_s is obtained.

* If σ_s is greater than the steel yield stress f_y, or if the layer has already yielded, the stresses are treated elasto-plastically.

6. The equivalent internal nodal forces $\underline{p}_i$ are evaluated using numerical integration, as,

$$\underline{p}_i = \int_V \underline{B}^T \underline{\sigma}_i \, dV \tag{29}$$

where $\underline{\sigma}_i$ are the total stress components corrected according to the constitutive equations.

7. The out of balance forces $\underline{\psi}_i$ are calculated,

$$\underline{\psi}_i = \underline{f} - \underline{p}_i \tag{30}$$

where $\underline{f}$ is the current external nodal force vector.

8. The convergence of the process is checked:

* If convergence has been achieved proceed to the next load

increment.

* If the convergence criterion has not been satisfied restart the iterative cycle from Step 1.

The size of the load increments, the solution algorithms and the convergence criterion necessary for efficient numerical solution are now discussed below.

(i) Load increment size

To establish the entire deformational response of structures up to collapse load necessitates the use of an incremental procedure in which a reasonable number of load increments must be used, since the solutions are generally path dependent. Comparative analyses have been performed varying the load increment size. It has been observed that the size of the load increments is not a critical parameter in tracing out the deformational response provided that an appropriate solution algorithm is employed.

In general, the size of the load increments used in the present work ranges from 0.1 to 0.2 of the structure cracking load. However, values of increment size about 0.5 times the cracking load have produced reasonable results for the cases analysed. In under-reinforced structures refined load increments should be employed during the first spread of cracking.

To obtain the collapse load within narrow limits the size of load increments must be refined when the structure is about to collapse. Use of the restart facility included in the code is particularly useful in this respect.

(ii) Solution algorithm

A detailed reference to the incremental and iterative modified Newton-Raphson scheme employed in solution has been made in Section 3.1 of the previous chapter. In general, the tangential stiffness matrix is recalculated for the second iteration of each load increment. However, to improve the solution stability, the stiffness matrix is evaluated at stages within a load increment when the change of material characteristics implies a local increase of stiffness. This recalculation has to be performed more frequently during the first spread of cracked zones and also near the collapse load.

(iii) Convergence Criterion

A convergence criterion is required in order to terminate the iterative cycle when the solution is considered to be sufficiently accurate. In the previous chapter a criterion based on the unbalanced nodal forces was adopted. In the analysis of reinforced concrete structures, however, a force convergence criterion may not be the most suitable as the unbalanced forces often form equilibrium groups [4] which do not have much influence on the overall structural response and the solution is not monotonically convergent. Convergence criteria both in terms of out of balance nodal forces and in terms of incremental nodal displacements are therefore more appropriate. However the sole use of a displacement convergence criterion, in which displacements and rotations are separately checked, has been favoured in most of the numerical examples analysed to date.

Together with the convergence criterion, a convergence tolerance must be specified to indicate when the results have converged. Higher solution cost and needless accuracy may result from too tight tolerances and on the other hand inaccurate solutions may result if the tolerance adopted is too coarse. Convergence tolerances varying from 0.1 to 1.0 percent have been employed in the problems analysed. Comparative studies performed on some examples have indicated that a tolerance of 2.5 percent is generally sufficient for practical purposes.

A maximum number of iterations for each increment of load is specified to stop the nonlinear solution if meanwhile the convergence limit has not been achieved. It has been observed that a maximum number of about 15 iterations is generally sufficient to detect the solution divergence or collapse, provided the most appropriate solution algorithm has been utilized. Obviously, this maximum number of iterations also depends on the problem and on the specified tolerance, but a maximum number between 10 and 15 for a tolerance of 0.5 percent has been found adequate.

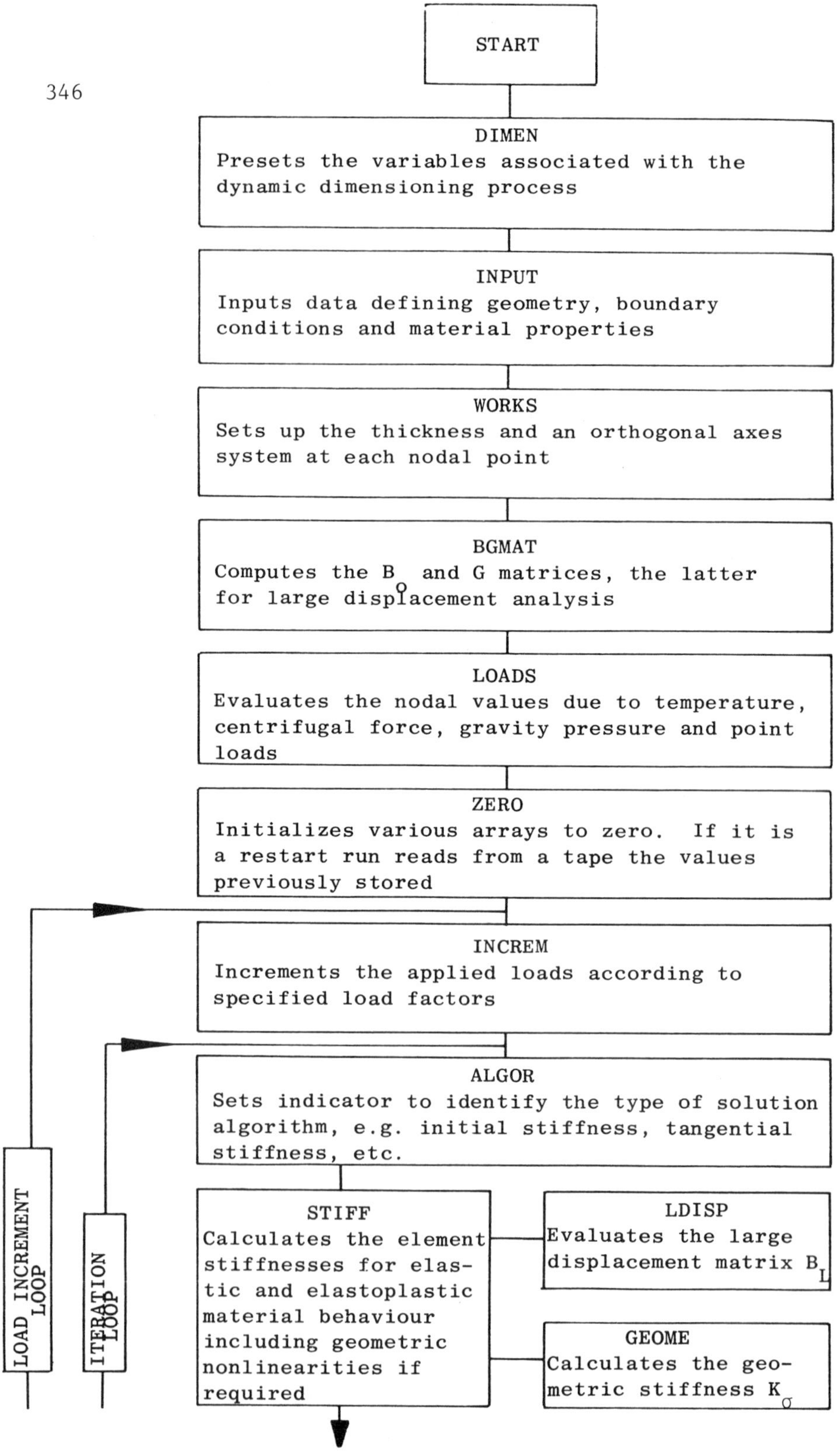

Fig. 5 Overall structure of Program CONSHELL

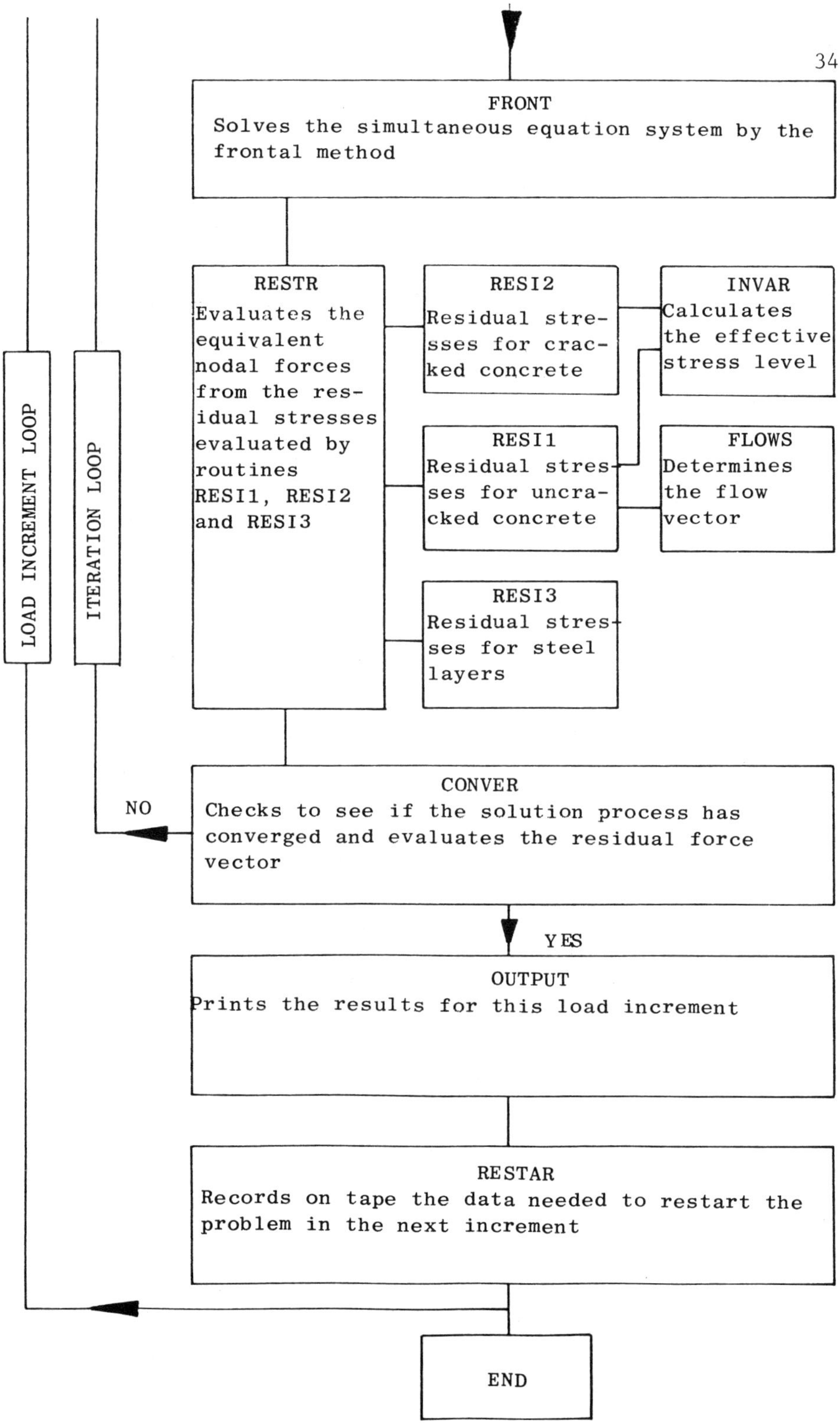

Fig. 5 (Continued)

4. FINITE ELEMENT COMPUTER CODE

In this section a computer code is presented which undertakes the ultimate load analysis of reinforced concrete plates and shells. A layered approach is adopted and geometric nonlinear effects may also be accounted for.

Most of the subroutines which comprise the package have already been presented in Section 4 of the previous chapter. It now remains only to describe the additional FORTRAN subroutines required and to assemble them to form a working program. User instructions for the program are provided in the appendix following this chapter. The structure of the program is shown schematically in Fig. 5. Where, in particular, the subroutines required to form the complete program are indicated. The function of each subroutine (not already defined in the previous chapter) is now explained and a FORTRAN listing provided.

4.1 Subroutine CONVRD

This subroutine monitors convergence of the nonlinear solution iteration process. For convergence both the norm of the incremental displacements and the norm of the incremental rotations are required to be less than a specified percentage, TOLER, of the norm of the total displacements and rotations respectively

```
      SUBROUTINE CONVRD(ASDIS,ELOAD,TDISP,TLOAD,IITER,MELEM,
     .                  MEVAB,MTOTV,NCHEK,NDOFN,NELEM,NEVAB,
     .                  NPOIN,TOLER)
C***********************************************************************
C
C*** THIS SUBROUTINE CHECKS FOR CONVERGENCE USING DISPLACEMENT NORMS
C
C***********************************************************************
      DIMENSION ASDIS(MTOTV),TDISP(MTOTV),ELOAD(MELEM,MEVAB),
     .          TLOAD(MELEM,MEVAB)
      KCHEK=1
      IF(IITER.EQ.1) GO TO 50
      KCHEK=0
      NCHEK=0
      RESID=0.0
      RETOT=0.0
      DO 10 IDOFN=1,3
      DO 10 IPOIN=1,NPOIN
      ITOTV=(IPOIN-1)*NDOFN+IDOFN
      RESID=RESID+ASDIS(ITOTV)*ASDIS(ITOTV)
   10 RETOT=RETOT+TDISP(ITOTV)*TDISP(ITOTV)
      RESID=SQRT(RESID)
      RETOT=SQRT(RETOT)
      RATIO=100.0*RESID/RETOT
      IF(RATIO.LE.TOLER) GO TO 20
      NCHEK=1
      KCHEK=KCHEK+1
```

```
   20 WRITE(6,900) NCHEK,RATIO
  900 FORMAT(1H0,3X,13HDISPLACEMENTS,3X,18HCONVERGENCE CODE =,I4,
     .       3X,28HNORM OF RESIDUAL SUM RATIO =,E14.6)
      NCHEK=0
      RESID=0.0
      RETOT=0.0
      DO 30 IDOFN=4,5
      DO 30 IPOIN=1,NPOIN
      ITOTV=(IPOIN-1)*NDOFN+IDOFN
      RESID=RESID+ASDIS(ITOTV)*ASDIS(ITOTV)
   30 RETOT=RETOT+TDISP(ITOTV)*TDISP(ITOTV)
      RESID=SQRT(RESID)
      RETOT=SQRT(RETOT)
      IF(RETOT.EQ.RESID) GO TO 40
      RATIO=100.0*RESID/RETOT
      COEFE=1.0
      TOLEC=TOLER*COEFE
      IF(RATIO.LE.TOLEC) GO TO 40
      NCHEK=1
      KCHEK=KCHEK+1
   40 WRITE(6,910) NCHEK,RATIO
  910 FORMAT(4X,13HROTATIONS    ,3X,18HCONVERGENCE CODE =,I4,
     .       3X,28HNORM OF RESIDUAL SUM RATIO =,E14.6)
   50 NCHEK=1
      IF(KCHEK.EQ.0) NCHEK=0
      DO 60 IELEM=1,NELEM
      DO 60 IEVAB=1,NEVAB
   60 ELOAD(IELEM,IEVAB)=TLOAD(IELEM,IEVAB)-ELOAD(IELEM,IEVAB)
      RETURN
      END
```

4.2 Subroutine FLOWS

The function of this subroutine is to evaluate the flow vector $\underline{a}$ in (8) as well as the denominator of the last term of the elasto-plastic matrix $\underline{D}_{ep}$ (see expression (68) previous chapter).

```
      SUBROUTINE FLOWS(ABETA,AVECT,DVECT,LPROP,
     .                 MMATS,NPROP,PROPS,SG)
C***********************************************************************
C
C*** THIS SUBROUTINE CALCULATES THE FLOW VECTOR -AVECT- AND
C    COMPUTES  -DVECT- AND -ABETA-
C
C***********************************************************************
      DIMENSION AVECT(5),DMATX(5,5),DVECT(5),
     .          PROPS(MMATS,NPROP),SG(5)
C
C*** SET UP MATERIAL PROPERTIES
C
      HARDS=0.0
C
C*** COMPUTES THE VECTOR  AVECT
C
      CONS1=0.355*0.5
      CONS2=CONS1*CONS1+1.355
      CONS3=2.0*CONS1*CONS1-1.355
      CONS4=1.355*3.0
      AFUNC=(CONS2*(SG(1)*SG(1)+SG(2)*SG(2))+CONS3*SG(1)*SG(2)+
     .      CONS4*(SG(3)*SG(3)+SG(4)*SG(4)+SG(5)*SG(5)))**0.5
      AFUNC=2.0*AFUNC
      AVECT(1)=CONS1+(2.0*CONS2*SG(1)+CONS3*SG(2))/AFUNC
      AVECT(2)=CONS1+(2.0*CONS2*SG(2)+CONS3*SG(1))/AFUNC
      AVECT(3)=2.0*CONS4*SG(3)/AFUNC
      AVECT(4)=2.0*CONS4*SG(4)/AFUNC
      AVECT(5)=2.0*CONS4*SG(5)/AFUNC
```

```
C
C*** COMPUTE  DVECT=DMATX*AVECT
C
      CALL       MODUL(DMATX,PROPS,NPROP,MMATS,LPROP,     1,0.0,
     .                STRA1,STRA2,  0.0,  0.0)
      DO 10 I=1,5
      DVECT(I)=0.0
      DO 10 J=1,5
   10 DVECT(I)=DVECT(I)+DMATX(I,J)*AVECT(J)
C
C*** COMPUTE  ABETA=1.0/(H+ATDA)
C
      DENOM=HARDS
      DO 20 ISTRE=1,5
   20 DENOM=DENOM+AVECT(ISTRE)*DVECT(ISTRE)
      ABETA=1.0/DENOM
      RETURN
      END
```

4.3 Subroutine INVAR

This subroutine merely evaluates the current value of the yield function given by (5).

```
      SUBROUTINE INVAR(ST,YIELD)
C***********************************************************************
C
C***THIS SUBROUTINE EVALUATES THE CURRENT VALUE OF THE
C    YIELD FUNCTION
C
C***********************************************************************
      DIMENSION ST(5)
      BEETA=1.355
      ALPHA=0.355
      GJNV2=BEETA*(ST(1)*ST(1)+ST(2)*ST(2)-ST(1)*ST(2)+3.0*
     .      (ST(3)*ST(3)+ST(4)*ST(4)+ST(5)*ST(5)))
      GINV1=ALPHA*(ST(1)+ST(2))*0.5
      GASH1=SQRT(GINV1*GINV1+GJNV2)
      YIELD=GINV1+GASH1
      RETURN
      END
```

4.4 Subroutine INVA2

This subroutine evaluates the current value of the equivalent compressive strain (crushing condition) given by (13).

```
      SUBROUTINE INVA2(SN,CRUSH)
C***********************************************************************
C
C*** THIS SUBROUTINE EVALUATES THE CURRENT VALUE OF THE COMPRESSIVE
C    FRACTURE STRAIN FUNCTION
C
C***********************************************************************
      DIMENSION SN(5)
      BEETA=1.355
      ALPHA=0.355
      GASH1=BEETA*(SN(1)*SN(1)+SN(2)*SN(2)-SN(1)*SN(2)+
     .      0.75*(SN(3)*SN(3)+SN(4)*SN(4)+SN(5)*SN(5)))
      GESH1=ALPHA*(SN(1)+SN(2))*0.5
      GISH1=SQRT(GESH1*GESH1+GASH1)
      CRUSH=GESH1+GISH1
      RETURN
      END
```

4.5 Subroutine MODUL

This subroutine evaluates the elasticity matrix D for uncracked and cracked concrete defined by (14), (15) and for elastic and elasto-plastic steel. This matrix is first calculated for the material axes and then transformed to the reference axes (local axes).

```
      SUBROUTINE MODUL(DMATX,PROPS,NPROP,MMATS,LPROP,NSTAT,ANGLE,
     .                 STRA1,STRA2,YOUNA,YOUNB)
C*********************************************************************
C
C***CALCULATES THE ELASTICITY MATRIX -DMATX- IN THE LOCAL AXIS SYSTEM
C
C*********************************************************************
      DIMENSION DMATX(5,5),PROPS(MMATS,NPROP),GASHM(5,5),
     .          TRAMX(5,5)
      DO 10 IGASH=1,5
      DO 10 JGASH=1,5
   10 DMATX(IGASH,JGASH)=0.0
      YOUNG=PROPS(LPROP,1)
      GO TO (1,2,2,1,5,6,7),NSTAT
    1 POISS=PROPS(LPROP,2)
      COEF1=5.0/6.0
      CONS1=YOUNG/(1.0-POISS*POISS)
      CONS2=YOUNG/(2.0*(1.0+POISS))
      DMATX(1,1)=CONS1
      DMATX(2,2)=CONS1
      DMATX(1,2)=POISS*CONS1
      DMATX(2,1)=POISS*CONS1
      DMATX(3,3)=CONS2
      DMATX(4,4)=COEF1*CONS2
      DMATX(5,5)=COEF1*CONS2
      GO TO 100
    2 POISS=PROPS(LPROP,2)
      GMODU=YOUNG/(2.0*(1.0+POISS))
      CONSG=0.25
      GTENS=0.004
      YOUN2=YOUNG
      GMOD2=GMODU*5.0/6.0
      YOUN1=YOUNA
      GMOD1=CONSG*GMODU*(1.0-ABS(STRA1)/GTENS)
      IF(GMOD1.LE.0.0) GMOD1=0.0
      IF(NSTAT.EQ.2) GO TO 35
      YOUN2=YOUNB
      GMOD2=CONSG*GMODU*(1.0-ABS(STRA2)/GTENS)
      IF(GMOD2.LE.0.0) GMOD2=0.0
      GMOD3=GMOD1
      IF(GMOD2.LT.GMOD1) GMOD3=GMOD2
   35 DMATX(1,1)=YOUN1
      DMATX(2,2)=YOUN2
      DMATX(3,3)=GMOD1
      IF(NSTAT.EQ.3) DMATX(3,3)=0.5*GMOD3
      DMATX(4,4)=GMOD1
      DMATX(5,5)=GMOD2
      GO TO 50
    5 DMATX(1,1)=YOUNG
      GO TO 50
    6 DMATX(1,1)=PROPS(LPROP,2)
   50 IF(ABS(ANGLE).LT.0.001) GO TO 100
      CALL      TRANS(TRAMX,ANGLE)
C
C*** CALCULATE DMATX IN THE LOCAL AXIS SYSTEM
C
      DO 60 IGASH=1,5
```

```
      DO 60 JGASH=1,5
      GASHM(IGASH,JGASH)=0.0
      DO 60 KGASH=1,5
   60 GASHM(IGASH,JGASH)=GASHM(IGASH,JGASH)+DMATX(IGASH,KGASH)*
     .                   TRAMX(KGASH,JGASH)
      DO 70 IGASH=1,5
      DO 70 JGASH=1,5
      DMATX(IGASH,JGASH)=0.0
      DO 70 KGASH=1,5
   70 DMATX(IGASH,JGASH)=DMATX(IGASH,JGASH)+TRAMX(KGASH,IGASH)*
     .                   GASHM(KGASH,JGASH)
    7 CONTINUE
  100 RETURN
      END
```

4.6 Subroutine PRIST

This subroutine evaluates the two principal stresses in the structural plane (x'y'). The maximum principal stress is employed in defining the tensile fracture (cracking) plane.

```
      SUBROUTINE PRIST(S,SGMAX,SGMIN)
C***********************************************************************
C
C*** THIS SUBROUTINE CALCULATES THE  PRINCIPAL STRESSES IN STRUCTURAL PL
C
C***********************************************************************
      DIMENSION S(5)
      GASH1=(S(1)+S(2))*0.5
      GASH2=(S(1)-S(2))*0.5
      GASH3=SQRT(GASH2*GASH2+S(3)*S(3))
      SGMAX=GASH1+GASH3
      SGMIN=S(1)+S(2)-SGMAX
      RETURN
      END
```

4.7 Subroutine RESI1

This subroutine calculates the residual stresses for compression concrete to satisfy the yield condition.

```
      SUBROUTINE RESI1(STRSG,MSTAT,EFFST,EPSTN,PROPS,STRES,
     .                 SIGMA,TSTRN,NSTRE,MELGP,NPROP,MMATS,
     .                 LPROP,KGAUS,KUNLO,DESIG,DIREC,IELEM)
C***********************************************************************
C
C*** THIS SUBROUTINE CALCULATES THE RESIDUAL STRESSES FOR
C    NON-CRACKED CONCRETE
C
C***********************************************************************
      DIMENSION STRSG(NSTRE,MELGP),MSTAT(MELGP),EFFST(MELGP),
     .          EPSTN(MELGP,2),PROPS(MMATS,NPROP),STRES(5),
     .          SIGMA(5),TSTRN(5),SGTOT(5),AVECT(5),DVECT(5),
     .          DESIG(5),DIREC(2,MELGP)
      NSTAT=MSTAT(KGAUS)
      IF(NSTAT.NE.2) GO TO 10
      GASH1=EPSTN(KGAUS,1)
      EPSTN(KGAUS,1)=EPSTN(KGAUS,2)
      GASH2=EFFST(KGAUS)
      EFFST(KGAUS)=DIREC(2,KGAUS)
   10 CONTINUE
```

```
      IF(NSTAT.EQ.7) GO TO 160
      UNIAX=PROPS(LPROP,6)
      UNSTN=PROPS(LPROP,7)
      HARDS=0.0
      YOUNG=PROPS(LPROP,1)
      EPSTN(KGAUS,1)=ABS(EPSTN(KGAUS,1))
      CALL       INVA2(TSTRN,CRUSH)
      IF(NSTAT.EQ.2) CRUSH=EFFST(KGAUS)/YOUNG+EPSTN(KGAUS,1)
      IF(CRUSH.GE.UNSTN) GO TO 160
C
C*** REDUCE STRESSES TO THE YIELD SURFACE FOR YIELDED GAUSS POINTS
C
      PREYS=UNIAX+EPSTN(KGAUS,1)*HARDS
      CALL       INVAR(SIGMA,YIELD)
      ESPRE=EFFST(KGAUS)-PREYS
      IF(ESPRE.GE.0.0) GO TO 55
      ESCUR=YIELD-PREYS
      IF(ESCUR.LE.0.0) GO TO 60
      RFACT=ESCUR/(YIELD-EFFST(KGAUS))
      GO TO 70
   55 ESCUR=YIELD-EFFST(KGAUS)
      IF(ESCUR.LE.0.0) GO TO 60
      RFACT=1.0
   70 MSTEP=ESCUR*8.0/UNIAX+1.0
      ASTEP=MSTEP
      REDUC=1.0-RFACT
      DO 80 ISTR1=1,NSTRE
      SGTOT(ISTR1)=STRSG(ISTR1,KGAUS)+REDUC*STRES(ISTR1)
   80 STRES(ISTR1)=RFACT*STRES(ISTR1)/ASTEP
      DO 90 ISTEP=1,MSTEP
      CALL       INVAR(SGTOT,YIELD)
      CALL       FLOWS(ABETA,AVECT,DVECT,LPROP,
     .                 MMATS,NPROP,PROPS,SGTOT)
      AGASH=0.0
      DO 100 ISTR1=1,NSTRE
  100 AGASH=AGASH+AVECT(ISTR1)*STRES(ISTR1)
      DLAMD=AGASH*ABETA
      IF(DLAMD.LT.0.0) DLAMD=0.0
      BGASH=0.0
      DO 110 ISTR1=1,NSTRE
      BGASH=BGASH+AVECT(ISTR1)*SGTOT(ISTR1)
  110 SGTOT(ISTR1)=SGTOT(ISTR1)+STRES(ISTR1)-DLAMD*DVECT(ISTR1)
      EPSTN(KGAUS,1)=EPSTN(KGAUS,1)+DLAMD*BGASH/YIELD
   90 CONTINUE
      CALL       INVAR(SGTOT,YIELD)
      CURYS=UNIAX+EPSTN(KGAUS,1)*HARDS
      BRING=CURYS/YIELD
      DO 130 ISTR1=1,NSTRE
  130 STRSG(ISTR1,KGAUS)=BRING*SGTOT(ISTR1)
      EFFST(KGAUS)=CURYS
      MSTAT(KGAUS)=4
C*** ALTERNATIVE LOCATION OF STRESS REDUCTION LOOP TERMINATION CARD
C  90 CONTINUE
C***
      GO TO 190
  160 MSTAT(KGAUS)=7
      WRITE(6,900) IELEM,KGAUS,(TSTRN(I),I=1,5)
  900 FORMAT(//,10H CRUSHED  ,8H IELEM =,I5,5X,8H KGAUS =,I5,/,
     .        10X,8HSTRAINS ,5E16.8)
      DO 165 ISTRE=1,NSTRE
  165 STRSG(ISTRE,KGAUS)=0.0
      EFFST(KGAUS)=0.0
      GO TO 190
   60 DO 180 ISTR1=1,NSTRE
  180 STRSG(ISTR1,KGAUS)=STRSG(ISTR1,KGAUS)+DESIG(ISTR1)
      EFFST(KGAUS)=YIELD
      IF(EPSTN(KGAUS,1).EQ.0.0.OR.ESCUR.EQ.0.0) GO TO 190
      EPSTN(KGAUS,1)=-EPSTN(KGAUS,1)
      KUNLO=KUNLO+1
  190 CONTINUE
      IF(NSTAT.NE.2) GO TO 200
      MSTAT(KGAUS)=NSTAT
      EPSTN(KGAUS,2)=EPSTN(KGAUS,1)
      EPSTN(KGAUS,1)=GASH1
      DIREC(2,KGAUS)=EFFST(KGAUS)
      EFFST(KGAUS)=GASH2
  200 CONTINUE
      RETURN
      END
```

4.8 Subroutine RESI2

This subroutine calculates the residual stresses for cracked concrete according to the tension stiffening model.

```
      SUBROUTINE RESI2(PROPS,MSTAT,DIREC,EPSTN,EFFST,                   RES2
     .                 SIG,TSTRN,KGAUS,LPROP,NPROP,                     RES2
     .                 MMATS,MELGP,NSTRE,SGMAX,STRSG,                   RES2
     .                 GRTST,KUNLO,YIELD)                               RES2
C***********************************************************************RES2
C                                                                       RES2
C*** THIS SUBROUTINE CALCULATES THE RESIDUAL STRESSES FOR CRACKED CONCRERES2
C                                                                       RES2
C***********************************************************************RES2
      DIMENSION PROPS(MMATS,NPROP),MSTAT(MELGP),DIREC(2,MELGP),         RES2
     .          EPSTN(MELGP,2),EFFST(MELGP),SIG(5),TSTRN(5),STRNP(5),   RES2
     .          SGTOT(5),STRSP(2),DMATX(5,5),TRAMX(5,5),                RES2
     .          STRSG(NSTRE,MELGP),GRTST(MELGP,2)                       RES2
      NSTAT=MSTAT(KGAUS)                                                RES2
      IF(NSTAT.EQ.2.OR.NSTAT.EQ.3) GO TO 30                             RES2
      IF(NSTAT.EQ.1) GO TO 5                                            RES2
C**   NSTAT=4 AND FOR THE FIRT TIME SGMAX.GT.TENST                      RES2
      EPSTN(KGAUS,2)=EPSTN(KGAUS,1)                                     RES2
      DIREC(2,KGAUS)=EFFST(KGAUS)                                       RES2
    5 EPSTN(KGAUS,1)=0.0                                                RES2
      IF(DIREC(1,KGAUS).NE.0.0) GO TO 25                                RES2
C**   SET UP THE ANGLE -DIREC(1,-)- BETWEEN THE MAXIMUM PRINCIPAL DIRECTIRES2
C     AND X-AXIS,AND THE ANGLE -DIREC(2,-)- BETWEEN THE CRACKED PLANE ANDRES2
      GASH1=SGMAX-SIG(2)                                                RES2
      IF(GASH1.EQ.0.0) GASH1=0.1E-20                                    RES2
      DIREC(1,KGAUS)=ATAN(SIG(3)/GASH1)                                 RES2
   25 NSTAT=2                                                           RES2
   30 ANGLE=DIREC(1,KGAUS)                                              RES2
      CALL        TRANS(TRAMX,ANGLE)                                    RES2
C**   CALCULATE THE TWO NORMAL STRESSES IN THE CRACKING SYSTEM          RES2
      C=COS(ANGLE)                                                      RES2
      S=SIN(ANGLE)                                                      RES2
      STRSP(1)=C*C*SIG(1)+S*S*SIG(2)+2.0*C*S*SIG(3)                     RES2
      STRSP(2)=SIG(1)+SIG(2)-STRSP(1)                                   RES2
      TENST=PROPS(LPROP,5)                                              RES2
      UNIAX=PROPS(LPROP,6)                                              RES2
      IF(NSTAT.EQ.2.AND.STRSP(2).GE.TENST) NSTAT=3                      RES2
C**   CALCULATE THE STRAIN COMPONENTS IN CRACKED SYSTEM                 RES2
      DO 40 ISTRE=1,NSTRE                                               RES2
      STRNP(ISTRE)=0.0                                                  RES2
      DO 40 JSTRE=1,NSTRE                                               RES2
   40 STRNP(ISTRE)=STRNP(ISTRE)+TRAMX(ISTRE,JSTRE)*TSTRN(JSTRE)         RES2
C**   EVALUATE THE -DMATX- IN CRACKED SYSTEM                            RES2
      CONSE=PROPS(LPROP,9)                                              RES2
      TENSN=PROPS(LPROP,8)                                              RES2
      STRA1=STRNP(1)                                                    RES2
      STRA2=STRNP(2)                                                    RES2
      THETA=0.0                                                         RES2
      CALL        MODUL(DMATX,PROPS,NPROP,MMATS,LPROP,NSTAT,THETA,      RES2
     .                  STRA1,STRA2,  0.0,  0.0)                        RES2
C**   CALCULATE THE STRESS COMPONENTS FOR CRACKED SYSTEM                RES2
      DO 50 ISTRE=1,NSTRE                                               RES2
      SGTOT(ISTRE)=0.0                                                  RES2
      DO 50 JSTRE=1,NSTRE                                               RES2
   50 SGTOT(ISTRE)=SGTOT(ISTRE)+DMATX(ISTRE,JSTRE)*STRNP(JSTRE)         RES2
C**   CALCULATE NORMAL STRESSES ACCOUNTING FOR TENSION STIFFENING       RES2
      IF(STRA1.GE.TENSN) GO TO 52                                       RES2
      IF(ABS(STRA1).GE.GRTST(KGAUS,1)) GO TO 51                         RES2
      GASH1=CONSE*TENST*(1.0-GRTST(KGAUS,1)/TENSN)                      RES2
      IF(GASH1.LT.0.0) GASH1=0.0                                        RES2
      SGTOT(1)=GASH1*STRA1/GRTST(KGAUS,1)                               RES2
      GO TO 52                                                          RES2
   51 SGTOT(1)=CONSE*TENST*(1.0-STRA1/TENSN)                            RES2
   52 IF(NSTAT.EQ.2) GO TO 55                                           RES2
      IF(STRA2.GE.TENSN) GO TO 55                                       RES2
      IF(ABS(STRA2).GE.GRTST(KGAUS,2)) GO TO 54                         RES2
      GASH2=CONSE*TENST*(1.0-GRTST(KGAUS,2)/TENSN)                      RES2
      IF(GASH2.LT.0.0) GASH2=0.0                                        RES2
      SGTOT(2)=GASH2*STRA2/GRTST(KGAUS,2)                               RES2
      GO TO 55                                                          RES2
```

```
   54 SGTOT(2)=CONSE*TENST*(1.0-STRA2/TENSN)
   55 CONTINUE
C**  ANALYSE IF THE STIFFNESS MATRIX MUST BE RECALCULATED
      IF(NSTAT.EQ.2) GO TO 57
      IF(EPSTN(KGAUS,2).LT.GRTST(KGAUS,2)) GO TO 57
      IF(STRA2.LT.EPSTN(KGAUS,2)) KUNLO=KUNLO+1
   57 IF(EPSTN(KGAUS,1).LT.GRTST(KGAUS,1)) GO TO 60
      IF(STRA1.LT.EPSTN(KGAUS,1)) KUNLO=KUNLO+1
   60 CONTINUE
C**  SET UP THE MATRIX OF MAXIMUM TENSILE STRAINS
      IF(STRA1.GT.GRTST(KGAUS,1)) GRTST(KGAUS,1)=STRA1
      IF(NSTAT.EQ.3.AND.STRA2.GT.GRTST(KGAUS,2)) GRTST(KGAUS,2)=STRA2
C**  STORE THE ACTUAL CRACKING STRAINS
      EPSTN(KGAUS,1)=STRA1
      IF(NSTAT.EQ.3) EPSTN(KGAUS,2)=STRA2
      EFFST(KGAUS)=SGTOT(1)
C
C**  IF THE CRACK CLOSES CHANGE THE MATERIAL STATE NUMBER
C
      IF(NSTAT.EQ.2) GO TO 72
      IF(STRA2.GT.0.0) GO TO 72
C**  NSTAT=3 AND CRACK CLOSES IN THE 2-DIRECTION
      EPSTN(KGAUS,2)=0.0
      NSTAT=2
      KUNLO=KUNLO+1
   72 IF(STRA1.GT.0.0) GO TO 76
      IF(NSTAT.EQ.2) GO TO 74
C**  NSTAT=3 AND CRACK CLOSES IN THE 1-DIR.;CHANGE DIRECTIONS 1 AND 2
      EPSTN(KGAUS,1)=EPSTN(KGAUS,2)
      EPSTN(KGAUS,2)=0.0
      GAMMA=1.570796327
      IF(DIREC(1,KGAUS).GE.0.0) GAMMA=-GAMMA
      DIREC(1,KGAUS)=DIREC(1,KGAUS)+GAMMA
      NSTAT=2
      KUNLO=KUNLO+1
      GO TO 76
   74 EPSTN(KGAUS,1)=0.0
      NSTAT=1
      KUNLO=KUNLO+1
      IF(EPSTN(KGAUS,2).EQ.0.0) GO TO 76
C**  THEN IT IS A YIELD POINT;SEE LABEL 10 OF RESI1
      EPSTN(KGAUS,1)=EPSTN(KGAUS,2)
      EFFST(KGAUS)=DIREC(2,KGAUS)
   76 CONTINUE
      MSTAT(KGAUS)=NSTAT
C**  EVALUATE STRESS COMPONENTS IN LOCAL SYSTEM
      DO 70 ISTRE=1,NSTRE
      SIG(ISTRE)=0.0
      DO 70 JSTRE=1,NSTRE
   70 SIG(ISTRE)=SIG(ISTRE)+TRAMX(JSTRE,ISTRE)*SGTOT(JSTRE)
      CALL      INVAR(SIG,YIELD)
      IF(YIELD.GT.UNIAX) GO TO 90
      DO 80 ISTRE=1,NSTRE
   80 STRSG(ISTRE,KGAUS)=SIG(ISTRE)
   90 CONTINUE
      RETURN
      END
```

4.9 Subroutine RESI3

This subroutine evaluates the residual stresses for steel layers.

```
      SUBROUTINE RESI3(STRSG,MSTAT,DIREC,EFFST,EPSTN,PROPS,
     .                 STRES,KGAUS,MELGP,NSTRE,NPROP,MMATS,
     .                 LPROP,KUNLO)
C*********************************************************************
C
C***THIS SUBROUTINE EVALUATES THE RESIDUAL STRESSES FOR STEEL LAYERS
C
C*********************************************************************
      DIMENSION STRSG(NSTRE,MELGP),MSTAT(MELGP),DIREC(2,MELGP),
     .          EFFST(MELGP),EPSTN(MELGP,2),PROPS(MMATS,NPROP),
     .          STRES(5)
      ANGLE=DIREC(1,KGAUS)
      YOUNG=PROPS(LPROP,1)
      YOUN2=PROPS(LPROP,2)
      HARDS=YOUNG*YOUN2/(YOUNG-YOUN2)
      YIELD=PROPS(LPROP,5)
```

```
      CONS1=COS(ANGLE)*COS(ANGLE)
      CONS2=SIN(ANGLE)*SIN(ANGLE)
      CONS3=2.0*COS(ANGLE)*SIN(ANGLE)
      STREP=CONS1*STRES(1)+CONS2*STRES(2)+CONS3*STRES(3)
      XGASH=STREP
      STRPP=CONS1*STRSG(1,KGAUS)+CONS2*STRSG(2,KGAUS)+
     .      CONS3*STRSG(3,KGAUS)
      STOTP=STRPP+STREP
      PREYS=YIELD+HARDS*ABS(EPSTN(KGAUS,1))
      IF(ABS(STRPP).GE.PREYS) GO TO 20
      ESCUR=ABS(STOTP)-PREYS
      IF(ESCUR.LE.0.0) GO TO 40
      RFACT=ESCUR/ABS(STREP)
      GO TO 30
   20 IF(STOTP.GT.0.0.AND.STREP.LE.0.0) GO TO 35
      IF(STOTP.LT.0.0.AND.STREP.GT.0.0) GO TO 35
      RFACT=1.0
   30 REDUC=1.0-RFACT
      STRPP=STRPP+REDUC*STREP+RFACT*XGASH*(1.0-YOUNG/
     .      (YOUNG+HARDS))
      EPSTN(KGAUS,1)=EPSTN(KGAUS,1)+RFACT*XGASH/(YOUNG+HARDS)
      MSTAT(KGAUS)=6
      EPSTN(KGAUS,2)=0.0
      GO TO 50
   35 MSTAT(KGAUS)=5
      KUNLO=KUNLO+1
   40 STRPP=STRPP+STREP
   50 STRSG(1,KGAUS)=CONS1*STRPP
      STRSG(2,KGAUS)=CONS2*STRPP
      STRSG(3,KGAUS)=0.5*CONS3*STRPP
      STRSG(4,KGAUS)=0.0
      STRSG(5,KGAUS)=0.0
      EFFST(KGAUS)=STRPP
      RETURN
      END
```

4.10 Subroutine RESTR

This subroutine monitors the calling of the three previous subroutines (Sections 4.7, 4.8 and 4.9) and evaluates the nodal forces which are statically equivalent to the (residual) stress field.

```
      SUBROUTINE RESTR(ASDIS,EFFST,ELOAD,LNODS,
     .                 MATNO,MELEM,MMATS,MPOIN,MELGP,MTOTV,
     .                 NDOFN,NELEM,NEVAB,NGAUS,NNODE,
     .                 NPROP,NSTRE,POSGP,PROPS,STRSG,
     .                 TDISP,WEIGP,EPSTN,KUNLO,MACON,MASTE,
     .                 THICK,MLAYR,NCLAY,NSLAY,DIREC,
     .                 MSTAT,NRTAP,NWTAP,LARGE,GRTST)
C**********************************************************************
C
C*** THIS SUBROUTINE REDUCES THE STRESSES AND
C    EVALUATES THE EQUIVALENTE NODAL FORCES
C
C**********************************************************************
      DIMENSION ASDIS(MTOTV),NSLAY(MLAYR),BMATX(5,45),
     .          DMATX(5,5),NCLAY(MLAYR),EFFST(MELGP),ELDIS(45),
     .          ELOAD(MELEM,NEVAB),GVECT(5),LNODS(MELEM,9),
     .          MATNO(MELEM),POSGP(5),PROPS(MMATS,NPROP),
     .          DESIG(5),SIGMA(5),GMATX(2,45),
     .          STRES(5),EPSTN(MELGP,2),TDISP(MTOTV),THICK(MPOIN),
     .          STRSG(5,MELGP),WEIGP(5),ETDIS(45),
     .          MACON(MLAYR,10),TSTRN(5),
     .          MASTE(MLAYR,4),DIREC(2,MELGP),MSTAT(MELGP),
     .          GRTST(MELGP,2),BLARG(5,45)
      REWIND NRTAP
      REWIND NWTAP
      REWIND 9
```

```
      DO 5 IELEM=1,NELEM
      DO 5 IEVAB=1,NEVAB
    5 ELOAD(IELEM,IEVAB)=0.0
      KUNLO=0
C
C*** LOOP OVER EACH ELEMENT
C
      DO 210 IELEM=1,NELEM
      LLAYR=MATNO(IELEM)
      NCONL=NCLAY(LLAYR)
      NSTEL=NSLAY(LLAYR)
      READ(NRTAP) MSTAT,DIREC,EFFST,EPSTN,STRSG,GRTST
C
C*** IDENTIFY THE DISPLACEMENTS OF THE ELEMENT NODAL POINTS
C
      JPOSI=0
      DO 10 INODE=1,NNODE
      LNODE=IABS(LNODS(IELEM,INODE))
      NPOSN=(LNODE-1)*NDOFN
      DO 10 IDOFN=1,NDOFN
      NPOSN=NPOSN+1
      JPOSI=JPOSI+1
      ELDIS(JPOSI)=ASDIS(NPOSN)
      ETDIS(JPOSI)=TDISP(NPOSN)
   10 CONTINUE
      KGAUS=0
      DO 15 I=4,5
      DO 15 IEVAB=1,NEVAB
   15 BLARG(I,IEVAB)=0.0
C
C*** ENTER LOOPS OVER EACH SAMPLING POINTS
C
      DO 205 IGAUS=1,NGAUS
      DO 205 JGAUS=1,NGAUS
C
C*** EVALUATE THE STRESSES ON CONCRETE LAYERS
C
      DO 200 ICONL=1,NCONL
      LPROP=MACON(LLAYR,ICONL)
      KGAUS=KGAUS+1
      READ(9) BMATX,GMATX,DVOLU
C
C*** CALL SUBROUTINE WHICH SETS UP -BMATX- TAKING INTO ACCOUT
C    THE GEOMETRIC NONLINEARITY
C
      IF(LARGE.EQ.1)
     .CALL        LDISP(BMATX,GMATX,ETDIS,NEVAB,BLARG)
C
C*** NOW PROCEED TO CALCULATE STRESSES FROM STRES=DMATX*BMATX*ELDIS
C    FIRST STORE IN GASH VECTOR GVECT THE PRODUCT BMATX*ELDIS
C
      DO 30 IDOFN=1,NDOFN
      GASH=0.0
      TSTRN(IDOFN)=0.0
      DO 25 IEVAB=1,NEVAB
      TSTRN(IDOFN)=TSTRN(IDOFN)+(BMATX(IDOFN,IEVAB)-0.5*
     .              BLARG(IDOFN,IEVAB))*ETDIS(IEVAB)
   25 GASH=GASH+BMATX(IDOFN,IEVAB)*ELDIS(IEVAB)
   30 GVECT(IDOFN)=GASH
C
C*** CALCULATE THE FIVE LOCAL STRESSES IN THE ORDER XX,YY,XY,XZ,YZ
C
      NSTAT=MSTAT(KGAUS)
      ANGLE=DIREC(1,KGAUS)
      STRA1=EPSTN(KGAUS,1)
      STRA2=EPSTN(KGAUS,2)
      CALL        MODUL(DMATX,PROPS,NPROP,MMATS,LPROP,NSTAT,ANGLE,
     .                  STRA1,STRA2,  0.0,  0.0)
      DO 50 ISTRE=1,NSTRE
      GASH=0.0
      DO 45 JSTRE=1,NSTRE
   45 GASH=GASH+DMATX(ISTRE,JSTRE)*GVECT(JSTRE)
   50 STRES(ISTRE)=GASH
      DO 60 ISTRE=1,NSTRE
      DESIG(ISTRE)=STRES(ISTRE)
   60 SIGMA(ISTRE)=STRSG(ISTRE,KGAUS)+STRES(ISTRE)
      CALL        PRIST(SIGMA,SGMAX,SGMIN)
      TENST=PROPS(LPROP,5)
      UNIAX=PROPS(LPROP,6)
```

```
      IF(SGMAX.GT.TENST) GO TO 90
      IF(NSTAT.EQ.1.OR.NSTAT.EQ.4) GO TO 95
      IF(NSTAT.EQ.7) GO TO 95
   90 CALL        RESI2(PROPS,MSTAT,DIREC,EPSTN,EFFST,
     .                  SIGMA,TSTRN,KGAUS,LPROP,NPROP,
     .                  MMATS,MELGP,NSTRE,SGMAX,STRSG,
     .                  GRTST,KUNLO,YIELD)
      IF(YIELD.LE.UNIAX) GO TO 100
      DO 92 ISTRE=1,NSTRE
      STRES(ISTRE)=SIGMA(ISTRE)-STRSG(ISTRE,KGAUS)
   92 DESIG(ISTRE)=STRES(ISTRE)
   95 CALL        RESI1(STRSG,MSTAT,EFFST,EPSTN,PROPS,STRES,
     .                  SIGMA,TSTRN,NSTRE,MELGP,NPROP,MMATS,
     .                  LPROP,KGAUS,KUNLO,DESIG,DIREC,IELEM)
  100 CONTINUE
C
C*** CALCULATE THE EQUIVALENT NODAL FORCES AND ASSOCIATE WITH THE
C    ELEMENT NODES
      MGASH=0
      DO 140 INODE=1,NNODE
      DO 140 IDOFN=1,NDOFN
      MGASH=MGASH+1
      DO 140 ISTRE=1,NSTRE
  140 ELOAD(IELEM,MGASH)=ELOAD(IELEM,MGASH)+BMATX(ISTRE,MGASH)*
     .STRSG(ISTRE,KGAUS)*DVOLU
  200 CONTINUE
C
C*** EVALUATE THE STRESSES IN STEEL LAYERS
C
      DO 180 ISTEL=1,NSTEL
      KGAUS=KGAUS+1
      LPROP=MASTE(LLAYR,ISTEL)
      NSTAT=MSTAT(KGAUS)
      ANGLE=DIREC(1,KGAUS)
      READ(9) BMATX,GMATX,DVOLU
C
C*** CALL SUBROUTINE WHICH SETS UP -BMATX- TAKING INTO ACCOUT
C    THE GEOMETRIC NONLINEARITY
C
      IF(LARGE.EQ.1)
     .CALL        LDISP(BMATX,GMATX,ETDIS,NEVAB,BLARG)
C
C*** NOW PROCEED TO CALCULATE STRESSES FROM STRES=DMATX*BMATX*ELDIS
C    FIRST STORE IN GASH VECTOR GVECT THE PRODUCT BMATX*ELDIS
C
      DO 70 IDOFN=1,NDOFN
      GASH=0.0
      DO 65 IEVAB=1,NEVAB
   65 GASH=GASH+BMATX(IDOFN,IEVAB)*ELDIS(IEVAB)
   70 GVECT(IDOFN)=GASH
C
C*** CALCULATE THE FIVE LOCAL STRESSES IN THE ORDER XX,YY,XY,XZ,YZ
C
      CALL        MODUL(DMATX,PROPS,NPROP,MMATS,LPROP,    5,ANGLE,
     .                  STRA1,STRA2,  0.0,  0.0)
      DO 80 ISTRE=1,NSTRE
      GASH=0.0
      DO 75 JSTRE=1,NSTRE
   75 GASH=GASH+DMATX(ISTRE,JSTRE)*GVECT(JSTRE)
   80 STRES(ISTRE)=GASH
  915 FORMAT(8H KGAUS =,I5,5X,I5,E12.4,5X,5E12.4)
      CALL        RESI3(STRSG,MSTAT,DIREC,EFFST,EPSTN,PROPS,
     .                  STRES,KGAUS,MELGP,NSTRE,NPROP,MMATS,
     .                  LPROP,KUNLO)
C
C*** CALCULATE THE EQUIVALENT NODAL FORCES AND ASSOCIATE WITH THE
C    ELEMENT NODES
      MGASH=0
      DO 150 INODE=1,NNODE
      DO 150 IDOFN=1,NDOFN
      MGASH=MGASH+1
      DO 150 ISTRE=1,NSTRE
  150 ELOAD(IELEM,MGASH)=ELOAD(IELEM,MGASH)+BMATX(ISTRE,MGASH)*
     .STRSG(ISTRE,KGAUS)*DVOLU
  180 CONTINUE
  205 CONTINUE
      WRITE(NWTAP) MSTAT,DIREC,EFFST,EPSTN,STRSG,GRTST
  210 CONTINUE
```

```
      NTAPE=NRTAP
      NRTAP=NWTAP
      NWTAP=NTAPE
      RETURN
      END
```

4.11 Subroutine STIFF

This subroutine evaluates the stiffness matrix for each element in turn. Large deformation effects can be considered.

```
      SUBROUTINE STIFF(EPSTN,ESTIF,KITER,LNODS,MATNO,
     .                 MELEM,MEVAB,MMATS,MPOIN,MELGP,NDOFN,
     .                 NELEM,NEVAB,NGAUS,NNODE,NPROP,
     .                 NSTRE,POSGP,PROPS,STRSG,WEIGP,MACON,
     .                 NCLAY,MLAYR,NSLAY,THICK,
     .                 MASTE,DIREC,MSTAT,EFFST,NRTAP,
     .                 TDISP,MTOTV,LARGE,GRTST)
C**********************************************************************
C
C     THIS SUBROUTINE EVALUATES THE STIFFNESS MATRIX
C     FOR EACH ELEMENT IN TURN
C
C**********************************************************************
      DIMENSION BMATX(5,45),DBMAT(5,45),DMATX(5,5),
     .          ESTIF(MEVAB,MEVAB),LNODS(MELEM,9),MATNO(MELEM),
     .          POSGP(5),PROPS(MMATS,NPROP),MACON(MLAYR,10),
     .          STRES(5),THICK(MPOIN),WEIGP(5),EPSTN(MELGP,2),
     .          STRSG(5,MELGP),AVECT(5),DVECT(5),MASTE(MLAYR,4),
     .          DIREC(2,MELGP),MSTAT(MELGP),GMATX(2,45),
     .          EFFST(MELGP),NCLAY(MLAYR),NSLAY(MLAYR),TDISP(MTOTV),
     .          ETDIS(45),GRTST(MELGP,2),BLARG(5,45)
      REWIND 1
      REWIND NRTAP
      REWIND 9
C
C*** LOOP OVER EACH ELEMENT
C
      DO 110 IELEM=1,NELEM
      LLAYR=MATNO(IELEM)
      NCONL=NCLAY(LLAYR)
      NSTEL=NSLAY(LLAYR)
      READ(NRTAP) MSTAT,DIREC,EFFST,EPSTN,STRSG,GRTST
C**  SET UP THE ELEMENT DISPLACEMENT VECTOR -ETDIS-
      JPOSI=0
      DO 10 INODE=1,NNODE
      LNODE=IABS(LNODS(IELEM,INODE))
      NPOSN=(LNODE-1)*NDOFN
      DO 10 IDOFN=1,NDOFN
      NPOSN=NPOSN+1
      JPOSI=JPOSI+1
   10 ETDIS(JPOSI)=TDISP(NPOSN)
C
C*** INITIALIZE THE ELEMENT STIFFNESS MATRIX
C
      DO 20 IEVAB=1,NEVAB
      DO 20 JEVAB=1,NEVAB
   20 ESTIF(IEVAB,JEVAB)=0.0
      KGAUS=0
C
C*** ENTER LOOPS OVER GAUSS POINTS FOR NUMERICAL INTEGRATION
C
      DO 105 IGAUS=1,NGAUS
      DO 105 JGAUS=1,NGAUS
C
C*** CALCULATE THE CONCRETE CONTRIBUTION TO THE STIFFENESS MATRIX
C
      DO 100 ICONL=1,NCONL
      LPROP=MACON(LLAYR,ICONL)
      KGAUS=KGAUS+1
      NSTAT=MSTAT(KGAUS)
      ANGLE=DIREC(1,KGAUS)
      CONSE=PROPS(LPROP,4)
      TENST=PROPS(LPROP,5)
```

```
      TENSN=PROPS(LPROP,8)
      READ(9) BMATX,GMATX,DVOLU
C**   CALL SUBROUTINE WHICH SETS UP -BMATX- TAKING INTO ACCOUT
C     THE LARGE DISPLACEMENTS
      IF(LARGE.EQ.1.AND.KITER.GT.2)
     .CALL        LDISP(BMATX,GMATX,ETDIS,NEVAB,BLARG)
      STRA1=EPSTN(KGAUS,1)
      STRA2=EPSTN(KGAUS,2)
      IF(NSTAT.NE.4) GO TO 80
      IF(EPSTN(KGAUS,1).LE.0.0) GO TO 80
C
C*** CALCULATE THE ELASTO-PLASTIC -D- MATRIX
C
      CALL        MODUL(DMATX,PROPS,NPROP,MMATS,LPROP,NSTAT,ANGLE,
     .                  STRA1,STRA2,  0.0,  0.0)
      DO 50 ISTRE=1,NSTRE
   50 STRES(ISTRE)=STRSG(ISTRE,KGAUS)
      CALL        FLOWS(ABETA,AVECT,DVECT,LPROP,
     .                  MMATS,NPROP,PROPS,STRES)
      DO 70 ISTRE=1,NSTRE
      DO 70 JSTRE=1,NSTRE
   70 DMATX(ISTRE,JSTRE)=DMATX(ISTRE,JSTRE)-ABETA*
     .                    DVECT(ISTRE)*DVECT(JSTRE)
C
C*** CALCULATE THE PRODUCT OF  D MATRIX BY  B MATRIX
C
      DO 35 ISTRE=1,NSTRE
      DO 35 IEVAB=1,NEVAB
      DBMAT(ISTRE,IEVAB)=0.0
      DO 35 JSTRE=1,NSTRE
      DBMAT(ISTRE,IEVAB)=DBMAT(ISTRE,IEVAB)+DMATX(ISTRE,JSTRE)*
     .                    BMATX(JSTRE,IEVAB)
   35 CONTINUE
      GO TO 90
   80 CONTINUE
      IF(NSTAT.LT.2.OR.NSTAT.GT.3) GO TO 83
      YOUNA=0.0
      YOUNB=0.0
      IF(NSTAT.EQ.2) GO TO 81
      IF(STRA2.EQ.GRTST(KGAUS,2)) GO TO 81
      IF(GRTST(KGAUS,2).GE.TENSN) GO TO 81
      IF(GRTST(KGAUS,2).EQ.0.0) GO TO 81
      YOUNB=CONSE*TENST*(1.0-GRTST(KGAUS,2)/TENSN)/GRTST(KGAUS,2)
   81 IF(STRA1.EQ.GRTST(KGAUS,1)) GO TO 83
      IF(GRTST(KGAUS,1).GE.TENSN) GO TO 83
      IF(GRTST(KGAUS,1).EQ.0.0) GO TO 83
      YOUNA=CONSE*TENST*(1.0-GRTST(KGAUS,1)/TENSN)/GRTST(KGAUS,1)
   83 CALL        MODUL(DMATX,PROPS,NPROP,MMATS,LPROP,NSTAT,ANGLE,
     .                  STRA1,STRA2,YOUNA,YOUNB)
      DO 85 ISTRE=1,NSTRE
      DO 85 IEVAB=1,NEVAB
      DBMAT(ISTRE,IEVAB)=0.0
      DO 85 JSTRE=1,NSTRE
   85 DBMAT(ISTRE,IEVAB)=DBMAT(ISTRE,IEVAB)+
     .        DMATX(ISTRE,JSTRE)*BMATX(JSTRE,IEVAB)
C
C*** CALCULATE THE ELEMENT STIFFNESS
C
   90 DO 40 IEVAB=1,NEVAB
      DO 40 JEVAB=IEVAB,NEVAB
      DO 40 ISTRE=1,NSTRE
   40 ESTIF(IEVAB,JEVAB)=ESTIF(IEVAB,JEVAB)+BMATX(ISTRE,IEVAB)*
     .                    DBMAT(ISTRE,JEVAB)*DVOLU
      IF(LARGE.EQ.0.OR.KITER.EQ.2) GO TO 98
C
C*** CALL SUBROUTINE WHICH CALCULATES THE GEOMETRIC MATRIX -GEMTX-
C
      CALL        GEOME(ESTIF,GMATX,STRSG,MEVAB,NEVAB,MELGP,
     .                  KGAUS,DVOLU)
   98 CONTINUE
  100 CONTINUE
C
C*** CALCULATE THE STEEL CONTRIBUTION TO THE STIFFENESS MATRIX
C
      DO 130 ISTEL=1,NSTEL
      KGAUS=KGAUS+1
      LPROP=MASTE(LLAYR,ISTEL)
      ANGLE=DIREC(1,KGAUS)
      NSTAT=MSTAT(KGAUS)
```

```
      READ(9) BMATX,GMATX,DVOLU
C**   CALL SUBROUTINE WHICH SETS UP -BMATX- TAKING INTO ACCOUT
C     THE LARGE DISPLACEMENTS
      IF(LARGE.EQ.1.AND.KITER.GT.2)
     .CALL        LDISP(BMATX,GMATX,ETDIS,NEVAB,BLARG)
      CALL        MODUL(DMATX,PROPS,NPROP,MMATS,LPROP,NSTAT,ANGLE,
     .                  STRA1,STRA2,  0.0,  0.0)
      DO 115 ISTRE=1,NSTRE
      DO 115 IEVAB=1,NEVAB
      DBMAT(ISTRE,IEVAB)=0.0
      DO 115 JSTRE=1,NSTRE
  115 DBMAT(ISTRE,IEVAB)=DBMAT(ISTRE,IEVAB)+
     .       DMATX(ISTRE,JSTRE)*BMATX(JSTRE,IEVAB)
      DO 120 IEVAB=1,NEVAB
      DO 120 JEVAB=IEVAB,NEVAB
      DO 120 ISTRE=1,NSTRE
  120 ESTIF(IEVAB,JEVAB)=ESTIF(IEVAB,JEVAB)+BMATX(ISTRE,IEVAB)*
     .                        DBMAT(ISTRE,JEVAB)*DVOLU
      IF(LARGE.EQ.0.OR.KITER.EQ.2) GO TO 97
C
C*** CALL SUBROUTINE WHICH CALCULATES THE GEOMETRIC MATRIX -GEMTX-
C
      CALL        GEOME(ESTIF,GMATX,STRSG,MEVAB,NEVAB,MELGP,
     .                  KGAUS,DVOLU)
   97 CONTINUE
  130 CONTINUE
  105 CONTINUE
C
C*** CONSTRACT THE LOWER TRIANGLE OF THE STIFFNESS MATRIX
C
      DO 60 IEVAB=1,NEVAB
      DO 60 JEVAB=1,NEVAB
      ESTIF(JEVAB,IEVAB)=ESTIF(IEVAB,JEVAB)
   60 CONTINUE
C
C*** STORE THE STIFFNESS MATRIX FOR EACH
C    ELEMENT ON DISC FILE
C
      WRITE(1) ESTIF
  110 CONTINUE
      RETURN
      END
```

4.12 Subroutine TRANS

This subroutine evaluates the transformation matrix which performs the transformation of quantities from the material system of axes to the local (reference) system of axis.

```
      SUBROUTINE TRANS(TRAMX,ANGLE)
C*********************************************************************
C
C**   SET UP THE TRANSFORMATION MATRIX -TRAMX-
C
C*********************************************************************

      DIMENSION TRAMX(5,5)
      DO 10 IGASH=1,5
      DO 10 JGASH=1,5
   10 TRAMX(IGASH,JGASH)=0.0
      C=COS(ANGLE)
      S=SIN(ANGLE)
      TRAMX(1,1)=C*C
      TRAMX(2,2)=C*C
      TRAMX(1,2)=S*S
      TRAMX(2,1)=S*S
      TRAMX(1,3)=C*S
      TRAMX(3,1)=-2.0*C*S
      TRAMX(2,3)=-1.0*C*S
```

```
      TRAMX(3,2)=2.0*C*S
      TRAMX(3,3)=C*C-S*S
      TRAMX(4,4)=C
      TRAMX(4,5)=S
      TRAMX(5,4)=-S
      TRAMX(5,5)=C
      RETURN
      END
```

4.13 Main or Master segment

This segment controls the calling, in order, of the other subroutines. Its other function is to control the iterative process and the incrementing of the applied loads.

The following channel numbers are employed by the program: 5 (input channel), 6 (output channel), 1,2,3,4,7,8,9 (scratch files), 12 (permanent file for problem restart).

```
      PROGRAM CONCR1(INPUT,OUTPUT,TAPE5=INPUT,TAPE6=OUTPUT,
     .              TAPE1,TAPE2,TAPE3,TAPE4,TAPE7,TAPE8,TAPE9,
     .              TAPE12)
C**********************************************************************
C
C     PROGRAM FOR ELASTO-PLASTIC-FRACTURE ANALYSIS OF REINFORCED CONCRETE
C     SHELLS.THIS LAYERED APPROACH USES THE -8- -9- AND HETEROSIS ELEMENT
C     A DRUCKER-PRAGER TYPE YIELD FUNCTION PLUS A MAXIMUM TENSILE STRESS
C     CRITERION IS USED.ACCOUNTING FOR LARGE DISPLACEMENTS AND SELECTIVE INT.
C     RESTART FACILITIES ARE INCLUDED.
C
C**********************************************************************
      DIMENSION ASDIS(500),COORD(100,8),ELOAD(20,45),EQRHS(10),
     .          EQUAT(75,10),EFFST(126),EPSTN(126,2),ESTIF(45,45),
     .          FIXED(500),GLOAD(75),GSTIF(2850),GRAVI(3),
     .          IFFIX(500),LOCEL(45),LNODS(20,9),MATNO(20),
     .          NACVA(75),NAMEV(10),NDEST(45),NDFRO(20),
     .          NOFIX(36),NOUTP(2),NPIVO(10),POSGP(5),THICK(100),
     .          PRESC(36,5),PROPS(10,9),RLOAD(20,45),STFOR(500),
     .          STRSG(5,126),TDISP(500),TLOAD(20,45),TOFOR(500),
     .          TREAC(36,5),VECRV(75),WEIGP(5),DICOS(3,300),
     .          MSTAT(126),DIREC(2,126),MACON(5,10),MASTE(5,4),
     .          NCLAY(5),NSLAY(5),GRTST(126,2)
C
C*** PRESET VARIABLES ASSOCIATED WITH DYNAMIC DIMENSIONING
C
      CALL      DIMEN(MBUFA,MELEM,MEVAB,MFRON,MMATS,MPOIN,MSTIF,
     .                MELGP,MTOTV,MVFIX,NDOFN,NPROP,NSTRE,M3POI,
     .                MLAYR)
C
C*** CALL THE SUBROUTINE WHICH READS MOST OF THE PROBLEM DATA
C
      CALL      INPUT(ANVEL,COORD,GRAVI,IFFIX,LNODS,
     .                MATNO,MFRON,MELEM,MMATS,MPOIN,
     .                MTOTV,MVFIX,NDFRO,NDOFN,NELEM,NCOLA,
     .                NEVAB,NGAUS,NGAUZ,NNODE,MLAYR,
     .                NOFIX,NPOIN,NPROP,NSLAY,NREST,LARGE,
     .                NTOTV,NVFIX,POSGP,PRESC,PROPS,WEIGP,
     .                NALGO,NINCS,MACON,MASTE,NCLAY)
C
C*** CREATE THE THICKNESS AND A LOCAL ORTHOGONAL SET AT EACH NODAL POINT
C
      CALL      WORKS(COORD,DICOS,LNODS,THICK,MELEM,MPOIN,
     .                NPOIN,M3POI)
C
C*** CALL SUBROUTINE WHICH COMPUTES BMATX AND GMATX.THESE MATRICES
C    ARE STORED ON TAPE 9 FOR LATER USAGE
C
```

```
      CALL       BGMAT(COORD,DICOS,LNODS,MATNO,MELEM,
     .                 MLAYR,MMATS,MPOIN,M3POI,NELEM,
     .                 NEVAB,NGAUS,NGAUZ,NCLAY,NNODE,NPROP,
     .                 POSGP,PROPS,THICK,WEIGP,NSLAY,MACON,
     .                 MASTE)
C
C*** CALL THE SUBROUTINE WHICH COMPUTES THE APPLIED LOADS
C    AFTER READING SOME LOADING DATA
C
      CALL       LOADS(ANVEL,COORD,RLOAD,GRAVI,LNODS,
     .                 MATNO,MELEM,MEVAB,MMATS,MPOIN,DICOS,
     .                 NELEM,NEVAB,NGAUS,THICK,
     .                 NNODE,NPROP,NSTRE,POSGP,M3POI,NSLAY,
     .                 PROPS,WEIGP,MLAYR,NCLAY,MACON,MASTE)
C
C*** INITIALISE CERTAIN ARRAYS
C
      CALL       ZERO(EFFST,ELOAD,EPSTN,MELEM,MEVAB,
     .                MELGP,MTOTV,NDOFN,NELEM,NEVAB,NGAUS,
     .                NSTRE,NTOTV,NVFIX,MVFIX,STRSG,MSTAT,
     .                TDISP,TFACT,TLOAD,TREAC,DIREC,MATNO,
     .                NCLAY,NSLAY,MASTE,PROPS,NPROP,MMATS,
     .                MLAYR,NRTAP,NWTAP,GRTST,NREST,KINCS)
C
C*** LOOP OVER EACH INCREMENT
C
      DO 100 IINCS=1,NINCS
C
C*** READ DATA FOR CORRENT INCREMENT
C
      CALL       INCREM(ELOAD,FIXED,IINCS,MELEM,MEVAB,MITER,
     .                  MTOTV,MVFIX,NDOFN,NELEM,NEVAB,NOUTP,
     .                  NOFIX,NTOTV,NVFIX,PRESC,RLOAD,TFACT,
     .                  TLOAD,TOLER,LNODS,IFFIX,NNODE,NCOLA,
     .                  NREST,KINCS)
C
C*** LOOP OVER EACH ITERATION
C
      KSTOP=0
      KUNLO=0
      DO 50 IITER=1,MITER
      KITER=IINCS+IITER
      JINCS=IINCS-KINCS
C
C*** CALL SUBROUTINE WHICH SELECTS SOLUTION ALGORITHM VARIABLE KRESL
C
      CALL       ALGOR(FIXED,KITER,IITER,KRESL,MTOTV,NALGO,
     .                 NTOTV,KUNLO,KINCS)
C
C*** CHECK WHETHER A NEW EVALUATION OF THE STIFFNESS MATRIX
C    IS REQUIRED
C
      IF(KRESL.EQ.1)
     .CALL       STIFF(EPSTN,ESTIF,KITER,LNODS,MATNO,
     .                 MELEM,MEVAB,MMATS,MPOIN,MELGP,NDOFN,
     .                 NELEM,NEVAB,NGAUS,NNODE,NPROP,
     .                 NSTRE,POSGP,PROPS,STRSG,WEIGP,MACON,
     .                 NCLAY,MLAYR,NSLAY,THICK,
     .                 MASTE,DIREC,MSTAT,EFFST,NRTAP,
     .                 TDISP,MTOTV,LARGE,GRTST)
C
C*** MERGE AND SOLVE THE RESULTING EQUATIONS BY THE FRONTAL SOLVER
C
      CALL       FRONT(ASDIS,ELOAD,EQRHS,EQUAT,ESTIF,FIXED,
     .                 GLOAD,GSTIF,IFFIX,JINCS,IITER,KRESL,
     .                 LOCEL,LNODS,MBUFA,MELEM,MEVAB,MFRON,
     .                 MSTIF,MTOTV,MVFIX,NACVA,NAMEV,NDEST,
     .                 NDOFN,NELEM,NEVAB,NNODE,NOFIX,NPIVO,
     .                 NPOIN,NTOTV,TDISP,TLOAD,TREAC,VECRV)
C
C*** CALCULATE RESIDUAL FORCES
C
      CALL       RESTR(ASDIS,EFFST,ELOAD,LNODS,
     .                 MATNO,MELEM,MMATS,MPOIN,MELGP,MTOTV,
     .                 NDOFN,NELEM,NEVAB,NGAUS,NNODE,
     .                 NPROP,NSTRE,POSGP,PROPS,STRSG,
     .                 TDISP,WEIGP,EPSTN,KUNLO,MACON,MASTE,
     .                 THICK,MLAYR,NCLAY,NSLAY,DIREC,
     .                 MSTAT,NRTAP,NWTAP,LARGE,GRTST)
```

```
C
C*** CHECK FOR CONVERGENCE
C
C      CALL       CONVER(ELOAD,IITER,LNODS,MELEM,MEVAB,MTOTV,
C     .                  NCHEK,NDOFN,NELEM,NEVAB,NNODE,NTOTV,
C     .                  STFOR,TLOAD,TOFOR,TOLER)
       CALL       CONVRD(ASDIS,ELOAD,TDISP,TLOAD,IITER,MELEM,
      .                  MEVAB,MTOTV,NCHEK,NDOFN,NELEM,NEVAB,
      .                  NPOIN,TOLER)
C
C*** OUTPUT RESULTS IF REQUIRED
C
       IF(IITER.EQ.1.AND.NOUTP(1).GT.0)
      .CALL       OUTPUT(IITER,MELGP,MTOTV,MVFIX,NCHEK,NELEM,
      .                  NGAUS,NOFIX,NOUTP,NPOIN,
      .                  NSTRE,NVFIX,STRSG,TDISP,TREAC,EPSTN,
      .                  POSGP,EFFST,MATNO,MMATS,PROPS,
      .                  NPROP,MELEM,THICK,
      .                  MPOIN,LNODS,MLAYR,NCLAY,NSLAY,MACON,
      .                  MASTE,MSTAT,DIREC,NRTAP,GRTST)
C
C*** IF SOLUTION HAS CONVERGED STOP ITERATING AND OUTPUT RESULTS
C
       IF(IITER.EQ.1.AND.NCHEK.EQ.0) GO TO 100
       IF(NCHEK.EQ.0) GO TO 75
    50 CONTINUE
C
C***
C
       KSTOP=1
    75 CALL       OUTPUT(IITER,MELGP,MTOTV,MVFIX,NCHEK,NELEM,
      .                  NGAUS,NOFIX,NOUTP,NPOIN,
      .                  NSTRE,NVFIX,STRSG,TDISP,TREAC,EPSTN,
      .                  POSGP,EFFST,MATNO,MMATS,PROPS,
      .                  NPROP,MELEM,THICK,
      .                  MPOIN,LNODS,MLAYR,NCLAY,NSLAY,MACON,
      .                  MASTE,MSTAT,DIREC,NRTAP,GRTST)
       IF(KSTOP.EQ.1) STOP
C
C*** RECORD ONTO TAPE 12 THE DATA NEEDED FOR RESTART THE PROBLEM
C    IN NEXT INCREMENT
C
       CALL       RESTAR(DIREC,EFFST,ELOAD,EPSTN,GRTST,MELEM,
      .                  MELGP,MEVAB,MTOTV,MVFIX,NELEM,MSTAT,
      .                  STRSG,TDISP,TFACT,TLOAD,TREAC,NRTAP,
      .                  IINCS)
   100 CONTINUE
       STOP
       END
```

4.14 Subroutines INPUT, LOADS and OUTPUT

The role of these subroutines has been described in Sections 4.15, 4.18 and 4.22 of the previous Chapters respectively. However, for the present reinforced concrete application small differences exist between the versions presented previously and those required for the current case. Therefore, for completeness, these modified subroutines are listed below.

```
      SUBROUTINE INPUT(ANVEL,COORD,GRAVI,IFFIX,LNODS,
     .                 MATNO,MFRON,MELEM,MMATS,MPOIN,
     .                 MTOTV,MVFIX,NDFRO,NDOFN,NELEM,NCOLA,
     .                 NEVAB,NGAUS,NGAUZ,NNODE,MLAYR,
     .                 NOFIX,NPOIN,NPROP,NSLAY,NREST,LARGE,
     .                 NTOTV,NVFIX,POSGP,PRESC,PROPS,WEIGP,
     .                 NALGO,NINCS,MACON,MASTE,NCLAY)
C*********************************************************************
C
C**** THIS SUBROUTINE ACCEPTS MOST OF THE INPUT DATA
C
C*********************************************************************
      DIMENSION COORD(MPOIN,8),IFFIX(MTOTV),LNODS(MELEM,9),
     .          MATNO(MELEM),NDFRO(MELEM),GRAVI(3),
     .          NOFIX(MVFIX),POSGP(5),PRESC(MVFIX,NDOFN),
```

```
     .         PROPS(MMATS,NPROP),TITLE(12),WEIGP(5),NSLAY(MLAYR),
     .         MACON(MLAYR,10),MASTE(MLAYR,4),NCLAY(MLAYR)
      READ(5,920)  TITLE
      WRITE(6,920) TITLE
  920 FORMAT(12A6)
C
C*** READ THE FIRST DATA CARD, AND ECHO IT IMMEDIATELY.
C
      READ(5,900) NPOIN,NELEM,NVFIX,NNODE,NMATS,NGAUS,NGAUZ,NCOLA,
     .NALGO,NINCS,NLAYR,LARGE,NREST
  900 FORMAT(16I5,/5X,15I5)
      NEVAB=NDOFN*NNODE
      NTOTV=NPOIN*NDOFN
      WRITE(6,901)NPOIN,NELEM,NVFIX,NNODE,NMATS,NGAUS,NGAUZ,
     .NEVAB,NCOLA,NALGO,NINCS,NLAYR,LARGE,NREST
  901 FORMAT(///,5X,8H NPOIN =,I5/5X,8H NELEM =,I5/5X,8H NVFIX =,I5/
     .        5X,8H NNODE =,I5/5X,8H NMATS =,I5/5X,8H NGAUS =,I5/
     .        5X,8H NGAUZ =,I5/5X,8H NEVAB =,I5/5X,8H NCOLA =,I5/5X,
     .        8H NALGO =,I5/5X,8H NINCS =,I5,/,
     .        5X,8H NLAYR =,I5,/,5X,8H LARGE =,I5,/,5X,8H NREST =,I5)
      WRITE (6,912)
      READ (5,913) GRAVI(1),GRAVI(2),GRAVI(3),ANVEL
      WRITE (6,913) GRAVI(1),GRAVI(2),GRAVI(3),ANVEL
  912 FORMAT(//41H X-GRAVITY  Y-GRAVITY  Z-GRAVITY  ANG VEL/)
  913 FORMAT(4F10.5)
      CALL       CHECK1(NDOFN,NELEM,NGAUS,NMATS,NNODE,NPOIN,
     .                  MMATS,NVFIX,NGAUZ,NLAYR)
C
C*** READ THE ELEMENT NODAL CONNECTIONS, AND THE PROPERTY NUMBERS.
C
      WRITE(6,902)
  902 FORMAT(//8H ELEMENT,5X,8HPROPERTY,10X,12HNODE NUMBERS)
      DO 2 IELEM=1,NELEM
      READ(5,900) NUMEL,MATNO(NUMEL),
     .            (LNODS(NUMEL,INODE),INODE=1,NNODE)
    2 WRITE(6,903) NUMEL,MATNO(NUMEL),(LNODS(NUMEL,INODE),INODE=1,NNODE)
  903 FORMAT(1X,I5,4X,I5,4X,9I5)
C
C*** ZERO ALL THE NODAL COORDINATES, PRIOR TO READING SOME OF THEM.
C
      DO 4 IPOIN=1,NPOIN
      DO 4 IDIME=1,8
    4 COORD(IPOIN,IDIME)=0.0
C
C*** READ SOME NODAL COORDINATES, FINISHING WITH THE LAST NODE OF ALL.
C
      WRITE(6,904)
  904 FORMAT(//5H NODE,8X,1HX,14X,1HY,14X,1HZ,13X,5HPRESS)
    6 READ(5,905) IPOIN,(COORD(IPOIN,IDIME),IDIME=1,8)
  905 FORMAT(I5,4F15.10/5X,4F15.10)
      IF(IPOIN.NE.NPOIN) GO TO 6
C
C*** INTERPOLATE COORDINATES OF MID-SIDE NODES
C
      CALL       NODEX(COORD,LNODS,MELEM,MPOIN,NELEM,NNODE)
      DO 10 IPOIN=1,NPOIN
   10 WRITE(6,906) IPOIN,(COORD(IPOIN,IDIME),IDIME=1,8)
  906 FORMAT(I5,4F15.10/5X,4F15.10)
C
C*** READ THE FIXED VALUES.
C
      WRITE(6,907)
  907 FORMAT(//5H NODE,6X,4HCODE,15X,12HFIXED VALUES)
      DO 8 IVFIX=1,NVFIX
      READ(5,908) NOFIX(IVFIX),IFPRE,(PRESC(IVFIX,IDOFN),IDOFN=1,NDOFN)
      WRITE(6,908) NOFIX(IVFIX),IFPRE,(PRESC(IVFIX,IDOFN),IDOFN=1,NDOFN)
      NLOCA=(NOFIX(IVFIX)-1)*NDOFN
      IFDOF=10**(NDOFN-1)
      DO 8 IDOFN=1,NDOFN
      NGASH=NLOCA+IDOFN
      IF(IFPRE.LT.IFDOF) GO TO 8
      IFFIX(NGASH)=1
      IFPRE=IFPRE-IFDOF
    8 IFDOF=IFDOF/10
  908 FORMAT(1X,I4,5X,I5,5X,5F10.6)
C
C*** READ THE CONCRETE AND THE STEEL DISCRETIZATION PATTERN
C
      WRITE(6,935)
      DO 20 ILAYR=1,NLAYR
      READ(5,900) NCLAY(ILAYR),NSLAY(ILAYR)
      NCONL=NCLAY(ILAYR)
      NSTEL=NSLAY(ILAYR)
```

```
      READ(5,900) (MACON(ILAYR,ICONL),ICONL=1,NCONL),
     .            (MASTE(ILAYR,ISTEL),ISTEL=1,NSTEL)
   20 WRITE(6,925) ILAYR,NCONL,NSTEL,(MACON(ILAYR,ICONL),ICONL
     .             =1,NCONL),(MASTE(ILAYR,ISTEL),ISTEL=1,NSTEL)
  925 FORMAT(I6,5X,I5,5X,I5,5X,15I5)
  935 FORMAT(//11H LAYER PAT.,8H   NCONL,5X,5HNSTEL,5X,
     .        44HMATERIAL NUMBER PER CONCRETE AND STEEL LAYER/)
C
C*** READ THE AVAILABLE SELECTION OF ELEMENT PROPERTIES.
C
      WRITE(6,910)
  910 FORMAT(//7H  NUMAT,10X,18HELEMENT PROPERTIES)
      DO 18 IMATS=1,NMATS
      READ(5,900) NUMAT
      READ(5,930) (PROPS(NUMAT,IPROP),IPROP=1,7)
      READ(5,930) (PROPS(NUMAT,IPROP),IPROP=8,NPROP)
  930 FORMAT(7F10.5)
   18 WRITE(6,911) NUMAT,(PROPS(NUMAT,IPROP),IPROP=1,NPROP)
  911 FORMAT(1X,I4,3X,9E12.4)
C
C*** SET UP GAUSSIAN INTEGRATION CONSTANTS
C
      CALL      GAUSSQ(NGAUS,POSGP,WEIGP)
      CALL      CHECK2(COORD,IFFIX,LNODS,MATNO,MELEM,MFRON,MPOIN,MTOTV,
     .                 MVFIX,NDFRO,NDOFN,NELEM,NNODE,NOFIX,NPOIN,
     .                 NVFIX,NLAYR)
      RETURN
      END

      SUBROUTINE LOADS(ANVEL,COORD,ELOAD,GRAVI,LNODS,
     .                 MATNO,MELEM,MEVAB,MMATS,MPOIN,DICOS,
     .                 NELEM,NEVAB,NGAUS,THICK,
     .                 NNODE,NPROP,NSTRE,POSGP,M3POI,NSLAY,
     .                 PROPS,WEIGP,MLAYR,NCLAY,MACON,MASTE)
C**********************************************************************
C
C*** THIS SUBROUTINE EVALUATES THE NODAL LOADS DUE TO TEMPERATURE,
C    CENTRIFUGAL FORCE, GRAVITY,PRESSURE,AND POINT LOADS
C
C**********************************************************************
      COMMON WORMX(3,24),QVALU,DJACB
      DIMENSION BMATX(5,45),COORD(MPOIN,8),
     .          ELOAD(MELEM,MEVAB),GRAVI(3),LNODS(MELEM,9),
     .          MATNO(MELEM),POSGP(5),DICOS(3,M3POI),
     .          PROPS(MMATS,NPROP),SHAPE(3,9),STREN(5),
     .          THICK(MPOIN),WEIGP(5),
     .          NCLAY(MLAYR),MACON(MLAYR,10),NSLAY(MLAYR),
     .          MASTE(MLAYR,4),GMATX(2,45)
      REWIND 9
C
C*** LOOP OVER EACH ELEMENT
C
      DO 150 IELEM=1,NELEM
      LLAYR=MATNO(IELEM)
      NCONL=NCLAY(LLAYR)
      NSTEL=NSLAY(LLAYR)
      NLAYT=NCONL+NSTEL
C
C*** READ THE CHARACTERISTICS OF THE  U. D. AND POINT LOADS
C
      READ(5,900) NPRES,NUCLO,NBODY
      WRITE(6,901) NPRES,NUCLO,NBODY
  900 FORMAT(5I5)
  901 FORMAT(8H NPRES =,I5,5X,8H NUCLO =,I5,5X,8H NBODY =,I5)
      IF(NPRES.EQ.0) GO TO 3
      READ(5,902) KPRES,CFACE,PREVA,SURFA
      WRITE(6,902) KPRES,CFACE,PREVA,SURFA
  902 FORMAT(I5,F5.1,2F15.5)
C**  CFACE IS +1.0 OR -1.0 ACCORDING AS PRESSURE IS ON TOP OR BOTTOM SUR
    3 CONTINUE
C
C*** INITIALIZE THE LOAD MATRIX ELOAD  ONE COLUMN AT EACH TIME
C
      DO 4 IEVAB=1,NEVAB
    4 ELOAD(IELEM,IEVAB)=0.0
C
C*** ENTER LOOPS OVER GAUSS POINTS FOR NUMERICAL INTEGRATION
C
      DO 145 IGAUS=1,NGAUS
      DO 145 JGAUS=1,NGAUS
      EXISP=POSGP(IGAUS)
      ETASP=POSGP(JGAUS)
```

```
      CALL       SFR1(SHAPE,EXISP,ETASP)
      IF(NBODY.EQ.0) GO TO 141
      ZETSP=-1.0
      DO 140 ILAYT=1,NLAYT
      IF(ILAYT.GT.NCONL) GO TO 6
      LPROP=MACON(LLAYR,ILAYT)
      DZETA=PROPS(LPROP,3)
      ZETSP=ZETSP+DZETA/2.0
      NSTAT=1
      ANGLE=0.0
      GO TO 8
    6 LPROP=MASTE(LLAYR,ILAYT-NCONL)
      ZETSP=PROPS(LPROP,6)
      DZETA=PROPS(LPROP,3)
      NSTAT=5
      ANGLE=PROPS(LPROP,7)
    8 READ(9) BMATX,GMATX,DVOLU
C
C*** CALCULATE THE CENTRIFUGAL,GRAVITATIONAL ,PRESSURE AND POINT LOADS
C
C
C*** CENTRIFUGAL FORCE
C
      IF(ANVEL.EQ.0.0) GO TO 70
      NBORP = 2
      CALL       FUNC(BMATX,SHAPE,THICK,NBORP,NNODE,ZETSP,
     .                MELEM,COORD,DICOS,LNODS,IELEM,MPOIN,
     .                M3POI,GMATX)
      GASH=PROPS(LPROP,4)*ANVEL*ANVEL*DVOLU
      DO 45 I=1,2
   45 STREN(I)=GASH*WORMX(I,1)
      STREN(3)=0.0
      DO 65 INODE=1,NNODE
C**  FIND THE POSITION OF THE  V-1 AND  V-2 VECTORS
      IPOIN=IABS(LNODS(IELEM,INODE))
      JPOSI=(IPOIN-1)*3
      DO 65 ISTRE=1,NSTRE
      IEVAB=(INODE-1)*5+ISTRE
      IF(ISTRE.GT.3) GO TO 50
      ELOAD(IELEM,IEVAB)=ELOAD(IELEM,IEVAB)+STREN(ISTRE)*
     .                          SHAPE(1,INODE)
      GO TO 65
   50 JPOSI=JPOSI+1
      GASH=SHAPE(1,INODE)*(THICK(IPOIN)/2.0)*ZETSP
      IF(ISTRE.NE.5) GO TO 55
      GASH=-GASH
   55 DO 60 I=1,2
   60 ELOAD(IELEM,IEVAB)=ELOAD(IELEM,IEVAB)+STREN(I)*
     .                          DICOS(I,JPOSI)*GASH
   65 CONTINUE
   70 CONTINUE
C
C*** GRAVITY
C
      GASH=PROPS(LPROP,4)*DVOLU
      DO 75 I=1,3
   75 STREN(I)=GRAVI(I)*GASH
      DO 95 INODE=1,NNODE
      IPOIN=IABS(LNODS(IELEM,INODE))
      JPOSI=(IPOIN-1)*3
      DO 95 ISTRE=1,NSTRE
      IEVAB=(INODE-1)*5+ISTRE
      IF(ISTRE.GT.3) GO TO 80
      ELOAD(IELEM,IEVAB)=ELOAD(IELEM,IEVAB)+STREN(ISTRE)*
     .                          SHAPE(1,INODE)
      GO TO 95
   80 JPOSI=JPOSI+1
      GASH=SHAPE(1,INODE)*(THICK(IPOIN)/2.0)*ZETSP
      IF(ISTRE.NE.5) GO TO 85
      GASH=-GASH
   85 DO 90 I=1,3
   90 ELOAD(IELEM,IEVAB)=ELOAD(IELEM,IEVAB)+STREN(I)*
     .                          DICOS(I,JPOSI)*GASH
   95 CONTINUE
      ZETSP=ZETSP+DZETA/2.0
  140 CONTINUE
C
C*** CALCULATE THE NODAL LOADS DUE TO PRESSURE
C
  141 IF(NPRES.EQ.0) GO TO 142
      CALL       PRES(COORD,ELOAD,LNODS,POSGP,SHAPE,THICK,
     .                WEIGP,IELEM,IGAUS,JGAUS,MELEM,MPOIN,NNODE,
     .                NEVAB,KPRES,CFACE,PREVA,SURFA,DICOS,M3POI)
```

```
  142 CONTINUE
  145 CONTINUE
C
C*** POINT LOADS
C
      IF(NUCLO.EQ.0) GO TO 150
C
C*** IS THE PRESENT ELEMENT A LOADED ELEMENT. IF IT IS
C    READ AND ACCUMULATE THE LOADS IN ELOAD
C
      DO 120 IGASP=1,NUCLO
      READ(5,950) LPOIN,LDOFN,CARGA
      WRITE(6,960) LPOIN,LDOFN,CARGA
      IEVAB=(LPOIN-1)*5+LDOFN
  120 ELOAD(IELEM,IEVAB)=ELOAD(IELEM,IEVAB)+CARGA
  150 CONTINUE
  950 FORMAT(2I5,F10.5)
  960 FORMAT(//8H LPOIN =,I5,8H LDOFN =,I5,7H LOAD =,F16.8/)
      RETURN
      END

      SUBROUTINE OUTPUT(IITER,MELGP,MTOTV,MVFIX,NCHEK,NELEM,NGAUS,
     .                  NOFIX,NOUTP,NPOIN,NSTRE,
     .                  NVFIX,STRSG,TDISP,TREAC,EPSTN,POSGP,
     .                  EFFST,MATNO,MMATS,PROPS,NPROP,MELEM,THICK,
     .                  MPOIN,LNODS,MLAYR,NCLAY,
     .                  NSLAY,MACON,MASTE,MSTAT,DIREC,NRTAP,GRTST)
C*********************************************************************
C
C**** THIS SUBROUTINE OUTPUTS DISPLACEMENTS,REACTIONS AND STRESSES
C
C*********************************************************************
      DIMENSION NOFIX(MVFIX),NOUTP(2),STRSG(5,MELGP),STRES(6),
     .          TDISP(MTOTV),TREAC(MVFIX,5),EPSTN(MELGP,2),
     .          POSGP(5),SHAPE(3,9),EFFST(MELGP),
     .          THICK(MPOIN),LNODS(MELEM,9),
     .          FORCE(8),MATNO(MELEM),
     .          PROPS(MMATS,NPROP),NCLAY(MLAYR),NSLAY(MLAYR),
     .          MACON(MLAYR,10),MASTE(MLAYR,4),MSTAT(MELGP),
     .          DIREC(2,MELGP),GRTST(MELGP,2)
      REWIND NRTAP
      KOUTP=NOUTP(1)
      IF(IITER.GT.1) KOUTP=NOUTP(2)
      IF(IITER.EQ.1.AND.NCHEK.EQ.0) KOUTP=NOUTP(2)
C
C*** OUTPUT DISPLACEMENTS
C
      IF(KOUTP.LT.1) GO TO 10
      WRITE(6,900)
  900 FORMAT(1H0,5X,13HDISPLACEMENTS)
      WRITE(6,905)
  905 FORMAT(1H0,6X,4HNODE,4X,6HX-DISP,8X,6HY-DISP,8X,6HZ-DISP,
     .       8X,6HAF-ROT,8X,6HBT-ROT)
      DO 20 IPOIN=1,NPOIN
      NGASH=IPOIN*5
      NGISH=NGASH-4
   20 WRITE(6,910) IPOIN,(TDISP(IGASH),IGASH=NGISH,NGASH)
  910 FORMAT(I10,5E14.6)
   10 CONTINUE
C
C*** OUTPUT REACTIONS
C
      IF(KOUTP.LT.2) GO TO 30
      WRITE(6,920)
  920 FORMAT(1H0,5X,9HREACTIONS)
      WRITE(6,925)
  925 FORMAT(1H0,6X,4HNODE,4X,6HX-REAC,8X,6HY-REAC,8X,6HZ-REAC,
     .       8X,6HAF-MOM,8X,6HBT-MOM)
      DO 40 IVFIX=1,NVFIX
   40 WRITE(6,910) NOFIX(IVFIX),(TREAC(IVFIX,IDOFN),IDOFN=1,5)
   30 CONTINUE
C
C*** OUTPUT STRESSES
C
      IF(KOUTP.LT.3) GO TO 120
      WRITE(6,927)
  927 FORMAT(1H0,5X,8HSTRESSES)
      WRITE(6,926)
  926 FORMAT(1H0,5HKLAYR,3X,6HXX-STR,6X,6HYY-STR,6X,6HXY-STR,
     .       6X,6HXZ-STR,6X,6HYZ-STR,4X,9HEFF.STRES,3X,
     .       8HEFF.STN1,3X,8HEFF.STN2,4X,5HDIR-1,2X,5HDIR-2,2X,4HSTAT)
      DO 110 IELEM=1,NELEM
      KGAUS=0
      KELGS=0
```

```
      READ(NRTAP) MSTAT,DIREC,EFFST,EPSTN,STRSG,GRTST
      LLAYR=MATNO(IELEM)
      NCONL=NCLAY(LLAYR)
      NSTEL=NSLAY(LLAYR)
      WRITE(6,940) IELEM
  940 FORMAT(1H0,50X,13HELEMENT NO. =,I5)
      DO 105 IGAUS=1,NGAUS
      DO 105 JGAUS=1,NGAUS
      EXISP=POSGP(IGAUS)
      ETASP=POSGP(JGAUS)
C
C*** SET TO ZERO THE STRESS RESULTANT VECTOR
C
      DO 70 JFORC=1,8
   70 FORCE(JFORC)=0.0
      KGASP=0
      KELGS=KELGS+1
      WRITE(6,945) KELGS
  945 FORMAT(11H G.P. NO. =,I5)
      CALL      SFR1(SHAPE,EXISP,ETASP)
C
C*** COMPUTE GAUSS POINT THICKNESS
C
      THIGP=0.0
      DO 65 INODE=1,8
      IPOIN=IABS(LNODS(IELEM,INODE))
   65 THIGP=THIGP+SHAPE(1,INODE)*THICK(IPOIN)
C
C*** CONCRETE LAYERS
C
      WRITE(6,947)
  947 FORMAT(/,16H CONCRETE LAYERS)
      ZETSP=-1.0
      DO 100 ICONL=1,NCONL
      LPROP=MACON(LLAYR,ICONL)
      DZETA=PROPS(LPROP,3)
      ZETSP=ZETSP+DZETA/2.0
      KGAUS=KGAUS+1
      KGASP=KGASP+1
      NSTAT=MSTAT(KGAUS)
      ALPHA=DIREC(1,KGAUS)*57.29577951
C     BEETA=DIREC(2,KGAUS)*57.29577951
      BEETA=DIREC(2,KGAUS)
C
C*** THE FIVE LOCAL STRESSES IN THE ORDER  XX,YY,XY,XZ,YZ
C
      DO 50 ISTRE=1,NSTRE
   50 STRES(ISTRE)=STRSG(ISTRE,KGAUS)
      WRITE(6,950) KGASP,(STRES(ISTRE),ISTRE=1,NSTRE),
     .             EFFST(KGAUS),EPSTN(KGAUS,1),EPSTN(KGAUS,2),
     .             ALPHA,BEETA,NSTAT
C
C*** SET UP THE STRESS RESULTANTS IN THE ORDER NX,NY,NXY,MX,MY,MXY,QX,QY
C
      DO 75 ISTRE=1,3
      FORCE(ISTRE)=FORCE(ISTRE)+STRES(ISTRE)*THIGP/2.0*DZETA
   75 FORCE(ISTRE+3)=FORCE(ISTRE+3)-STRES(ISTRE)*THIGP*THIGP*
     .                ZETSP*DZETA/4.0
      DO 80 ISTRE=4,5
   80 FORCE(ISTRE+3)=FORCE(ISTRE+3)+STRES(ISTRE)*THIGP/2.0*DZETA
  950 FORMAT(I4,8E12.4,2F7.2,I4)
      ZETSP=ZETSP+DZETA/2.0
  100 CONTINUE
C
C*** STEEL LAYERS
C
      WRITE(6,975)
  975 FORMAT(13H STEEL LAYERS)
      KGASP=0
      DO 130 ISTEL=1,NSTEL
      LPROP=MASTE(LLAYR,ISTEL)
      DZETA=PROPS(LPROP,3)
      ZETSP=PROPS(LPROP,6)
      KGASP=KGASP+1
      KGAUS=KGAUS+1
      NSTAT=MSTAT(KGAUS)
      ALPHA=DIREC(1,KGAUS)*57.29577951
      BEETA=DIREC(2,KGAUS)*57.29577951
C     BEETA=DIREC(2,KGAUS)
C
C*** THE FIVE LOCAL STRESSES IN THE ORDER  XX,YY,XY,XZ,YZ
C
```

```
      DO 51 ISTRE=1,NSTRE
   51 STRES(ISTRE)=STRSG(ISTRE,KGAUS)
      WRITE(6,950) KGASP,(STRES(ISTRE),ISTRE=1,NSTRE),
     .             EFFST(KGAUS),EPSTN(KGAUS,1),EPSTN(KGAUS,2),
     .             ALPHA,BEETA,NSTAT
C
C*** SET UP THE STRESS RESULTANTS IN THE ORDER NX,NY,NXY,MX,MY,MXY,QX,QY
C
      DO 71 ISTRE=1,3
      FORCE(ISTRE)=FORCE(ISTRE)+STRES(ISTRE)*THIGP/2.0*DZETA
   71 FORCE(ISTRE+3)=FORCE(ISTRE+3)-STRES(ISTRE)*THIGP*THIGP*
     .                 ZETSP*DZETA/4.0
  130 CONTINUE
      WRITE(6,960) FORCE(1),FORCE(4),FORCE(2),FORCE(5),FORCE(3),
     .             FORCE(6),FORCE(7),FORCE(8)
  960 FORMAT(/,22H STRESS RESULTANTS =  ,6HN-XX =,E12.5,5X,
     .       6HM-XX =,E12.5/22X,6HN-YY =,E12.5,5X,6HM-YY =,
     .       E12.5/22X,6HN-XY =,E12.5,5X,6HM-XY =,E12.5/,
     .       22X,6HQ-XZ =,E12.5,5X,6HQ-YZ =,E12.5)
  105 CONTINUE
  110 CONTINUE
  120 CONTINUE
      RETURN
      END
```

4.15 Subroutines HARDEN and YLSUF

These subroutines are only required to describe the strain hardening behaviour of concrete. In particular Subroutine HARDEN evaluates the position of the current strain point on the hardening curve and Subroutine YLSUF evaluates the corresponding stress according to (11) and also the strain-hardening parameter.

It should be noted that these subroutines must be called in subroutines RESI1 and STIFF whenever the current effective stress and the hardening value is required.

```
      SUBROUTINE HARDEN(ESTRN,ESTSS,PROPS,MMATS,NPROP,NPONT)
C**********************************************************************
C
C*** DEFINE THE STRAIN-HARDENING FUNCTION FOR CONCRETE
C
C**********************************************************************
      DIMENSION ESTRN(NPONT),ESTSS(NPONT),PROPS(MMATS,NPROP)
      CONST=0.3
      LPROP=1
      EMAXS=2.0*PROPS(LPROP,6)/PROPS(LPROP,1)
      EPMAX=EMAXS*0.5
      EPMIN=EMAXS*(1.0-CONST*0.5-SQRT(1.0-CONST))
      COEFE=EPMAX/(EPMAX-EPMIN)
      KPONT=NPONT-4
      WRITE(6,910) NPONT,EMAXS,EPMAX,EPMIN
  910 FORMAT(20H HARDENING CONSTANTS,I5,3E14.5)
      DO 20 IPONT=1,NPONT
      ESTSS(IPONT)=CONST*PROPS(LPROP,6)
      ESTRN(IPONT)=(EMAXS*(1.0-CONST*0.5-SQRT(ABS(1.0-CONST)))
     .             -EPMIN)*COEFE
      WRITE(6,920) IPONT,CONST,ESTSS(IPONT),ESTRN(IPONT)
  920 FORMAT(17H HARDENING VALUES,I5,3E14.5)
      IF(IPONT.LT.KPONT) CONST=CONST+0.05
      IF(IPONT.GE.KPONT) CONST=CONST+0.025
   20 CONTINUE
      ERROR=(PROPS(LPROP,6)-ESTSS(NPONT))/PROPS(LPROP,6)*100.0
      IF(ABS(ERROR).LT.0.01) GO TO 30
      WRITE(6,900) ESTSS(NPONT)
  900 FORMAT(50H ERROR IN THE STRESS-PL.STRAIN CURVE, MAX.STRESS =,
     .F10.5)
      STOP
   30 RETURN
      END
```

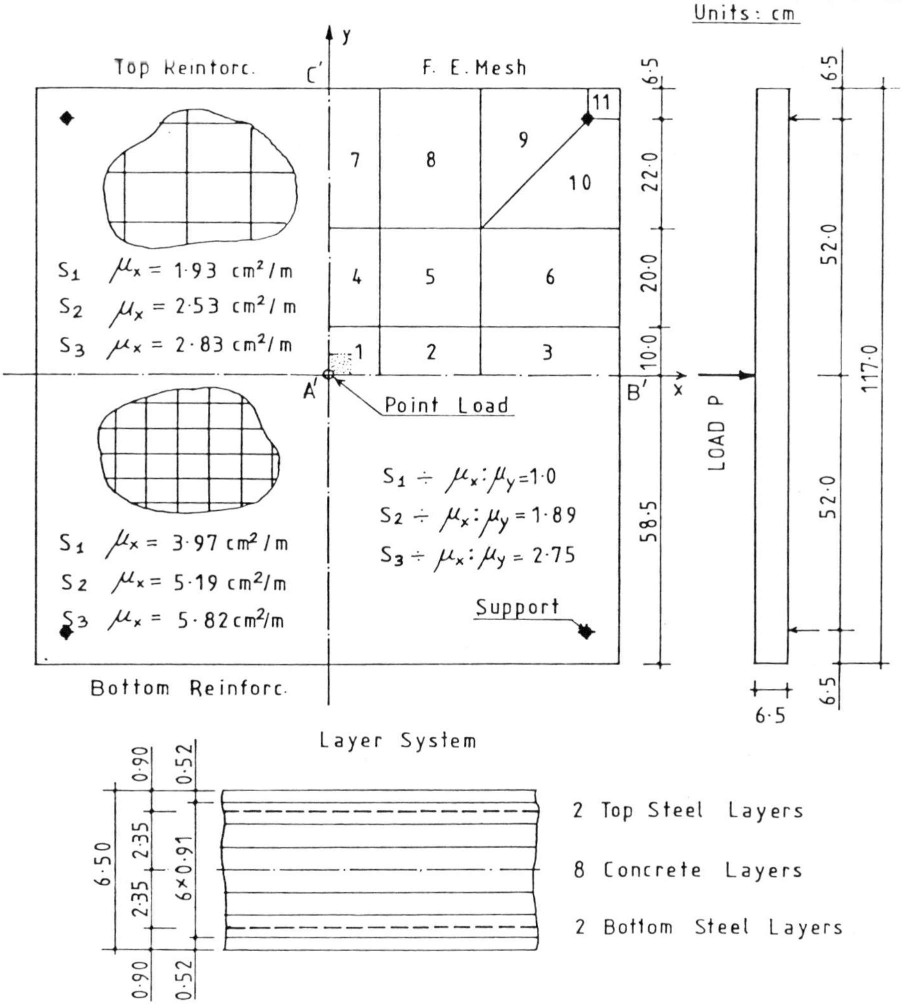

Fig. 6 **Corner supported centrally loaded plates: geometry, reinforcement and finite element idealisation (Example I).**

Table 1 **Problem parameters for Example I, Fig. 6.**

MATERIAL PROPERTIES [cm; KN]	
CONCRETE	STEEL
Youngs Modulus, E_c = 1640.0	Youngs Modulus, E_s = 20100.0
Poisson's Ratio, ν = 0.0	Yield stress, f_y = 60.0
Ult. Comp. st., f'_c = 4.30	Youngs Modulus, E'_s = 700.0
Ult. Tens. st., f_t = 0.30	Ult. stress, f'_y = 70.0
Ult. Comp. stn., ε_u = 0.0035	Youngs Modulus, E''_s = 0.0
Tension Stiff.Coeff. α = 0.7	Angle x-axis, θ_x = 0.0
Tension Stiff.Coeff. ε_m = 0.0020	Angle x-axis, θ_y = 1.57

```
      SUBROUTINE YLSUF(ESTRN,ESTSS,EPSTN,PROPS,KGAUS,MELGP,
     .                 MMATS,NPROP,NPONT,LPROP,HARDS,YLDST)
C***********************************************************************
C
C*** THIS SUBROUTINE CALCULATES THE CURRENT VALUES OF THE YIELD
C    SURFACE AND OF THE STRAIN HARDENING PARAMETER
C
C***********************************************************************
      DIMENSION ESTRN(NPONT),ESTSS(NPONT),EPSTN(MELGP,2),
     .          PROPS(MMATS,NPROP)
C     CONST=0.3
C     UNIAX=PROPS(LPROP,6)*CONST
C     YLDST=UNIAX
C     IF(EPSTN(KGAUS,1).EQ.0.0) RETURN
      DO 20 IPONT=2,NPONT
      IF(EPSTN(KGAUS,1).GT.ESTRN(IPONT)) GO TO 20
      HARDS=(ESTSS(IPONT)-ESTSS(IPONT-1))/(ESTRN(IPONT)-
     .      ESTRN(IPONT-1))
      YLDST=ESTSS(IPONT-1)+HARDS*(EPSTN(KGAUS,1)-ESTRN(IPONT-1))
      RETURN
   20 CONTINUE
      HARDS=0.0
      YLDST=PROPS(LPROP,6)
      RETURN
      END
```

5. NUMERICAL EXAMPLES

Two numerical examples are presented in this section which illustrate the applicability of the computer code described and also provide a comparison between numerical and experimental results.

5.1 Example I, corner supported square plate

A corner supported square plate loaded by a concentrated force at the centre, tested by Duddeck et al [52] and also reported in Ref. [12] is considered. This model, constructed from expanded slate lightweight concrete, has the advantage of well defined boundary conditions (horizontally free). Three plates were tested (S1, S2, S3) with different amounts of reinforcement in each direction, but with the total amount of reinforcement in the plates kept constant. The plate geometry, amount of reinforcement and finite element discretisation are shown in Fig. 6 and the material properties are specified in Table 1.

The load-deflection curves for the three plates analysed are compared in Fig. 7 with the corresponding yield line solutions. The plate S1, with isotropic reinforcement, shows smaller deflections and larger load carrying capacity, although the total amount of steel is equal for all three plates. In the analysis of plate S1 a smaller limit load (55.0 KN) is found

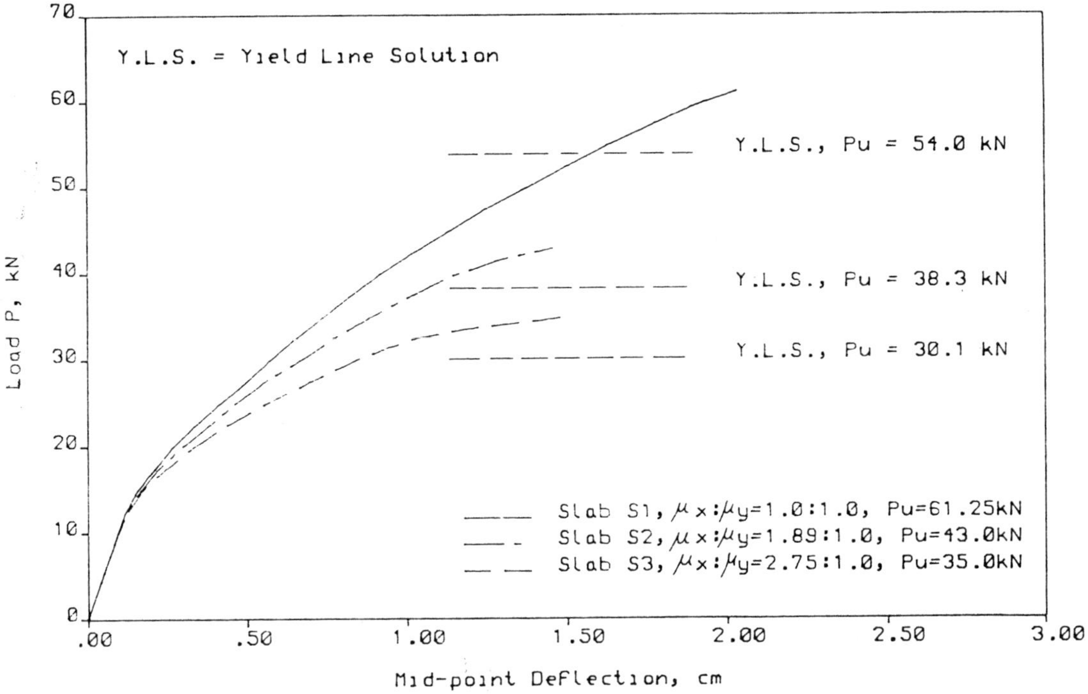

Fig. 7 Load-central deflection response for the three plates S1, S2 and S3 of Example I.

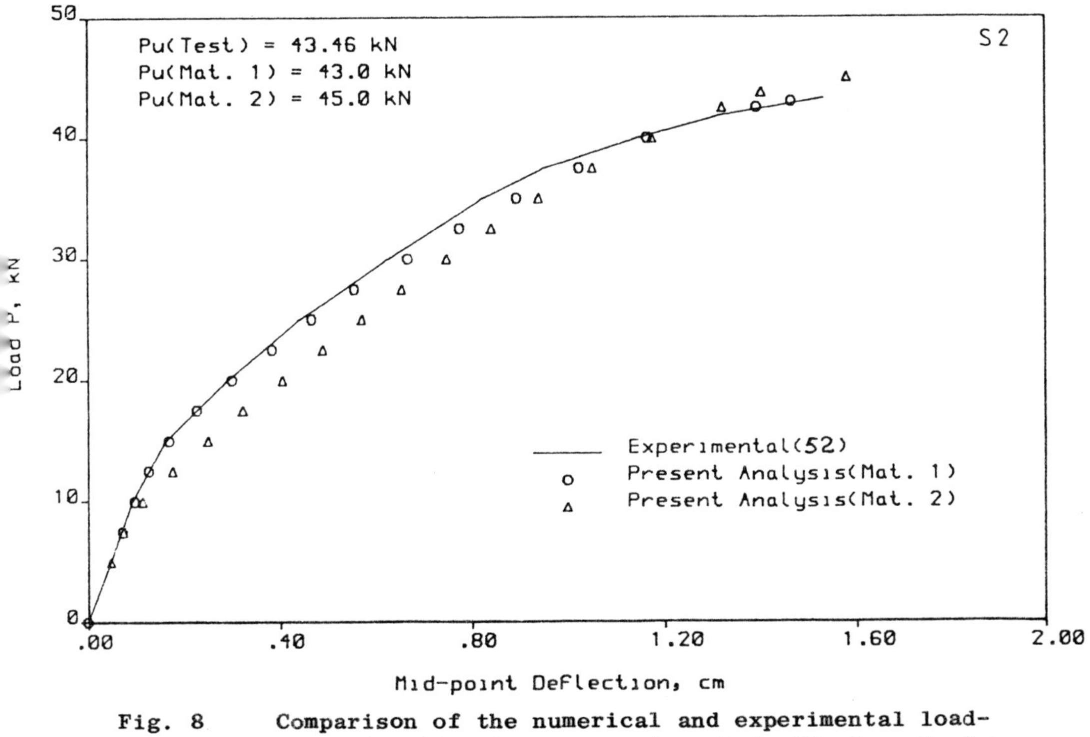

Fig. 8 Comparison of the numerical and experimental load-central deflection curves for plate S2, Example I.

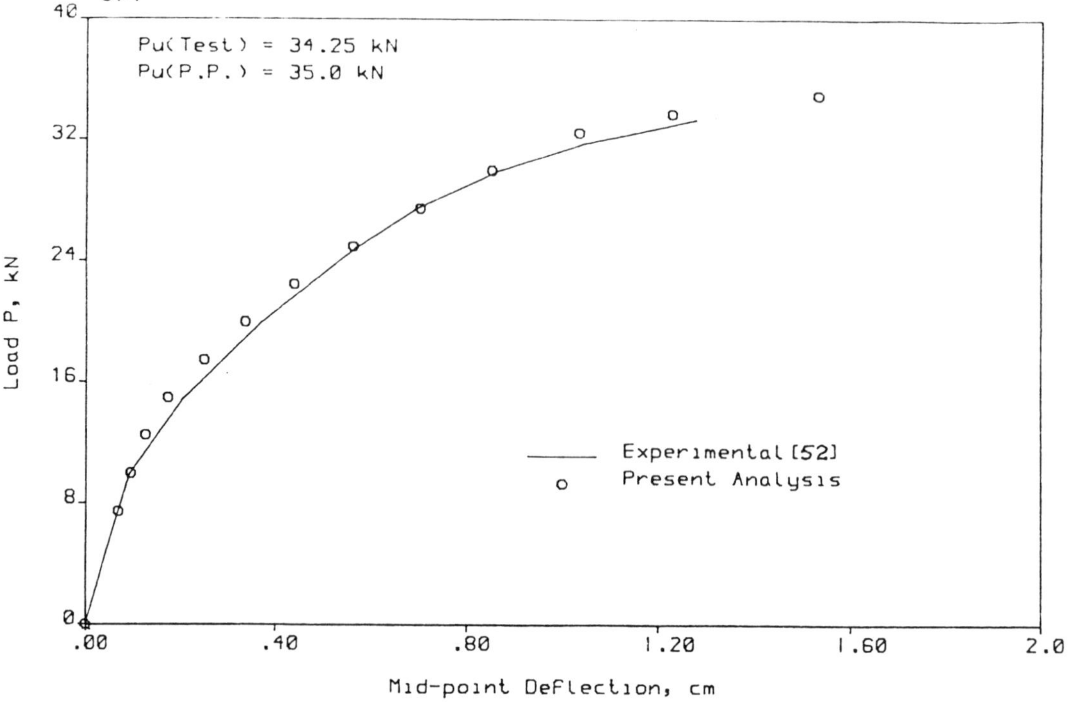

Fig. 9 Comparison of the numerical and experimental load-central deflection curve for plate S3, Example I.

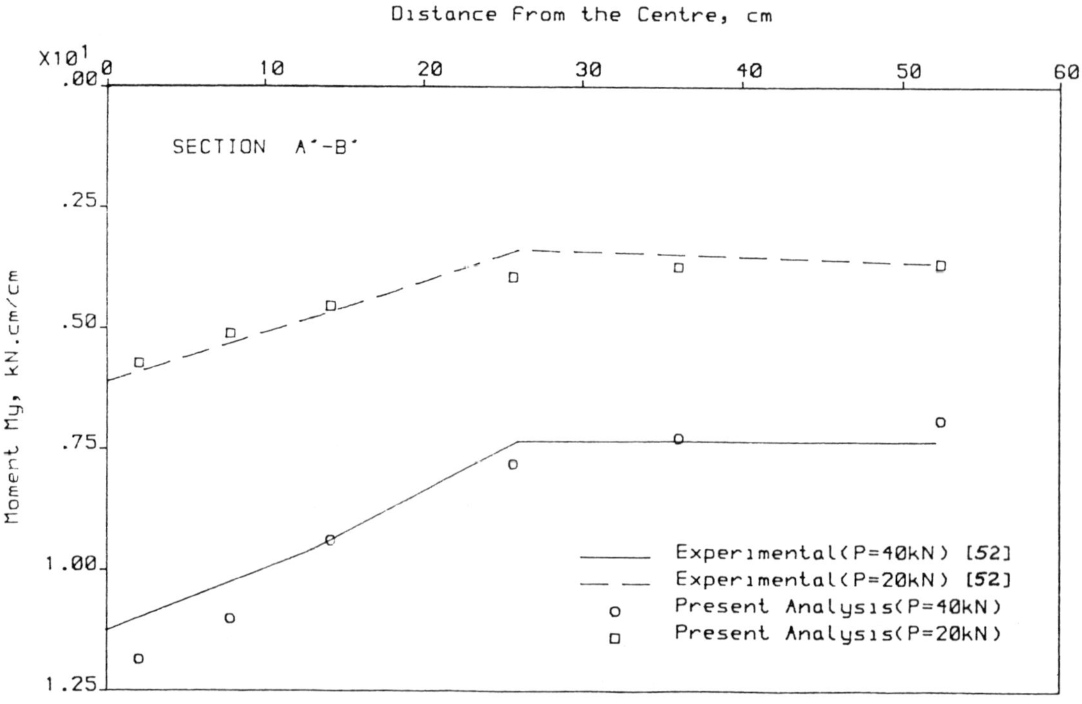

Fig. 10 Comparison of the numerical and experimental moments along the x axis for plate S1, Example I.

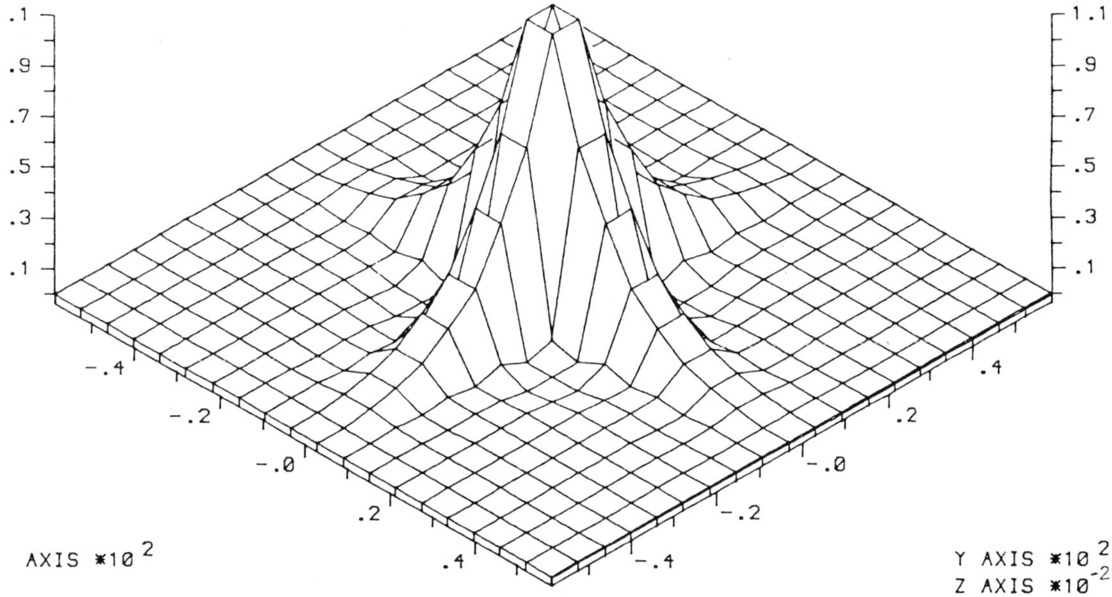

Fig. 11 Plastic strain in the steel at collapse for plate S1, Example I.

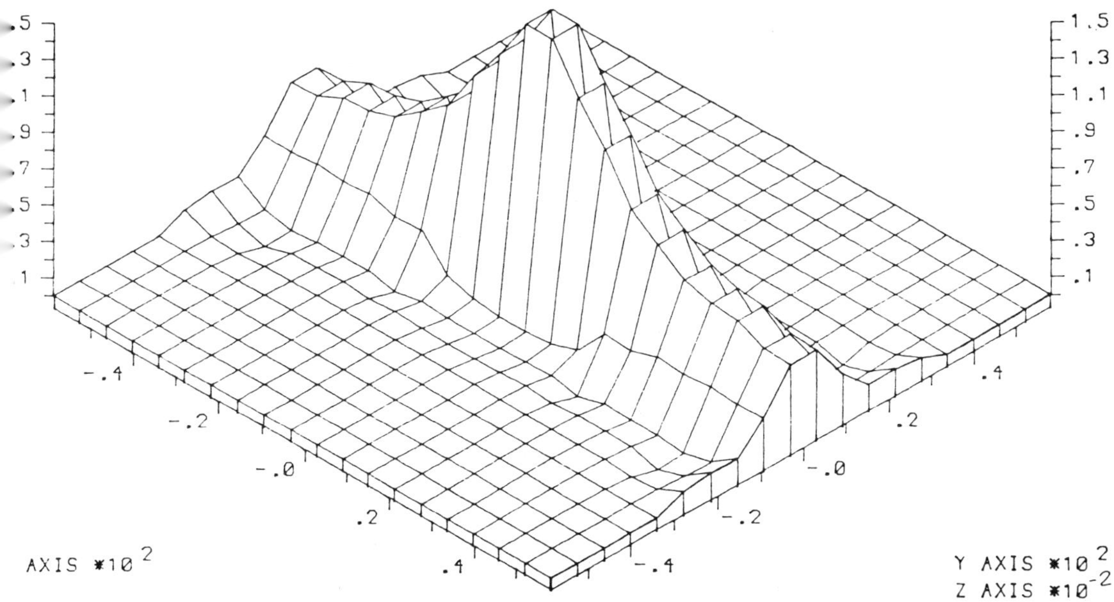

Fig. 12 Plastic strain in the steel at collapse for plate S3, Example I.

when a point load is applied. Failure occurs by crushing of the concrete in the vicinity of the load, where the transverse shear stress effects are dominant. The results presented for this plate have been obtained by applying the concentrated load on a small central area of 10 x 10 cm^2. Even then the collapse load (61.25 KN) is reached before a complete development of the yield lines (Fig. 11) for an ultimate strain $\varepsilon_u = 0.0035$. Figs. 8 and 9 compare the load-deflection curves and collapse loads of plates S2 and S3 (anisotropic reinforcement) with the experimental results and an excellent agreement is evident. Fig. 8 also shows the results obtained for plate S2 employing the same material properties used in Ref. [12,13]; which differ from the present ones (Mat.1) in the assumed tensile concrete strength ($f'_t = 0.20$ KN/cm^2) and in the prescription of an elastic-perfectly plastic diagram for the steel ($f_y = 67.0$ KN/cm^2). Fig. 10 shows the calculated and the experimentally measured moments M_y along the x axis for plate S1. The plastic steel strains at collapse are illustrated for plates S1 and S2 in Figs. 11 and 12 respectively. These diagrams show clearly the spread and geometry of the yield lines for the isotropic (S1) and anisotropic (S3) configurations.

5.2 Example II. Pressure loaded parabolic cylindrical shell

A parabolic cylindrical shell with variable thickness, subjected to uniformly distributed pressure, which was tested by Hedgren [54], is analysed with the present finite element code. The shell was tested with end support diaphragms and free edges. The shell geometry, finite element mesh, and the zones with different layer patterns are shown in Fig. 13. Thirty six finite elements are used to model one quarter of the shell, each of which is divided into 8 equal concrete layers.

The boundary conditions assumed to represent the effects of the end support diaphragms are free displacements in the x-direction and free rotations normal to the shell plane. The reinforcement characteristics for each layer pattern are

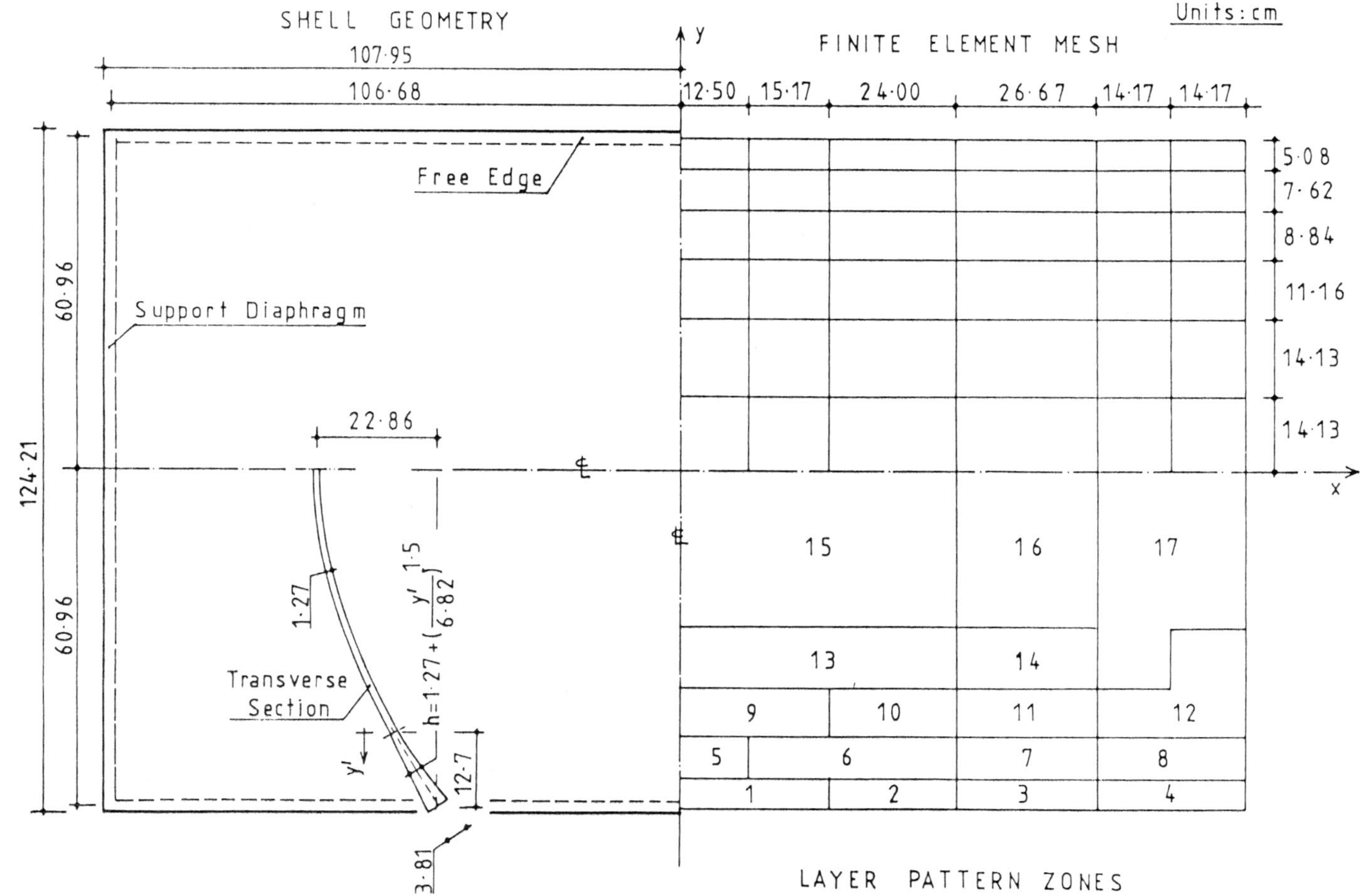

Fig. 13 Geometry and finite element idealisation for a parabolic cylindrical shell (Example II).

Table 2 Reinforcement details for each layer of parabolic cylindrical shell (Fig. 13).

REINFORCEMENT CHARACTERISTICS

LAYER PATTERN	LAYER 1 (1)	LAYER 1 (2)	LAYER 1 (3)	LAYER 1 (4)	LAYER 2 (1)	LAYER 2 (2)	LAYER 2 (3)	LAYER 2 (4)	LAYER 3 (1)	LAYER 3 (2)	LAYER 3 (3)	LAYER 3 (4)	LAYER 4 (1)	LAYER 4 (2)	LAYER 4 (3)	LAYER 4 (4)
1	4	0.602	-0.795	0.0	3	0.574	-0.640	90.0	9	2.996	-0.478	0.0	4	0.419	0.0	45.0
2	4	0.602	-0.795	0.0	3	0.574	-0.640	90.0	9	2.853	-0.478	0.0	4	0.419	0.0	45.0
3	4	0.602	-0.795	0.0	3	0.502	-0.640	90.0	9	2.853	-0.478	0.0	4	0.419	0.0	45.0
4	4	0.602	-0.795	0.0	3	0.302	-0.640	90.0	9	1.902	-0.478	0.0	4	0.629	0.0	45.0
5	3	0.691	-0.642	0.0	3	0.574	-0.640	90.0	3	0.251	0.625	0.0	4	0.479	0.740	90.0
6	3	0.691	-0.642	0.0	3	0.574	-0.640	90.0	4	0.419	0.0	45.0	3	0.251	0.625	0.0
7	3	0.691	-0.642	0.0	3	0.502	-0.640	90.0	4	0.419	0.0	45.0	3	0.251	0.625	0.0
8	3	0.691	-0.642	0.0	3	0.302	-0.640	90.0	4	0.629	0.0	45.0	3	0.251	0.625	0.0
9	3	0.483	-0.500	0.0	3	1.102	-0.308	90.0	3	0.482	0.280	0.0	4	0.920	0.500	90.0
10	3	0.483	-0.500	0.0	3	1.102	-0.308	90.0	4	0.511	0.0	45.0	3	0.482	0.280	90.0
11	3	0.483	-0.500	0.0	3	0.964	-0.308	90.0	4	0.511	0.0	45.0	3	0.482	0.280	90.0
12	3	0.483	-0.500	0.0	3	0.579	-0.308	90.0	4	1.208	0.0	45.0	3	0.482	0.280	90.0
13	3	0.483	-0.500	0.0	3	0.552	-0.308	90.0	3	0.482	0.280	0.0	4	0.920	0.500	90.0
14	3	0.483	-0.500	0.0	3	0.482	-0.308	90.0	3	0.482	0.280	0.0	4	0.805	0.500	90.0
15	3	0.483	-0.500	0.0	3	0.552	-0.308	90.0	3	0.482	0.280	0.0	4	1.841	0.500	90.0
16	3	0.483	-0.500	0.0	3	0.482	-0.308	90.0	3	0.482	0.280	0.0	4	1.611	0.500	90.0
17	3	0.483	-0.500	0.0	3	0.579	-0.308	90.0	3	0.482	0.280	0.0	4	0.614	0.500	90.0

LAYER PATTERN	LAYER 5 (1)	LAYER 5 (2)	LAYER 5 (3)	LAYER 5 (4)	LAYER 6 (1)	LAYER 6 (2)	LAYER 6 (3)	LAYER 6 (4)	LAYER 7 (1)	LAYER 7 (2)	LAYER 7 (3)	LAYER 7 (4)
1	9	2.996	0.478	0.0	3	0.250	0.625	0.0	4	0.479	0.740	90.0
2	9	2.853	0.478	0.0	3	0.250	0.625	0.0	4	0.479	0.740	90.0
3	9	2.853	0.478	0.0	3	0.250	0.625	0.0	4	0.419	0.740	90.0
4	9	1.702	0.478	0.0	3	0.250	0.625	0.0	4	0.503	0.740	90.0
5												
6	4	0.479	0.740	90.0								
7	4	0.419	0.740	90.0								
8	4	0.503	0.740	90.0								
9												
10	4	0.920	0.500	90.0								
11	4	0.805	0.500	90.0								

(1) REINFORCING BAR DESIGNATION (#3, #4, #9)
(2) LAYER THICKNESS IN NORMALIZED ζ-COORDINATE (x100)
(3) LAYER POSITION IN NORMALIZED ζ-COORDINATE (TOP: $\zeta=+1$; BOTTOM: $\zeta=-1$)
(4) ANGLE BETWEEN THE REINFORCEMENT AND THE LONGITUDINAL AXIS (DEGREES)

Table 3 Problem parameters for the parabolic cylindrical shell of Fig. 13.

MATERIAL PROPERTIES [cm, kN]	
CONCRETE	STEEL
Young's Modulus, E_c = 2069.0	Young's Modulus, E_s = 20000.0
Poisson's Ratio, ν = 0.145	Yield Stress, f_y (var.) (*)
Ult. Comp. St., f'_c = 3.02	Young's Modulus, E'_s = 4000.0
Ult. Tens. St., f'_t = 0.48	Ultimate Stress, f_u (var.)(*)
Ult. Comp. Stn., ε_u = 0.0035	Young's Modulus, E''_s = 0.0
Tens. Stiff. Coeff., α = 0.6	Angle with x-axis, θ (var.)
Tens.Stiff.Coeff., ε_m = 0.0020	(*) see table 4

Table 4 Reinforcement details for the parabolic cylindrical shell of Fig. 13.

REINFORCEMENT CHARACTERISTICS [cm, kN]				
DESIGNATION	DIAMETER	AREA	YIELD STRENGTH	ULT.STRENGTH
# 3	0.122	0.0117	25.29	36.42
# 4	0.157	0.0195	21.91	34.49
# 9	0.343	0.0923	30.66	42.00

summarised in Table 2 and the material properties are listed in Tables 3 and 4. The concrete and steel properties are well defined in Ref. [54], however the complete stress-strain diagram for the steel which could influence the ultimate behaviour of the shell is not provided. It is assumed for analysis that after initial yielding of the steel at stress level f_y, a linear hardening behaviour with $E'_s = 4000.0\ KN/cm^2$ follows up to the ultimate stress f_u after which a perfectly plastic behaviour occurs. In Fig. 14 the experimental load-vertical deflection curve at midspan of the free edge is compared with the present results using a linear and a nonlinear geometrical model. Geometric nonlinear analysis has an important effect in this shell problem as noted by Hedgren [54]. An increase of the shell stiffness and its load carrying capacity, due to the vertical uplift of the shell crown and the downward displacement of the free edges, is observed. An increase of the lever arm of the internal forces is consequently obtained, when the deformed geometry is taken into account.

The vertical deflections of the transverse section at midspan of the shell are presented in Fig. 15 for load factors of 0.8 and 2.4. The transverse variation of the stress resultant N_x at midspan is compared in Fig. 16 with theoretical and experimental results for two load levels. In all cases the numerical and experimental results are seen to be in good agreement provided that geometric nonlinear effects are included in the computations.

The plastic strains in the circumferential reinforcement are represented as contour plots in Fig. 16 for a load level, F = 4.35 which is close to ultimate load. Positive values correspond to the top layer and negative values refer to the bottom circumferential layer. The negative yield line (positive plastic strains) is completely formed, whereas the two positive yield lines are not yet completely developed along the edges.

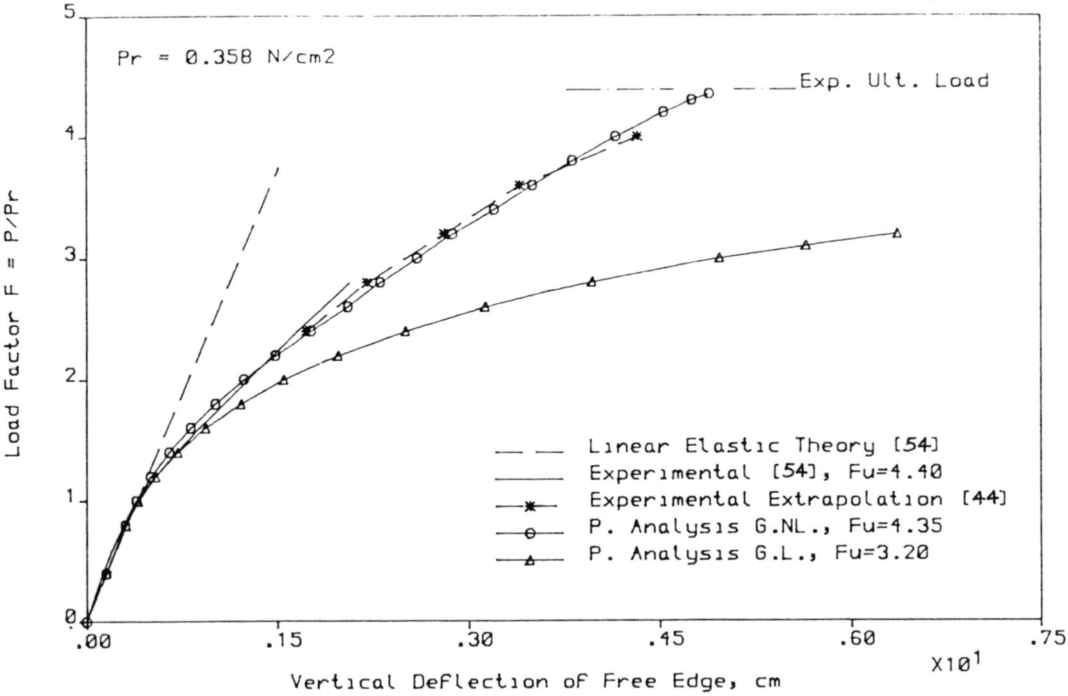

Fig. 14 Comparison of the experimental and numerical load-vertical deflection response for the cylindrical shell, Example II.

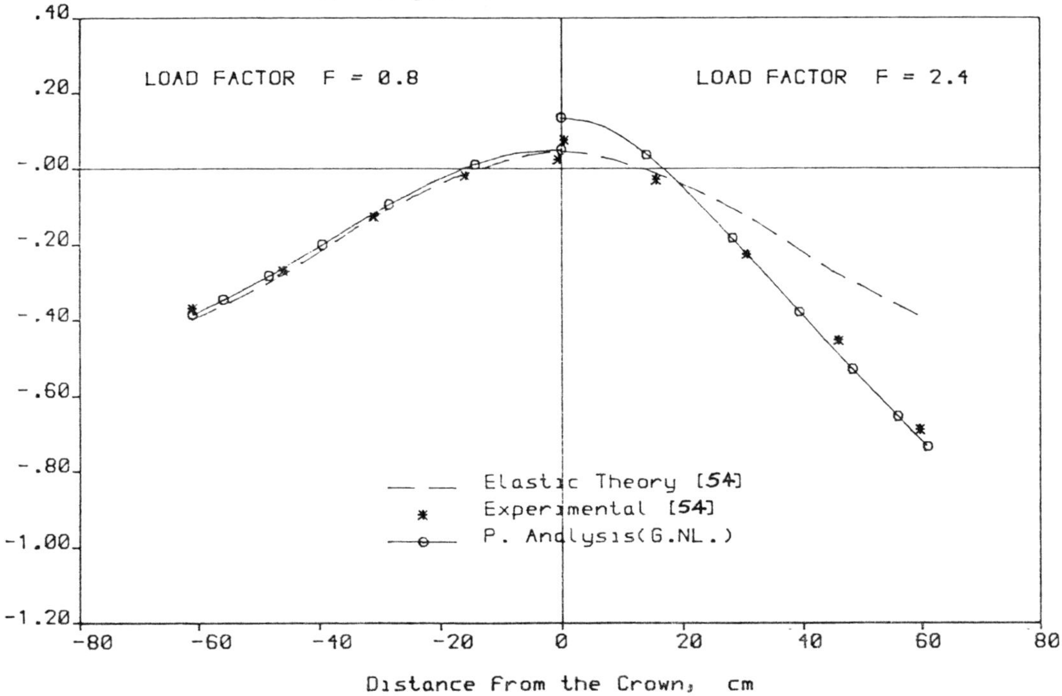

Fig. 15 Vertical deflections of the transverse section at mid-span (x=0) for Example II.

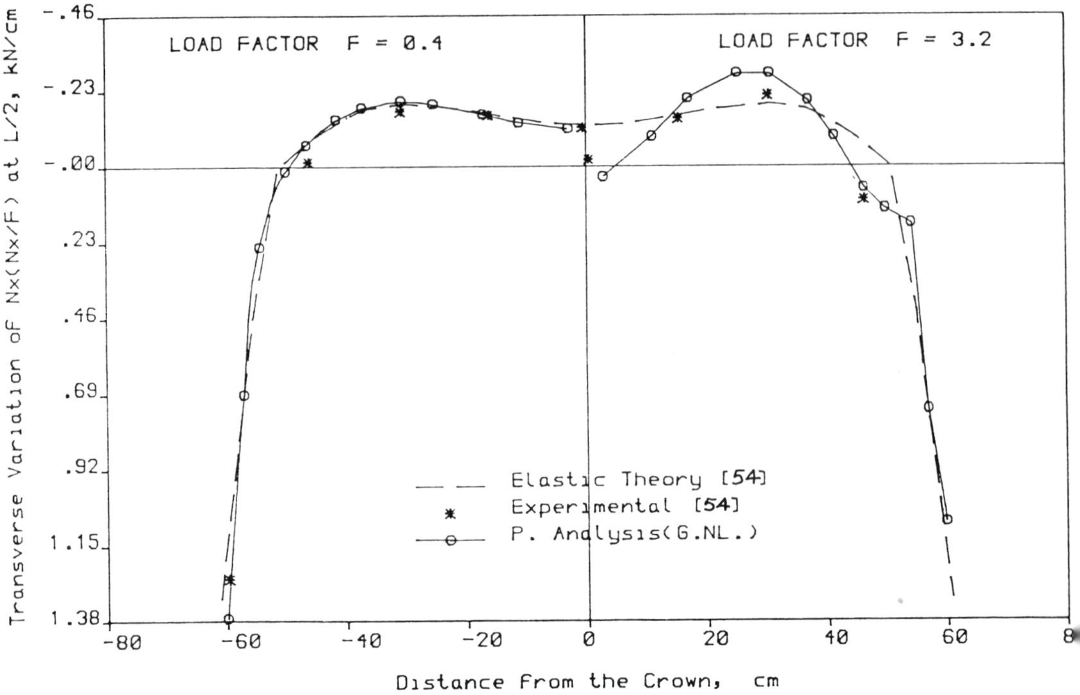

Fig. 16 Comparison of the transverse variation of the stress-resultant N_x at mid-span for Example II.

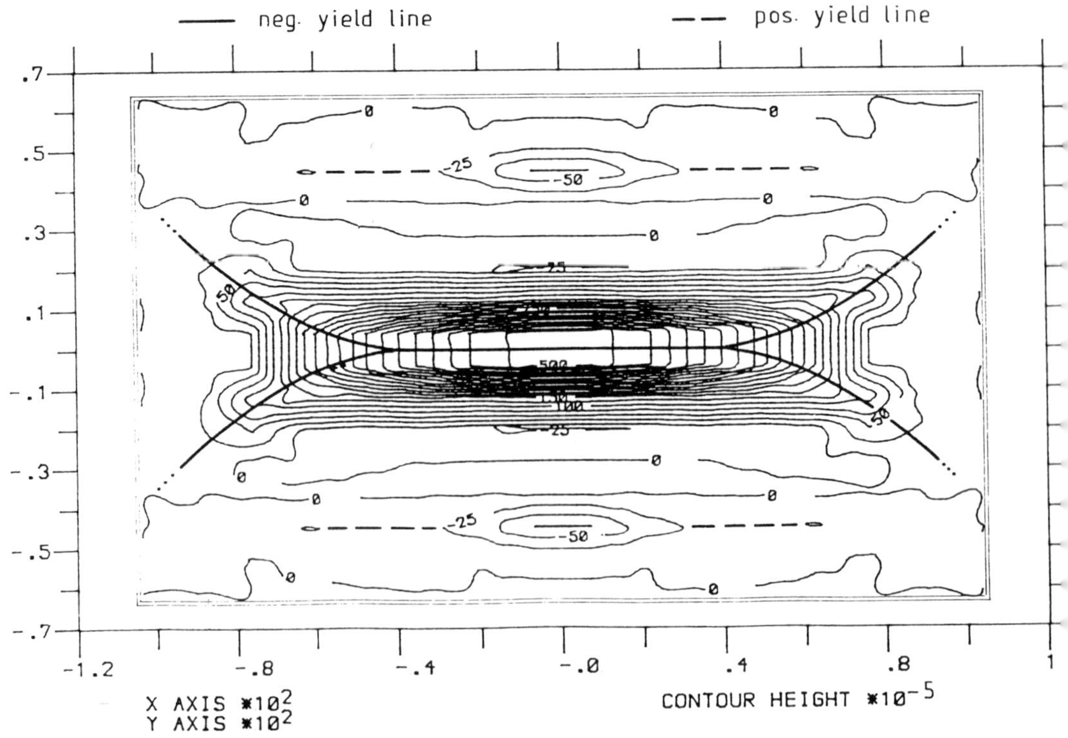

Fig. 17 Plastic strains in the circumferential reinforcement at failure for Example II.

6. REFERENCES

1. ASCE Committee on concrete and masonry structures, Task Committee on Finite Element Analysis of Reinforced Concrete Structures: A State-of-The-Art Report on Finite Element Analysis of Reinforced Concrete Structures, ASCE Spec. Pub. 1981.

2. W. C. SCHNOBRICH
Behaviour of reinforced concrete structures predicted by the finite element method, Computers & Structures, Vol. 7, pp. 365-376, 1977.

3. J. H. ARGYRIS, G. FAUST, J. SZIMMAT, E. P. WARNKE and K. J. WILLAM
Recent developments in the finite element analysis of prestressed concrete reactor vessels, Nuclear Engineering and Design, 28, pp. 42-75, 1974.

4. P. G. BERGAN and I. HOLAND
Nonlinear finite element analysis of concrete structures, Computer Methods in Applied Mechanics and Engineering, 17/18, pp. 443-467, 1979.

5. K. H. GERSTLE
Material modelling of reinforced concrete, Introductory Report, IABSE Colloquium on Advanced Mechanics of Reinforced Concrete, Delft, Volume-Band 33, pp. 41-63, 1981.

6. D. NGO and A. C. SCORDELIS
Finite element analysis of reinforced concrete beams, American Concrete Institute Journal, V.64, No.3, March 1967.

7. L. CEDOLIN and S. DEIPOLI
Finite element studies of shear-critical R/C beams, ASCE Journal of the Engineering Mechanics Division, Vol. 103, No.EM3, pp. 395-410, June 1977.

8. M. N. FARDIS and O. BUYUKOZTURK
Shear stiffness of concrete by finite elements, ASCE Journal of the Structural Division, Vol. 106, No.ST6, pp. 1311-1327, June 1980.

9. Z. P. BAZANT and L. CEDOLIN
Fracture mechanics of reinforced concrete, ASCE Journal of Eng. Mech. Division, Vol. 106, No.EM6, pp. 1287-1306, December 1980.

10. C. S. LIN
Nonlinear analysis of reinforced concrete slabs and shells, Ph.D. Thesis, UC-SESM 73-7, University of California, April 1973.

11. F. R. HAND, D. A. PECKNOLD and W. C. SCHNOBRICH
Nonlinear layered analysis of RC plates and shells, ASCE Journal of Structural Division Vol.99, No.ST7, pp. 1491-1505, July 1973.

12. G. MUELLER
Numerical problems in nonlinear analysis of reinforced concrete, Report No. UC-SESM 77-5, University of California, Berkeley, September 1977.

13. H. H. A. RAHMAN
Computational models for the nonlinear analysis of reinforced concrete flexural slab systems, Ph.D. Thesis, C/Ph/66/82, University of Wales, May 1982.

14. H. KUPFER, K. H. HILSDORF and H. RUSH
Behaviour of concrete under biaxial stresses, Proceedings, American Concrete Institute, Vol.66, No.8, pp. 656-666, August 1969.

15. L. L. MILLS and R. M. ZIMMERMAN
Compressive strength of plain concrete under multiaxial loading conditions. ACI Journal, Proc. Vol. 67, M-10, Oct. 1970, pp. 802-807.

16. K. H. GERSTLE et al.
Behaviour of concrete under multiaxial stress states, ASCE Journal of Eng. Mech. Div. Vol. 106, No. EM6, December 1980, pp. 1383-1403.

17. H. B. KUPFER and K. H. GERSTLE
Behaviour of concrete under biaxial stresses, ASCE Journal of the Eng. Mech. Div., Vol. 99, No. EM4, August 1973, pp. 853-866.

18. T. C. Y. LIU, A. H. NILSON and F. O. SLATE
Biaxial stress-strain relations for concrete, ASCE Journal of the Structural Division, Vol. 98, No. ST5, May 1972, pp. 1025-1034.

19. L. CEDOLIN, Y. R. J. CRUTZEN and S. DEIPOLI
Triaxial stress-strain relationship for concrete, ASCE J. of the Eng. Mech. Div. Vol. 103, No. EM3, June 1977, pp. 423-439.

20. K. H. GERSTLE
Simple formulation of biaxial concrete behaviour, ACI Journal, Vol. 78, Jan.-Feb. 1981, pp. 62-68.

21. C. S. LIN and A. C. SCORDELIS
Nonlinear analysis of RC shells of general form, ASCE Journal of the Structural Division, Vol. 101, No.ST3, March 1975, pp. 523-538.

22. A. C. T. CHEN and W. F. CHEN
Constitutive relations for concrete, ASCE Journal of the Eng. Mech. Div. Vol. 101, No. EM4, August 1975, pp. 465-481.

23. O. BUYUKOZTURK
Nonlinear analysis of reinforced concrete structures, Computers and Structures, Vol.7, pp. 149-156, 1977.

24. Z. P. BAZANT and P. D. BHAT
Endochronic theory of inelasticity and failure of concrete; ASCE Journal of the Eng. Mech. Div., Vol.102, No.EM4, August 1976, pp. 701-722.

25. Z. P. BAZANT and S. S. KIM
Plastic-Fracturing theory for concrete, ASCE Journal of the Eng. Mech. Div., Vol.105, No.EM3, June 1979, pp. 407-428.

26. J. C. JOFRIET and G. M. MCNEICE
Finite element analysis of reinforced concrete slabs, ASCE Journal of the Structural Division, Vol. 97, No. ST3, March 1971, pp. 785-806.

27. F. R. HAND, D. A. PECKNOLD, W. C. SCHNOBRICH
A layered finite element nonlinear analysis of reinforced concrete plates and shells, Civil Eng. Studies, SRS No.389, University of Illinois, Urbana, Illinois, August 1972.

28. E. HINTON, H. H. A. RAHMAN and O. C. ZIENKIEWICZ
Computational models for reinforced concrete slab systems, IABSE Colloquium on Advanced Mechanics of Reinforced Concrete, Final Report, Vol.-Band 34, Delft 1981, pp. 303-313.

29. N. S. OTTOSEN
A failure criterion for concrete, ASCE Journal of Eng. Mech. Div. Vol.103, No.EM4, August 1977, pp. 527-535.

30. J. WASTIELS
Behaviour of concrete under multiaxial stresses - a review. Cement and Concrete Research, Vol. 9, pp. 35-44, 1979.

31. J. WASTIELS
Failure criteria for concrete subjected to multiaxial stresses. Lecture held at University of Illinois at Chicago Circle, Department of Materials Engineering, May 27, 1981.

32. W. F. CHEN
Plasticity in Reinforced Concrete. McGraw-Hill Book Company, 1982.

33. E. ANDENAES, K. GERSTLE and H. Y. KO
Response of mortar and concrete to biaxial compression, ASCE Journal of Eng. Mech. Div., Vol.103, No.EM4, August 1977, pp.515-526.

34. D. R. J. OWEN and E. HINTON
Finite Elements in Plasticity - Theory and Practice. Pineridge Press, Swansea, U.K., 1980.

35. O. C. ZIENKIEWICZ
The Finite Element Method, McGraw-Hill, Third edition, 1977.

36. R. I. GILBERT and R. F. WARNER
Tension stiffening in reinforced concrete slabs, ASCE Journal of the Structural Division, Vol. 104, No.ST12, December 1978, pp. 1885-1900.

37. Y. GOTO
Cracks formed in concrete around deformed tension bars, ACI Journal, April 1971, pp. 244-251.

38. R. J. COPE, P. V. RAO, L. A. CLARK and P. NORRIS
Modelling of reinforced concrete behaviour for finite element analysis of bridge slabs. Numerical Meth. for Nonlinear Problems, Vol.1, Pineridge Press. Proceedings of the International Conference held at University College of Swansea, 2-5 September 1980, pp. 457-470.

39. R. C. FENWICK and T. PAULAY
Mechanics of shear resistance of concrete beams, ASCE Journal of the Structural Division, Vol.94, No.ST10, October 1968, pp. 2325-2350.

40. J. A. HOFBECK, I. O. IBRAHIM and A. H. MATTOCK
Shear transfer in reinforced concrete, ACI Journal, February 1969, pp. 119-128.

41. Y. D. HAMADI and P. E. REGAN
Behaviour in shear of beams with flexural cracks, Magazine of Concrete Research, Vol. 32, No. 111, June 1980, pp. 67-78.

42. S. AHMAD, B. M. IRONS and O. C. ZIENKIEWICZ
Analysis of thick and thin shell structures by curved finite elements. Int. J. Num. Meth. in Engng. Vol.2, pp. 419-451, 1970.

43. D. R. J. OWEN and J. A. FIGUEIRAS
Anisotropic elasto-plastic finite element analysis of thick and thin plates and shells, Int. J. Num. Meth. Engng., Vol.19, pp. 541-566, 1983.

44. H. FLOEGL and H. MANG
On tension stiffening in cracked reinforced concrete slabs and shells considering geometric and physical nonlinearity, Inginieur-Archiv 51 pp. 215-242, 1981.

45. G. MEHLHORN and D. KLEIN
Finite element analysis of reinforced concrete slabs and panels, Institut fur Massinbau, Technische Hochschule Darmstadt.

46. T. J. R. HUGHES
Unconditionally stable algorithms for nonlinear heat conduction, Comp. Meth. in Applied Mech. and Engng., Vol.10, pp. 135-139, 1977.

47. K. J. BATHE and E. L. WILSON
Numerical Methods in Finite Element Analysis, Prentice-Hall, New Jersey, USA, 1976.

48. M. A. HOGGE
Integration operators for first order linear matrix differential equations, Comp. Meth. in Applied Mech. and Engng., Vol.11, pp. 281-294, 1977.

49. D. R. J. OWEN and F. DAMJANIC
Reduced numerical integration in thermal transient finite element analysis. Computers and Structures (In press).

50. F. DAMJANIC and D. R. J. OWEN
Practical considerations for thermal transient finite element analysis using isoparametric elements. Nuc. Eng. & Design, Vol. 69, No.1, 109-126, 1982.

51. B. BRESLER and A. C. SCORDELIS
Shear strength of reinforced concrete beams, ACI Journal, Jan. 1963.

52. H. DUDDECK, G. GRIEBENOW and G. SHAPER
Material and time dependent nonlinear behaviour of cracked reinforced concrete slabs; in Nonlinear Behaviour of Reinforced Concrete Spatial Structures, Vol.1, Preliminary Report IASS Symposium held in Darmstadt, Eds. G. Mehlhorn, H. Ruhle and W. Zerna, Werner-Verlag Dusseldorf, pp. 101-113, July 1978.

53. R. TAYLOR, D. R. H. MAHER and B. HAYES
Effect of the arrangement of reinforcement on the behaviour of reinforced concrete slabs, Magazine of Concrete Research, Vol. 18, No.55, June 1966.

54. A. W. HEDGREN and D. P. BILLINGTON
Mortar model test on a cylindrical shell of varying curvature and thickness, ACI Journal, February 1967.

55. P. G. HODGE and G. N. WHITE
A quantitative comparison of flow and deformation theories of plasticity, J. Appl. Mech., Vol.17, pp. 180-184, June 1950.

56. A. ARNESEN, S. I. SORENSEN and P. G. BERGAN
Nonlinear analysis of reinforced concrete, Computers and Structures, Vol.12, 1980, pp. 571-579.

57. H. A. MANG and H. FLOEGL
Tension stiffening concept for reinforced concrete surface structures, IABSE Colloquium on Advanced Mechanics of Reinforced Concrete. Final Report, Delft, June 1981, pp. 351-369.

58. ACI Standard 318-77
Building Code Requirements for Reinforced Concrete, American Concrete Institute, Detroit, 1977.

APPENDIX

USER INSTRUCTIONS FOR PREPARING INPUT DATA FOR PROGRAMS PLASTOSHELL AND CONSHELL

In this appendix detailed user instructions are provided for the anisotropic elasto-plastic code (PLASTOSHELL) and the reinforced concrete code (CONSHELL) presented in the previous two chapters. A common set of instructions is provided in which the different data items required by the two programs are clearly indicated.

The programs undertake the ultimate load analysis of thick and thin plates and shells, employing:-

*Layered Approach

*Anisotropic material behaviour

*8 node Serendipity, 9 node Lagrangian and Heterosis elements

*Reduced, selective and full integration.

Nonlinear behaviour can be included due to:-

*Large geometric deformation, together with

*Elasto-plastic material response, or

*Reinforced concrete ultimate load characteristics.

<u>CARD SET 1</u> - TITLE CARD (12A6) One Card

<u>CARD SET 2</u> - CONTROL CARD (12I5) One Card

Cols. 1-5	NPOIN	Total number of nodal points
6-10	NELEM	Total number of elements
11-15	NVFIX	Total number of points where one or more degrees of freedom are prescribed
16-20	NNODE	Number of nodes per element 8 - for 8 node Serendipity 9 - for Heterosis and 9 node Lagrangian
21-25	NMATS	Total number of different materials
26-30	NGAUS	Number of Gauss points per element (Bending, Membrane)

31-35 NGAUZ — Number of Gauss points per element (Shear)
NGAUS=3, NGAUZ=3 - Normal integration rule
NGAUS=3, NGAUZ=2 - Selective integration rule
NGAUS=2, NGAUZ=2 - Reduced integration rule

36-40 NCOLA — Set the constraints for the Lagrangian 9 node element:
=0 9 node Lagrangian element (no constraints)
=1 Heterosis - constrain the 9th node displacements (u,v,w)

41-45 NALGO — Nonlinear solution process indicator:
=1 initial stiffness method is used
=2 tangential stiffness method is used
=3 stiffness matrix is recalculated in the first iteration of each increment
=4 stiffness matrix is recalculated in the second iteration of each increment and also when there are one or more unloaded integration points in the previous iteration

46-50 NINCS — Total number of load increments

51-55 NLAYR — (i) Total number of layers through the thickness (PLASTOSHELL)
(ii) Total number of layer patterns in the structure (CONSHELL)

56-60 LARGE — Large deformation parameter
=0 Geometrically linear analysis
=1 Geometrically nonlinear analysis

61-65 NREST — Restart facility parameter
=0 to start the analysis
=1 to restart the analysis from the last previously converged load increment

CARD SET 3 (5F10.5) One Card

Cols.	1-10	GRAVI(1)	Gravitational acceleration in the x-direction
	11-20	GRAVI(2)	Gravitational acceleration in the y-direction
	21-30	GRAVI(3)	Gravitational acceleration in the z-direction
	31-40	ANVEL	Angular velocity (referred to the z axis)

(i) PLASTOSHELL

CARD SET 4 - ELEMENT CARDS (16I5,/,5X,15I5) One or two Cards for each element

Cols.	1-5	NUMEL	Element number
	6-10	MATNO(NUMEL) (NUMEL,1)	Material property number for each layer, ILAYR from Bottom to Top
	⋮		
	56-60	MATNO(NUMEL, NLAYR)	(case of NLAYR = 10)
	61-65	LNODS(NUMEL, INODE)	Element node numbers (anticlockwise)
	⋮		
	106-110	LNODS(NUMEL, NNODE)	(Case of NNODE = 9)

(ii) CONSHELL

CARD SET 4 - ELEMENT CARDS (11I5) One Card for each element

Cols.	1-5	NUMEL	Element number
	6-10	MATNO(NUMEL)	Element layer pattern number
	11-15	LNODS(NUMEL,1)	Element node number (anticlockwise)
	16-20	LNODS(NUMEL,2)	
	⋮		
	46-50	LNODS(NUMEL,8)	
	51-55	LNODS(NUMEL,9)	(Case of NNODE = 9)

CARD SET 5 NODAL COORDINATE CARDS (I5,4F15.10/5X,4F15.10)

Two Cards for each node whose coordinates must be input - finishing with the last node. (Coordinates of the central 9th node and also mid-side nodes whose coordinates are obtained by a linear interpolation of the corresponding corner nodes need not be input).

First Card

Cols. 1-5 IPOIN Node number

6-20 COORD(IPOIN,1) Top x coordinate

21-35 COORD(IPOIN,2) Top y coordinate

36-50 COORD(IPOIN,3) Top z coordinate

51-65 COORD(IPOIN,4) Top pressure

Second Card

Cols. 6-20 COORD(IPOIN,5) Bottom x coordinate

21-35 COORD(IPOIN,6) Bottom y coordinate

36-50 COORD(IPOIN,7) Bottom z coordinate

51-65 COORD(IPOIN,8) Bottom pressure

CARD SET 6 RESTRAINED NODE CARDS (I5,5X,I5,5X,5F10.6) One Card for each restrained node. (Total of NVFIX Cards)

Cols. 1-5 NOFIX Restrained node number

11-15 IFPRE Condition of the degree of freedom:
restrained (=1)
otherwise (=0)
position 11 - u displacement (x-direction)
12 - v displacement (y-direction)
13 - w displacement (z-direction)
14 - β_1 rotation
15 - β_2 rotation

21-30 PRESC(IVFIX,1) - The prescribed value of the nodal variables (u,v,w,β_1 and β_2 respectively)

31-40 ⋮

41-50 ⋮

51-60 ⋮

61-70 PRESC(IVFIX,5)

(i) PLASTOSHELL

CARD SET 7 MATERIAL CARDS Four Cards for each different material (Total number of cards = 4*NMATS)

First card (I5)

Cols. 1-5 NUMAT Material identification number

Second card (7F10.5)

Cols. 1-10 PROPS(NUMAT,1) E_1 Young's modulus in 1 direction

Cols. 11-20 PROPS(NUMAT,2) ν_{12} Poisson's ratio ($\nu_{12}/E_1=\nu_{21}/E_2$)

21-30 PROPS(NUMAT,3) t_ζ Layer thickness expressed in the normalised ζ coordinate

31-40 PROPS(NUMAT,4) ρ Material density

41-50 PROPS(NUMAT,5) α Coefficient of thermal expansion

51-60 PROPS(NUMAT,6) σ_{01} Uniaxial yield stress (1 direction)

61-70 PROPS(NUMAT,7) H' Hardening parameter (1 direction)

Third card (7F10.5)

Cols. 1-10 PROPS(NUMAT,8) E_2 Young's modulus in 2-direction

11-20 PROPS(NUMAT,9) G_{12} Shear modulus in $\overline{12}$ plane

21-30 PROPS(NUMAT,10) G_{13} Shear modulus in $\overline{13}$ plane

31-40 PROPS(NUMAT,11) G_{23} Shear modulus in $\overline{23}$ plane

41-50 PROPS(NUMAT,12) σ_{02} Uniaxial yield stress (2 direction)

51-60 PROPS(NUMAT,13) σ_{03} Uniaxial yield stress (3 direction)

or σ_{θ_o} Uniaxial yield stress (At 45° to 1 direction)

61-70 PROPS(NUMAT,14) τ_{012} Shear yield stress ($\overline{12}$ plane)

Fourth card (7F10.5)

Cols. 1-10 PROPS(NUMAT,15) τ_{013} Shear yield stress ($\overline{13}$ plane)

11-20 PROPS(NUMAT,16) τ_{023} Shear yield stress ($\overline{23}$ plane)

21-30 PROPS(NUMAT,17) θ Angle between the reference system and the material system in the layer plane (anticlockwise - in radians)

NOTE: The 1,2,3 axes are the principal material axes, with 1,2 being in the plane of the layer.

(ii) CONSHELL

CARD SET 7-A CONCRETE AND STEEL DISCRETIZATION PATTERN Two Cards for each layer pattern

First card (16I5)

Cols. 1-5 NCLAY(ILAYR) Number of concrete layers

6-10 NSLAY(ILAYR) Number of steel layers (ILAYR - Layer pattern identification number)

Second card (16I5)

Cols. 1-5 MACON(ILAYR,ICONL)

ICONL - Material identification

number for each concrete layer from bottom to top

MASTE(ILAYR,ISTEL)

ISTEL - Material identification number for each steel layer

CARD SET 7-B MATERIAL CARDS - Three Cards for each different material

First card (I5)

Cols. 1-5 NUMAT Material identification number

Second card (7F10.5) FOR CONCRETE MATERIAL ONLY

Cols. 1-10	PROPS(NUMAT,1)	E_c	Young's Modulus
11-20	PROPS(NUMAT,2)	ν	Poisson's ratio
21-30	PROPS(NUMAT,3)	t_ζ	Layer thickness expressed in the normalized ζ coordinate
31-40	PROPS(NUMAT,4)	ρ	Material density
41-50	PROPS(NUMAT,5)	f'_t	Concrete ultimate tensile strength
51-60	PROPS(NUMAT,6)	f'_c	Concrete ultimate compression strength
61-70	PROPS(NUMAT,7)	ε_u	Concrete ultimate compressive strain

Second card (7F10.5) - FOR STEEL MATERIAL ONLY

Cols. 1-10	PROPS(NUMAT,1)	E_s	Young's Modulus
11-20	PROPS(NUMAT,2)	E'_s	Elasto-plastic Young's Modulus
21-30	PROPS(NUMAT,3)	t_ζ	Layer thickness expressed in terms of the normalized ζ coordinate
31-40	PROPS(NUMAT,4)	ρ	Material density
41-50	PROPS(NUMAT,5)	f_y	Steel yield stress
51-60	PROPS(NUMAT,6)	ζ_s	Layer position in terms of the normalized ζ coordinate
61-70	PROPS(NUMAT,7)	θ	Angle between the reinforcement and the x'-axis (measured anti-clockwise in radians with $\pi/2 < \theta \leq \pi/2$)

Third card (7F10.5) - FOR CONCRETE MATERIAL ONLY

Cols. 1-10 PROPS(NUMAT,8) ε_m Tension stiffening parameter ε_m

11-20 PROPS(NUMAT,9) α Tension stiffening parameter α

Third card (7F10.5) - FOR STEEL MATERIAL ONLY

Blank card

CARD SET 8 LOAD CARDS At least one card for each element

First card (3I5)

Cols. 1-5 NPRES Distributed load indicator

=0 no distributed loads on this element

=1 distributed loads to be input

6-10 NUCLO Number of concentrated loads on this element (=0, no concentrated loads)

11-15 NBODY Body load indicator (gravity and/or centrifugal

=0 no body loads on this element

=1 body loads to be input

Second card (I5,F5.1,2F15.5) [Only exists if NPRES=1]

Cols. 1-5 KPRES Distributed load type indicator

=0 Uniformly distributed load

=1 Hydrostatic load

=2 Load specified as nodal values (See Card Set 5)

6-10 CFACE = + 1.0 Pressure is on top surface

= - 1.0 Pressure is on bottom surface

11-25 PREVA Uniformly distributed load if KPRES = 0

Maximum value of hydrostatic load if KPRES = 1

26-40 SURFA z coordinate of zero pressure if KPRES = 1

Third set cards (2I5,F10.5) [Only exists if NUCLO > 0]

Numer of cards to be input equals NUCLO

Cols. 1-5 LPOIN Local node number (in the range 1-8) at which the load is applied

6-10 LDOFN Nodal variable number corresponding to the applied load

=1 - x displacement

=2 - y displacement

=3 - z displacement

=4 - β_1 rotation

=5 - β_2 rotation

11-20 CARGA Concentrated load value

<u>CARD SET 9</u> LOAD INCREMENT CONTROL CARDS (2F10.5,3I5) One Card for each load increment (total of NINCS cards)

Cols. 1-10 FACTO Applied load factor for the current increment

11-20 TOLER Convergence tolerance factor

21-25 MITER Maximum number of iterations allowed

26-30 NOUTP(1) Control output parameter of the unconverged results after the first iteration

=1 - Print the displacements only

=2 - Print displacements and nodal reactions

=3 - Print displacements, reactions and stresses

31-35 NOUTP(2) Control output parameter of the converged results

=1 Print the final displacements only

=2 Print displacements and nodal reactions

=3 Print displacements, reactions and stresses.

AUTHOR INDEX

SUBJECT INDEX